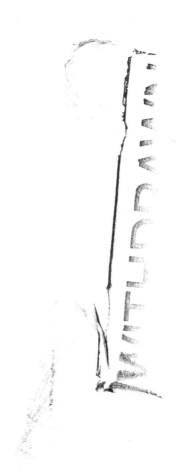

Atlas of Taphonomic Identifications

Vertebrate Paleobiology and Paleoanthropology Series

Edited by

Eric Delson
Vertebrate Paleontology, American Museum of Natural History
New York, NY 10024, USA
delson@amnh.org

Eric J. Sargis
Anthropology, Yale University
New Haven, CT 06520, USA
eric.sargis@yale.edu

Focal topics for volumes in the series will include systematic paleontology of all vertebrates (from agnathans to humans), phylogeny reconstruction, functional morphology, Paleolithic archaeology, taphonomy, geochronology, historical biogeography, and biostratigraphy. Other fields (e.g., paleoclimatology, paleoecology, ancient DNA, total organismal community structure) may be considered if the volume theme emphasizes paleobiology (or archaeology). Fields such as modeling of physical processes, genetic methodology, nonvertebrates or neontology are out of our scope.

Volumes in the series may either be monographic treatments (including unpublished but fully revised dissertations) or edited collections, especially those focusing on problem-oriented issues, with multidisciplinary coverage where possible.

More information about this series at http://www.springer.com/series/6978

Atlas of Taphonomic Identifications

1001+ Images of Fossil and Recent Mammal Bone Modification

Yolanda Fernández-Jalvo

Museo Nacional de Ciencias Naturales (CSIC), Madrid, Spain

Peter Andrews

Department of Paleontology, Natural History Museum, London, UK

 Springer

Yolanda Fernández-Jalvo
Museo Nacional de Ciencias Naturales (CSIC)
Madrid, Spain

Peter Andrews
Department of Paleontology
Natural History Museum
London, UK

ISSN 1877-9077   ISSN 1877-9085 (electronic)
Vertebrate Paleobiology and Paleoanthropology Series
ISBN 978-94-017-7430-7   ISBN 978-94-017-7432-1   (eBook)
DOI 10.1007/978-94-017-7432-1

Library of Congress Control Number: 2015954601

Springer Dordrecht Heidelberg New York London

Cover illustration:
Top left Scanning electron microphotograph of a human femur from Coldrum, Kent (UK). The sequence of modifications are, first, cut marks resulting from disarticulation and funerary processes; second, root marks produced by plant growth after burial; and, third, cracks resulting from sediment compaction (Atlas Text Figure 1.2).
Top middle Process of burial of a modern mandible of deer in Riofrío (Atlas Figure A.1074).
Top right Human primary burial at Çatalhöyük (Turkey), with the skeleton intact and all elements preserved and in articulation (Atlas Figure A.1070).
Bottom left Right human radius GC87-74 from Gough's Cave with extensive cut marks on the side of the shaft (courtesy of Silvia Bello and Chris Stringer). These marks were formerly interpreted as decorative engraving, but the 'groups' of incisions are each made by single strokes, with directionality towards the superior aspect of the shaft and interpreted as filleting of the arm muscles (Atlas Figure A.27 and Atlas Text Figure 3.5).
Bottom right Scanning electron microphotograph of a modern rodent incisor from the pellet of a long eared owl, *Asio otus*, raptor. Light digestion appears concentrated at the tip of the incisor (Atlas Figure A.822).

Printed on acid-free paper

Springer Science+Business Media B.V. Dordrecht is part of Springer Science+Business Media
(www.springer.com)

Preface

The object of this work is to provide sets of images of taphonomic modifications of vertebrate bones during their preservation in the archaeological and fossil record. The correct identification of taphonomic modifications is the first step in understanding the processes by which they are formed and the agents behind the processes (Weigelt 1927). The book is arranged in such a way as to facilitate comparisons of taphonomic modifications so that each modification produced by one process or agent can be compared with similar modifications produced by other processes and agents. *Modifications* are what are actually observed on recent and fossil bone, and modern simulations can reconstruct the *processes* by which they are formed. The third step is identification of the *agent* responsible. Thus, the modification of a rodent tooth showing the enamel partly dissolved away has undergone a process of solution which may be due to enzyme or acid attack; distinguishing the actual process requires experimentation to show which it is; and finally the agent responsible for the process can again be estimated by experimental work, comparing corrosive forces such as animal or bird digestion, soil corrosion or physical breakage.

In order to keep the book to a manageable size and yet illustrate it with high-quality images, we have opted for a minimum of text and text figures to organize the taphonomic modifications described and displayed here. We do not aim to update the literature published on vertebrate taphonomy. Many scientific papers have tackled specific taphonomic problems, and there are comprehensive books such as those of Shipman (1981), Brain (1981), Binford (1981), Lyman (1994a), Pickering et al. (2004), and Bell (2012) which describe and summarize the literature on vertebrate taphonomy. We aim here to show a broad range of images of most aspects of vertebrate taphonomy with identifications of causality which are largely the product of our own research, based on complete taphonomic analyses of a given site, long term field monitoring and experimental work. The text outlines the different modifications, processes and agents important in vertebrate taphonomy. Modifications are illustrated at the end of each chapter as Atlas images, and the essential parts of the book are the comparisons between images displayed in high resolution. The entire volume is available as a downloadable pdf file, as described on page 5 below.

Our interest in taphonomy goes back to 1971, when PA was introduced to the subject by Judy Van Couvering (now Judith Harris), an inspirational field worker and a good friend. This led to the establishment of a long-term monitoring project at Neuadd in Wales (see description in Chap. 2), the results of which have not yet been published. The taphonomy of small mammals complemented this with several projects on fossil faunas and on collections of modern predator assemblages (Andrews 1990; Andrews and Evans 1983), all of which were brought together in the book *Owls, Caves and Fossils* (Andrews 1990).

YFJ was encouraged to go into the field of taphonomy by Prof. Emiliano Aguirre in 1985 and was formally introduced to taphonomic modifications and interpretations of bone remodelling by T.G. Bromage, which was of great value for her training. She was also soon involved in the analysis of taphonomic and environmental information that small mammals could provide, and she started collaborating with PA on the fossil sites at Atapuerca and Olduvai (Fernández-Jalvo and Andrews 1992; Andrews and Fernández-Jalvo 1997; Fernández-Jalvo et al. 1998). She also recognized that the taphonomic modifications could be important in other areas of research, such as pollen analysis, and she has widened her approach to include diverse processes and agents of fossilization, including DNA preservation and experimental taphonomy. We have pursued together on our common interest in taphonomic studies of site formation and patterns in taphonomy, and on anthropological studies of human behaviour based on taphonomic evidence.

To a limited extent we have used images from the work of others, and we thank the following for permission to use their work: Luis Alcalá, Saleta Arcos, Graham Avery, Kay Behrensmeyer, Jill Cook, Isabel Cáceres, Arzu Demeril, Christiane Denys, Emma Jenkins, Peter Jones, Tania King, Dores Marin-Monfort, Theya Molleson, Dolores Pesquero, Ana Pinto, Tony Sutcliffe and Jim Williams. We also thank the following for providing specimens: E. Aguirre, G. Avery, A.K. Behrensmeyer, B. Brain, A. Cuadros, M. Domínguez-Rodrigo, E-M Geigl, G. Haynes, D. Fisher, C. Finlayson, P. Jones, J. Martínez, T. Molleson, S. Parfitt, S. Ripoll, A. Rosas, B. Sánchez, B. Sanchiz, C. Smith and D. Western. We are especially grateful to the late Anthony Sutcliffe, one of the early collectors of modern taphonomic specimens. He donated his extensive collection to the Natural History Museum and facilitated our taphonomic work. We are also grateful to the following for use of their figures: M. Antón, M. Bautista, S. Bello, I. Cáceres, J. Carrier, C. Denys, J. Fernández-Jalvo, G. Gómez, T. Jorstad, A. Louchart, F. Njau, Z. San Pedro, D. Pesquero, M. Salesa, C. Stringer, M. Wysocki and H. Tong.

Many people have helped in the collection of samples for modern analogue studies and in the excavation of fossil sites, particularly Libby Andrews, Miranda Armour-Chelu, Jill Cook, Isabel Cáceres, Montse Esteban, D. Marin-Monfort, D. Pesquero, V. Requejo, and Paul Richen. The DNA preservation project (i.e. DNA taphonomy) has been made possible thanks to collaboration with Eva-Maria Geigl, and pollen taphonomy thanks to collaboration with Louis Scott. We particularly thank the staff of the electron microscopy units and the photo studios at the Natural History Museum and at the Museo Nacional de Ciencias Naturales for their unstinting assistance and professional work. We thank four anonymous reviewers and especially Kay Behrensmeyer for significant suggestions on earlier versions of this book, and to L.S. Bell and T.G. Bromage for suggestions on specific subjects. We are especially grateful to Sylvia Hixson Andrews for comments, ideas and help, and to Eric Delson, co-editor of the Springer series Vertebrate Paleobiology and Paleoanthropology, for continued help and support. Thanks also to Sherestha Saini and Fermine Shaly and, in general, to the Publishing and Production editors/departments of Springer. Fernando Fernández helped in the final title of the book.

Yolanda Fernández-Jalvo
Peter Andrews

Contents

Part IV   Modification by Loss of Bone Tissue or Skeletal Elements

Part V   Conclusions

Chapter 1
Introduction and Rationale

Most simply put, taphonomy is the study of processes affecting the transition of the remains of past living organisms and their traces into the lithosphere as seen in the prehistoric record. It has many aspects, from processes affecting individual organisms to those affecting whole communities, but at its most basic level, the processes on which taphonomic interpretations are based are the same. Process is thus defined as the action of a taphonomic agent (Lyman 1994a), the agent being the immediate cause of modifications. The evidence by which process is identified in the fossil record is the effect it has on fossils, the biological, chemical and physical modifications preserved on the fossils. These modifications may be identified and interpreted through comparisons with observed processes produced by known agents acting on previously unmodified bones at the present time, either in experimentally controlled conditions or by naturalistic monitoring projects where agents and processes are known.

The taphonomic history of single specimens or of animal or plant fossil assemblages starts when living organisms die and are modified by decay, exposure, burial and later diagenetic effects. It is the sum of the effects (modifications) observed on the remains of the organisms that is the basis for interpreting the processes affecting the assemblages (Fig. 1.1). Similarly, groups of organisms individually pass through the same stages and are affected by the same or different processes, and it is the sum of these, the evidence of the variety of processes, that is the basis for the taphonomic history of the group. Whether working at the individual level or on biotic assemblages, therefore, the evidence needed to reconstruct taphonomic history begins at the level of the modifications observed on individual bones or skeletons, and these are combined to obtain general information about the fossil association.

Taphonomy provides evidence additional to taxonomic identification. Too often taphonomy is viewed as an unnecessary complication in the analysis of fossil assemblages, for example in the reconstruction of paleoecology, but it enhances the information content of fossils. Because modifications recorded on fossil bones are the response to organic or inorganic agents that may not fossilize, taphonomic modifications should be seen as additional information about these processes acting in the past rather than just as destructive processes (Fernández-López 1991). Figure 1.2 is a scanning electron micrograph of the proximal end of the shaft of a human femur from Coldrum, Kent (UK), and it shows several stages in the taphonomic history of this bone: first of all, two cuts marks were made by a left-handed person, and the cuts run diagonally across the shaft close to the junction with the head of the femur. The position of the cut marks relative to muscle insertions indicates that the object was to cut through the muscle at the top of the leg, perhaps so as to disarticulate the leg. Secondly, the femur shows no evidence of any weathering on the surface of the bone suggesting the bone was protected from direct sun exposure and changes in temperature and humidity, and this indicates that it was buried soon after disarticulation. Thirdly, the bone was buried in a biologically active soil and at shallow depth since there are root marks superimposed on the cut marks, and the root marks were probably made by woody herbs or bushes. The final event in the taphonomic history of this bone was that the bone was cracked and slightly displaced by soil pressure, displacing the line of both the cut marks and the root marks. This sequence of events, which is the taphonomic history of the bone, enhances the value of this specimen because it expands our knowledge of the past.

Building on the modifications and processes observed on the individual fossils preserved in a fossil assemblage, we may reconstruct the extent to which a prehistoric bone assemblage has been altered, for example by its mode of accumulation, the time taken for it to accumulate, and fossilization and/or diagenetic processes. The first might entail transport from one place to another or the intervention of predators, the second entails time averaging, the length of time it takes for an assemblage to accumulate and be buried and preserved, and the third depends on the sedimentary environment present at the place of preservation. All of these processes leave traces on the bones, and the correct identification of taphonomic processes depends on accurate

© Springer Science+Business Media Dordrecht 2016
Yolanda Fernández-Jalvo and Peter Andrews, *Atlas of Taphonomic Identifications: 1001+ Images of Fossil and Recent Mammal Bone Modification*, Vertebrate Paleobiology and Paleoanthropology, DOI 10.1007/978-94-017-7432-1_1

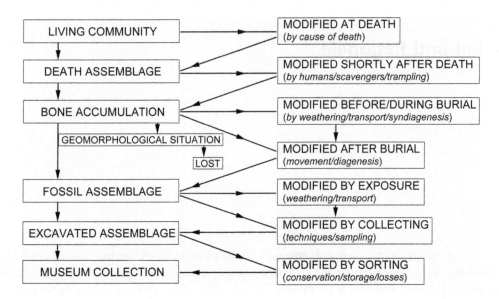

Fig. 1.1 Stages in the formation and modification of animal assemblages. The four stages are shown on the left, from living animals to fossil assemblage, and the modifications possible at each stage are on the right. The arrows indicate alternative pathways connecting both the stages themselves and the processes between stages

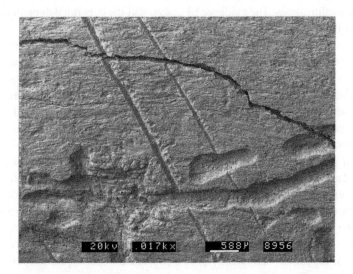

Fig. 1.2 Scanning electron micrograph of the proximal end of the shaft of a human femur from Coldrum, Kent (UK). The sequence of modifications on a fossilized bone is shown here. First to occur were slicing marks resulting from disarticulation; these are interrupted by root marks, which were made later and which indicate plant growth on the site after burial of the bone; and finally the bone has cracking superimposed on both cuts and root marks as a result of sediment compaction

description of the modifications resulting from these processes. There is much more that can be done, however, and in undertaking this compilation we have shown some of the strengths and weaknesses of the current state of taphonomic research. For example, we cannot at present reliably link tooth marks to the predators that made them, and often it is not possible to distinguish between predation and scavenging. Multiple modifications may further confuse the issue, such as the evidence left by bacterial attack on digested

bones. These and other issues will no doubt be the subject for future research for many years, and as our comparative database increases and becomes more sophisticated it will be possible to produce a more complete and accurate set of images.

We have produced a collection of images with the aim of providing images of taphonomic modifications on vertebrate bone, both recent and fossil. As stated earlier, the images we use, the photographs, SEM microphotographs and drawings, are largely drawn from our own work spanning the past three decades. We have investigated a range of sites from the Miocene to Holocene, but the majority of images come from the Pleistocene, from both cave deposits and open land surfaces. Our work has mainly been concentrated in Africa and Europe, and so again most images are from these two regions. This work, therefore, is a compilation of our ongoing research, some published and some not, and over the years it will be added to as additional work is done. We have restricted our use of work from other sources to subjects missing from our own work, and we are grateful to our colleagues who have given permission to use their images. We have made the work as comprehensive as possible with the evidence available, providing images of a wide range of modifications both from natural and experimental situations and from the fossil record. Unfortunately there are still many modifications that have not been studied fully, together with the agents that produce them and the processes by which they are formed, and some have not been studied at all. Therefore some images that we include have no positive identification as to agent. Images drawn from published work, both our own and from other authors, are all referenced in the main part of the book, and where there is no reference the images are from our unpublished taphonomy collections at the Natural History Museum, London, Blandford Museum, Blandford, or at the Museo Nacional de Ciencias Naturales, Madrid.

Layout

The primary organization of this book is based on *type of modification* rather than the more traditional *agent* of modification. Types of modification are the first evidence observed when paleontological or archaeological collections are made, and there are many processes producing similar types of modification (Table 1.1). For example, a groove or striation observed on a bone is a type of modification, but it can be produced by carnivore gnawing, by human butchery, by trampling damage, by root etching, by insects, by microorganisms (bacterial or fungal action), or by rodent gnawing. The organization of this book will make it possible to search through the section headed 'Linear marks', to make comparisons between different types of mark, and to use this to identify the process by which the observed feature was made. Identification is largely based on experience observing both fossil and modern sites, as well as experimental work (Fernández-Jalvo et al. 2010). A single agent may produce different modifications considered here. For instance human chewing will be considered in the chapters on linear marks, pits and perforations, and breakage because when humans chew bones, six types of damage occur:

- Bent ends (fraying)
- Curved shape at the very end of thin bones
- Crenulated edges
- Punctures on broken edges: double arch shaped
- Puncture marks on bone surfaces: triangular shaped
- Linear marks or grooves on bone surfaces: shallow, transverse, or oblique.

Surface shallow scratches may be associated with shallow crescent pits made by human incisors and located on bone surfaces and fragile anatomical edges.

The secondary organization of the book is based on the processes by which the different types of modification are formed. The broad classes are inorganic processes, such as action of stones or weather on bones, and organic processes, subdivided into the activities of animals, microorganisms and plants. Subdivisions of processes can usually be identified within these classes, such as weathering as opposed to corrosion, carnivore chewing versus herbivore chewing, microorganism corrosion versus vascular plant root marking, and so on. A specific process often can be linked with the agent responsible given more details of the modification features, allowing it to be distinguished from similar effects by other agents. For example, the type of stone tool used to make cut marks on a bone can sometimes be identified by the width, depth and irregularity of the mark. Similarly, the body size of a carnivore that has produced chewing marks can be identified, at least to a first approximation, by the size of the marks, and the type of plant producing root marks could potentially be identified by size and depth of the marks, although we lack comparative data for this at present. The aim of this book is to provide enough examples and descriptions of taphonomic modification traits to allow readers to match their samples to those depicted here and thereby to identify the processes and agents by which they were formed.

The sites from which we have taken samples, and the publications that may help users to understand the interpretations of causes of modification, are described in Chap. 2. Each figure in the book includes a reference to the site from which the specimen came and the nature of the photographic image, for example whether taken with the scanning electron microscope or camera. Short explanations of the nature of the modification depicted and the process by which it was formed are included in the figure captions.

This book is NOT intended to be a textbook on taphonomy and its importance in paleontological and archaeological studies. There is extensive published literature on this topic, and reference can be made for example to Lyman (1994a) or to Pickering et al. (2007) for a full review of past and current taphonomic research. It is intended rather as a guide for use in the field, in the laboratory, and in institutions that lack good comparative taphonomic collections. This

Table 1.1 Taphonomic modifications and the processes and agents by which they are produced

Modification	Agent 1	Agent 2	Agent 3	Agent 4	Agent 5
Abrasion	Sediment + water	Wind	Bioturbation	Trampling	Carnivores
Breakage	Humans	Carnivores	Sediment pressure	Trampling	Bioturbation
Cave corrosion	High humidity				
Chop marks	Human	Falling blocks			
Corrosion	Humic acids	Organic acids	Dense low plant cover	Cave humidity	Moss and lichen
Cracking	Weathering	Digestion	Organic acids	Humic acids	Plant roots
Curved ends	Human	Falling blocks			
Deformation/bending	Human	Sediment pressure			
Digestion	Carnivores	Birds	Crocodiles		
Disarticulation	Carnivores	Humans	Trampling	Bioturbation	Weathering
Discoloration: black	Burning	Manganese	Carbon stain	Fungi	
Discoloration: brown	Burning	Humic acids			
Discoloration: light	Burning	Gley soils	Leaching		
Discoloration: red	Burning	Oxydized soil			
Double arch punctures	Human molars				
Enamel loss	Digestion	Corrosion			
Gouges	Carnivores	Birds			
Linear mark U-shape	Carnivore chewing	Pland roots	Insects	Beak marks	Herbivore and rodent gnawing
Linear mark V-shape	Human cut marks	Trampling	Bone tool preparation		
Multiple scrapes	Humans	Trampling	Gnawing		
Notches	Human percussion marks	Carnivores			
Peeling	Human				
Pits	Carnivore	Birds	Insects	Trampling	Plant roots
Polishing	Wind + sand	Water + gravels	Licking	Digestion	
Punctures	Carnivore	Birds	Insects	Plants	Humans
Rounding	Sediment + water	Wind	Digestion	Carnivores	Trampling
Solution marks	Diatoms	Lichen	Algae		
Splitting	Weathering	Digestion			

book also does not cover changes in the chemical composition (organic or inorganic) during fossilization that cannot be illustrated in images. Fossils that appear identical may have different chemical composition and conversely fossils that appear different (e.g., in color) may have identical composition. These differences in chemical composition can only be known accurately and in detail after destructive analyses in the laboratory. Some exceptional cases may occur, but they are particular to a site history or site conditions.

There are two aspects of taphonomy that we consider of importance both now and in the future. The first is the recognition that the object of studying taphonomy is to gain information about past events, showing how fossils have come to be preserved in all the multitude of manifestations evident in the fossil record. The intention in this is not to limit the evidence of the fossil record but to enhance it by adding to the environmental background essential to a proper understanding of all fossil discoveries. As Behrensmeyer (1984) has put it, "the death of an organism is the beginning of a creative process of biological and physical transformation". The second aspect is to consider how the

possible quantification of the effects of taphonomic processes may be achieved in order both to obtain a quantitative estimate of the degree of bias introduced by taphonomic processes and to suggest corrections that could be applied to clearly biased fossil assemblages before they are used for further analysis, such as paleoecological reconstructions and analyses of paleogeography or paleoclimatology. We will come back to these issues in the final chapter of the book, Chap. 11.

How to Use the Book

Following the two introductory chapters, the book is organized into eight chapters targeting the major categories of taphonomic modification: Chap. 3 linear marks such as striations and grooves; Chap. 4 pits and perforations; Chap. 5 changes in color or discoloration; Chap. 6 surface abrasion producing rounding or polishing; Chap. 7 flaking and cracking of the surfaces of bones; Chap. 8 corrosion and digestion; Chap. 9 breakage; and Chap. 10 disarticulation and completeness/dispersal of skeletons. Each chapter has a short introductory text for each modification, with definitions, the agents that produce the modification, and brief descriptions of the characteristics and the processes that produce them. There are selected text figures and data (where available) for the range of morphologies seen for each modification type. The written text and text figures are followed at the end of each chapter by the Atlas figures relating to that chapter. Text figures are numbered according to the chapter, starting from 1 in every case. Thus, Chap. 2 has text Figs. 2.1, 2.2 and 2.3 and so on; Chap. 3 has text Figs. 3.1, 3.2 and 3.3 and so on. The comparative Atlas figures are numbered sequentially from A.1 to A.1080. They provide several examples of each modification type, and additional links to similar modifications enable the user to make comparisons among them.

In a pdf version of the work, links between text and figures, and from one figure to another, enable their comparison. The keyboard shortcuts Alt+Left Arrow can be used to return the reader to the last page viewed. At the end of the volume (pp. 341–353), there are pages with thumbnails of all of the Text Figures and Atlas Figures by chapter. These may help the reader to find the number of a figure seen previously.

The entire high-resolution pdf file of this book is available for download from the web site https://digital.csic.es/handle/ 10261/130147. At that page, you can click on the word "English" at the upper left to make sure the information is in English rather than Spanish. At the lower right, click on the button "Request a copy", which will bring you a new page from which you can send an email requesting access to the pdf. The book is also available from Springer as a pdf or e-book through institutional agreements.

To see the pdf with correctly facing paging reflecting the best organization of the Atlas figures, in a version of Adobe Acrobat™ or a similar pdf reader, click on View/Page Display/Two-Up. For chapters 3, 4, 6 and 9, it is necessary also to click on View/Page Display/Show Cover page during Two-Up. These features will result in facing pages correctly shown even though blank pages found in the print version are not included in the pdf.

Abbreviations

A numbers	Links to Atlas figures
BSE	Back-scattered electrons (Jones 2012) A.278
c.H.	Canal of Havers (Haversian canal)
ESEM	Environmental scanning electron microscope
DNA	Deoxyribonucleic acid
ECL	Endosteal cortical layer, Fig. 2.2, A.475
EDM	Electronic distance measuring
EDS	Energy Dispersive X-ray Spectroscopy mapping, Fig. 8.2, A.529
LVSEM	Low vacuum scanning electron microscope
MCL	Medial cortical layer, Fig. 2.2, A.475
MFD	Microscopic focal destruction A.297
MNCN	Museo National de Ciencias Naturales, Madrid
ND	Neuadd, Rhulen, Wales
NHM	Natural History Museum, London
PCL	Periosteal cortical layer, Fig. 2.2, A.475
SE	Secondary electrons (SEM) A.278
SEM	Scanning electron microscope, Fig. 2.5

Chapter 2
Methods in Taphonomy

The images in this book were produced using a variety of procedures and methods. This chapter provides a summary and discussion of these procedures and the potential effects they have on the interpretations of fossil assemblages. There are three main data sources: fossil faunas, studies of modern analogue faunas, and examining the impact of the methods used for investigating the modifications (such as DNA sampling). Data on fossil faunas can be collected directly from fossil sites, but in order to retrieve useful information on the taphonomy of fossil bones, basic information on the method of collection is needed. For example, the type of taphonomic information available from an excavated assemblage is different from that derived from surface collections, and the way the information is collected is also different. The interpretation of taphonomic modifications on fossil faunas is always based on inference, however, and the only way to obtain unequivocal data that links types of modification to specific taphonomic processes and agents, is through documentation of modern analogues. Such documentation can be achieved either by actualistic studies in the field or laboratory experiments under controlled conditions. The methods available for analyzing collections of fossils and analogue studies are similar, although some of the techniques differ slightly according to the type of analysis, especially during specimen preparation. In the analyses of fossils, non-destructive and non-invasive methods are preferred, although this is not always possible.

Collecting Methods for Fossils and Artifacts and Their Effects on Taphonomic Results

Surface Collection

Surface collections of fossils or human artifacts are generally organized according to some objective criteria. For example,

searches might be conducted following stratigraphic exposures, particularly if it is known that the source of the specimens is one particular horizon. In general, however, some additional control is needed if the fossil deposits are exposed over an unknown area. In this case the recommended method is to set out a linear grid that can be adapted to the nature and extent of the exposure. Each find can be photographed as it is found, and a record made of the type of sediment in which it was found. Criteria for collection of specimens depend on the type of collection being made. A frequently used practice is to collect all identifiable specimens, including all flake debitage and small mammals. Much taphonomic information can be lost by this procedure, and recommended practice is to collect in addition all specimens larger than 2 cm. These should be labeled with date of collection, numbered strip in the linear grid, distance along the grid (as stated above), and type of specimen. When available, location of finds can be made using a GPS or a total station. Where surface collections are able to locate specimens in place in the sediment, the same criteria apply for the recording of the specimens, but some additional criteria should be added (Fig. 2.1).

Excavation

The size and orientation of excavation grids are nearly always determined by the nature of the exposures. Meter grids in caves can often be set out above the surface of the deposits by running an aerial grid of lines from the side walls of the cave. Aerial grids are normally made with tensioned wire or metal cable rather than with string. Horizontal and vertical coordinates of fossils and artifacts, as well as sedimentary features, are recorded by measurements from one or more referential fixed points, or by digital total station using electronic distance measuring (EDM). The coordinates are used to make plans of excavations showing the 3-D spatial distribution of fossils and artifacts, with all finds and

© Springer Science+Business Media Dordrecht 2016
Yolanda Fernández-Jalvo and Peter Andrews, *Atlas of Taphonomic Identifications: 1001+ Images of Fossil and Recent Mammal Bone Modification*, Vertebrate Paleobiology and Paleoanthropology, DOI 10.1007/978-94-017-7432-1_2

No.	Material	Type	Identif.	L	W	T	X	Y	Z inf	orient.	tilt	Sediment	Remark
21	limestone	block	angular	300	210	70	314	4688	-274	NW-SE	SW	crumbly sediment with small dark pieces	rotten limestone
23	bone	II phalanx	equid	80	55	25	390	4687	-305	N-S	N	soft clayish sediment	DNA sample
24	stone tool	chert	BN2G	30	10	5	311	4613	-306	NW-SE	V	hard sediment	
25	obsidian	obsidian	BN2G	40	20	3	309	4611	-415	E-W	flat	silty clayish brown soft	smooth side facing bottom
26	charcoal			-	-	-	377	4654	-416	-	-	silty clayish brown soft	dating
27	coprolite			45	40	25	370	4655	-418	see remarks		silty clayish brown soft	round shape

Site Name: Azokh 1 — Unit: II (a) — Square: D46 — Page #
Diggers: — Date

Material	Type	Identification	IMPORTANT!!!		Sediment:	Remarks (e.g.)
bone/tooth Yellow	anat.elmt / frag.indet	bovid/cervid/ equid/bear	never forget to mark the ground side	Plot in map all coordinated finds.	silt/gravel/sand	Taken for dating, DNA
stone tool: Blue	obsidian/quartz /flint/chert	flake/debris/ or BN2G/BP	never forget to mark the ground side		crumbly/cemented/ soft	broken during extraction
limestone: Red	block/ stone	angular/rounded	bigger than 10 cms		heterogeneous/ homogeneous	consolidated in the field
charcoal: Black			wrap in foil		colour: red, brown, grey/yellow	uncertain coordinates
coprolites: Brown	isolated/several segments		wrap in paper			sample taken for geology

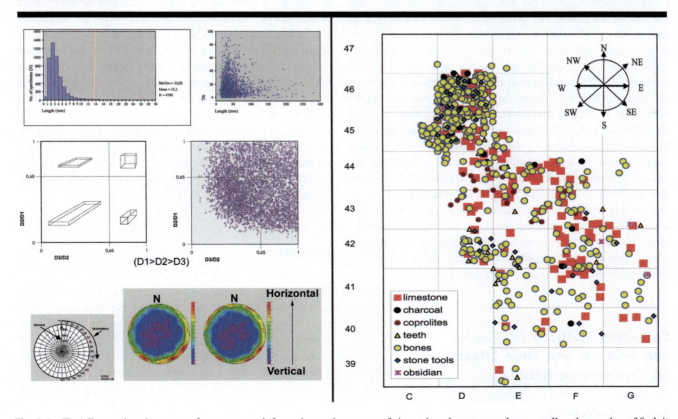

Fig. 2.1 (Top) Excavation sheet: apart from common information such as name of site, unit and square number, as well as the number of find, its dimensions and orientations, it is necessary also to identify the digger, write the date, and describe the color of the find according to a formerly established code. It is important taphonomic information to describe the sediment below and around the fossil, to collect samples of sediment and to distinguish clearly the purpose of the sample. Specimens should be marked with a permanent marker to identify the side in contact with the ground. Bottom right, the measured coordinates for all specimens are used to make plans of the excavation, in this case at Azokh Cave, Nagorno Karabakh. Bottom left, all these data allow the taphonomist to obtain diagrams of size, shape, tilt and orientation to discern the influence of processes that might select fossils and organize according to aerodynamic or hydrodynamic traits of the fossils

sediment described and labeled in detail. Collecting and screening for fragile small mammal specimens may increase fragmentation, but the advantage of this method is recovery of most small mammal fossil bones, both for taxonomy and taphonomy. Recommended protocols for dry screening are three nested screens with mesh sizes of 10, 5 and 1 mm; and for wet screening recommended mesh sizes are 2, 1 and 0.5 mm. Fine residues should be taken back to the laboratory for sorting using a mounted low magnification lens or binocular light microscope.

Collecting Samples for DNA Taphonomy

Sampling for DNA analyses is more specific than most other sampling in taphonomy. Washing and use of consolidants should be avoided, specimens should be kept away from changes in humidity and, in particular, from exposure to the sun. To avoid contamination, the specimen should never be exposed on the surface for longer than a few minutes before extraction under sterile conditions. The collector of DNA samples should wear latex or plastic gloves, face mask, and a disposable cap or clean handkerchief for the head. Tools to extract the fossil should have been previously treated to destroy DNA on the surface, either in the laboratory by exposure to UV-rays, or with bleach before extraction. Sample bags, boxes and foil paper for collection of samples must previously be sterilized or cleaned (nothing can be recycled!). When dealing with human remains, complete coverage in laboratory gowns is advisable, and as far as possible it is recommended that only one person should approach to the fossil and extract the specimens, and they should then provide a sample to characterize her/his DNA. After recovery, specimens are wrapped in foil and kept individually in labeled sterilized plastic bags recording them as DNA samples. Sediment beneath the fossil should be collected separately in foil and in a clean sample bag labeled in detail and noting it as a DNA sediment sample. A control sediment sample should also be collected from another part of the square or excavation level following identical procedures, noting it as a control sediment sample. All samples must be stored as soon as possible under cool conditions, either in a portable freezer, an icebox or buried deep enough into the ground to keep them at relatively low and stable temperatures. The samples should not be frozen before their arrival in the paleogenetic laboratory to prevent repeated freezing/thawing cycles that are harmful for DNA (Pruvost et al. 2007; Fortea et al. 2008). The main point to be considered is that fossil DNA is not modern DNA. The structure

is different and procedures to recover it are different, not only to prevent contamination, but because extraction and amplification of DNA from fossils are also different.

Histology and Preservation

Histological analyses to establish microorganism (bacterial or fungal) attack or other modification of the interior of the bone should be carried out on sectioned bones (Bell 2012). The bones are cut transversally and embedded in resin. The surface is polished with 1000 grit silicon carbide powder, finishing with 2500 grit, on a rotating lap using water as lubricant. The final polish is achieved using 3 and 1 micron diamond on a texmet 1000 polishing cloth, using low to medium weight and plenty of distilled water as a lubricant. This also helps to flush away any mineral laboratory contamination and minimize scratching.

In order to study histological traits of bone cross sections, particularly with regard to microorganisms and bone diagenesis modifications (Fig. 2.2) arbitrary divisions of the cortical bone are used: (PCL) periosteal cortical layer, (MCL) medial cortical layer and (ECL) endosteal cortical layer (Fig. 2.2 top left). These layers almost correspond to histological tissues in the bone cross sections: PCL is similar to the outer layer of periosteum and circumferential lamellae, and ECL is similar to the inner circumferential lamellae. Volkmann's canals run within osteons interconnecting canals of Havers (Fig. 2.2 top right). Osteons consist of concentric lacunae and lamellae around canals of Havers which contain the bone's nerve and blood supplies (Fig. 2.2 bottom right). Osteoblasts form the lamellae sequentially, from externally inwards toward the Haversian canal. Some of the osteoblasts develop into osteocytes, each living within its own small space, or lacuna. Osteocytes make contact with the cytoplasmic processes of their counterparts via a network of small canals named canaliculi (Fig. 2.2 bottom left).

The Oxford Histological Index established by Hedges et al. (1995) (see also Millard 2001) can be used to characterize the general preservation of the bone. Descriptions by Jans (2005), Jans et al. (2005), Bell (1990) and Bell et al. (1991, 1996) on the histological traits and destructive foci (MFD) provide further information on this process of modification and implications. Some of the histological modifications recently studied have been shown to provide valuable environmental information (e.g., Bell et al. 2009; Pesquero et al. 2010). Further research, however, is needed to obtain better identification of the origins of these modifications.

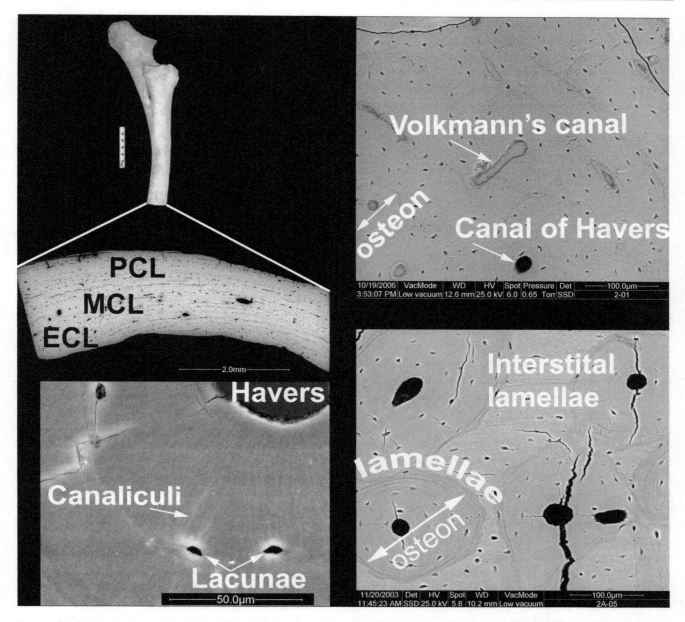

Fig. 2.2 Transverse long bone cross-section: (PCL) periosteal cortical layer, (MCL) medial cortical layer and (ECL) endosteal cortical layer (see text)

Modern Analogues and Experimental Field/Laboratory Projects

Taphonomic history of paleontological and archaeological collections is often complex and all variables need to be investigated in order to reconstruct the history of the assemblages. Interpreting taphonomic modifications depends on identifying the agents producing the modifications and the processes by which they are made, and this depends on actualistic studies providing comparable data from modern observations. These range from studies of the taxonomic composition of whole animal communities and the relative abundance of species within the fossil assemblages, to studies of the elements that make up the fossil assemblage (Behrensmeyer and Hill 1980). Dispersal and accumulation of faunal remains are other considerations, as are the agents responsible for all modifications. All these aspects contribute to the taphonomic history of fossil assemblages, and the modern analogues on which they are based are of several kinds, from single observations of specific types of modification to ranges of modification varying through space or time.

Long Term Monitoring Studies

Particularly valuable are studies monitoring animal remains over periods of time and over different types of landscape and/or climate. Large mammal monitoring has been in progress for many years in East Africa (Behrensmeyer 1984; Western and Behrensmeyer 2009) in UK and Abu Dhabi (Andrews and Armour-Chelu 1998; Andrews and Whybrow 2005) and in Spain (Cáceres et al. 2008). These experiments show how certain taphonomic modifications change through time and space (Behrensmeyer 1978). For example, to address the question of change through time, detailed monitoring by photographing, mapping, and measuring distances of skeletal dispersal are all necessary; while to address effects of the environment, quantifying environmental parameters and evaluating changes should be carried out on a continuing and regular basis, for example at least four times a year in different seasons. Environmental parameters to be measured are soil pH, vegetation type, temperature and humidity fluctuations, both in the air and beneath vegetation, altitude, precipitation, topographic relief, degree of slope, and qualitative assessments of the environment, e.g., damp versus dry, windy versus protected, exposed to the sun versus covered by vegetation, etc. Specimens can be collected periodically, but as much should be left in the field for continuing the long term monitoring. When no traces of specimens are left above ground, the areas should be excavated to determine the extent and depth of burial. Studies such as this, however, are in their infancy and need to be repeated in different environments and on different suites of animal. Figure 2.3 shows how the dispersal and weathering of bones of a camel skeleton in a desert environment changes over a period of 15 years (Andrews and Whybrow 2005): for example, there are episodic dispersal events due to the low and sporadic rainfall, and weathering progresses at less than half the rate observed by Behrensmeyer (1981) in tropical East Africa.

Long term monitoring should also have a spatial element, investigating taphonomic change in different environments. This is another facet of Behrensmeyer's project (Behrensmeyer 1993) in tropical Africa, and it is also incorporated into work in temperate environments both in England and in Spain (Andrews and Armour-Chelu 1998; Cáceres et al. 2007). Many differences have been observed in taphonomic trajectories between different environments even when they are within the same climatic zone, and these are attributed to variations in vegetation cover, type of soil, especially soil pH, topographic aspect of the ground and degree of exposure, degree of slope, existence of barriers, and biological variations due to preference of many animal species to specific parts of the environment.

Monitoring Studies of Small Mammals

Most of the actualistic monitoring studies are being conducted on large animal remains, but taphonomic monitoring of small vertebrate assemblages is also important (Andrews 1990; Montalvo et al. 2007, 2008a, b). Small vertebrate assemblages are usually the result of predator accumulations, and it is difficult to monitor small mammal remains in space and over long periods of time. We have attempted this on several occasions, but invariably the bone assemblages are damaged beyond recognition or disappear completely. Some controlled experiments have been done under field conditions (Korth 1979; Andrews 1990; Fernández-Jalvo and Andrews 2003) showing the long term effects of water movement on different types of bone and the long term effects of inorganic and organic acid soils. Scat collections from a species of mongoose in Kenya were compromised because on the nights following scat deposition many of the scats were eaten, probably by the same individuals that deposited them (Andrews and Evans 1983). The differential effects of the same predator hunting over an area over a period of years, or of different predators hunting over the same area within one time period (Fig. 2.4), have also been investigated (Andrews 1990). Laboratory experiments on insect damage, the effects of plant roots, dispersal of bones by earthworms, weathering and the effects of sediment abrasion need urgent attention.

Preparation Methods

Methods of preparation of animal remains in the laboratory can have marked effects on their preservation. For instance, for specific studies (such as DNA preservation) simply washing the fossils to get rid of the sediment is fully destructive. Similarly, consolidating the surfaces of fossils to preserve them makes it impossible to examine them subsequently with the scanning electron microscope. Some of the methods of bone preparation are not likely to compromise their surfaces or composition, but some of the methods produce modifications that are comparable to those observed in fossils. We have investigated the ones most frequently used by collectors, taxidermists, and museum osteological curators in order to record their effects on animal skeletons, for example boiling bodies to remove soft parts, treatment with enzymes, natural maceration in water, and treatment with insects such as dermestids. Some other methods of preparation have also been investigated, but these have no parallel in the fossil record. Laboratory experiments on the effects of acid during the preparation of fossil bones have

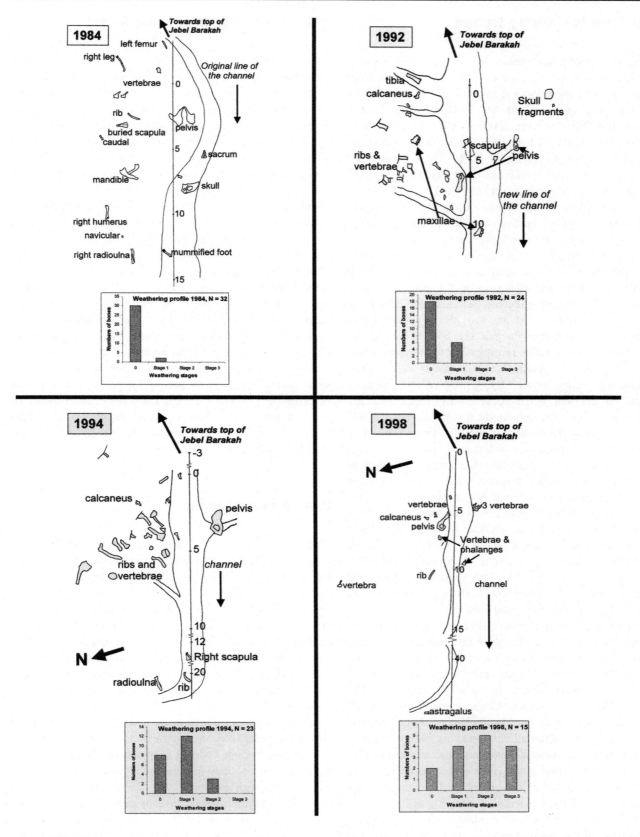

Fig. 2.3 Monitoring the decay and dispersal of a camel skeleton over a 15 year period at Jebel Barakah, UAE. The four plans show the skeleton along the same linear grid and from the same perspective, measured during 1984, 1992, 1994 and 1998. Also shown at the bottom of each figure are the weathering profiles and directionality of all bones visible on the surface at each period

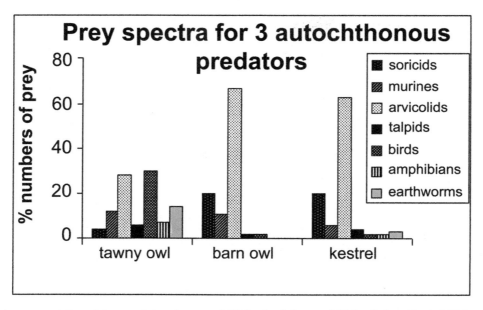

Fig. 2.4 Prey species representation of three predators, tawny owl (*Strix aluco*), barn owl (*Tyto alba*) and kestrel (*Falco tinnunculus*), which occupied different parts of the same barn at the same time over a period of five weeks at Neuadd, Wales

been undertaken, and the effects of weathering and transport, again on fossil bone, have also been investigated (Fernández-Jalvo and Marin-Monfort 2008).

Equipment Used in Taphonomic Research

For surface collections the tools needed for collecting fossils are minimal. Excavation grids are more complex, but at their most basic level no more equipment is needed for setting out a meter grid than is needed for a linear grid. Total station records distances and angles from the instrument to points to be surveyed, so that it records heights as well as horizontal coordinates, but a less expensive way of recording heights is with a laser beam forming a horizontal plane at a referential known height. Equipment needed for field observations on taphonomy comprise a tape measure and hand lens (magnification ×10). Collection of specimens in place in sediment requires in addition a compass for measurement of direction and an inclinometer for measurement of tilt. Magnifying glasses are adequate for a first inspection of the sample, but a binocular microscope is an asset in the field laboratory, although not a requirement. Analyzing fossils in the laboratory is mostly done with a binocular microscope, and magnification up to ×40 is more than adequate. A color chart is used to record color.

Further analyses with more sophisticated equipment may be needed in some cases to distinguish and conclusively identify some taphonomic modifications and the agents that produced them. Scanning electron microscopes (SEM) can be used at higher magnifications and with better resolution than light microscopes, and they also have greater depth of field allowing better three-dimensional images. This method works by directing electron beams at a specimen and collecting electrons emitted from the specimen, either from its surface as secondary electrons (SE) or from beneath the surface as backscattered electrons (BSE) (Fig. 2.5). Conventional SEMs require a metallic surface, and, therefore, fossil or modern organic samples have to be coated by gold or carbon to reflect electrons, which otherwise would be absorbed into the specimen both preventing observation and potentially with disastrous effects. The new generation of SEMs operates in a series of vacuum gradients that absorb some of the electrons and allow the examination of specimens without a metallic covering. These are known as environmental scanning electron microscopes (ESEM) or Low Vacuum (LVSEM) (Bromage 1987) and can operate in both secondary and backscattered mode (Fig. 2.5).

Chemical analysis linked with SEMs can be achieved in two ways. A conventional SEM can be fitted with detectors for X-rays (energy dispersive spectroscopy or EDS) which provides an approximate breakdown of the chemical composition of specimens (Fig. 2.6), but more accurate results

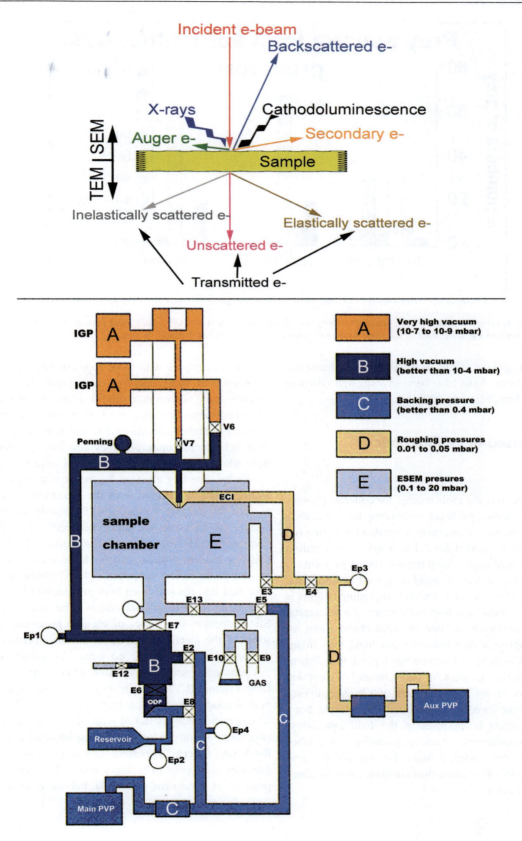

Fig. 2.5 Scanning electron microscope – the incident electron beam reflects off specimens to produce back scattered electrons (BSE), secondary electrons (SE) and X-rays, each of which requires separate detectors fitted to an SEM. Also shown here are the transmitted electrons used in a transmitted electron microscope (TEM); and an environmental scanning electron microscope showing the variations in vacuum in the different chambers

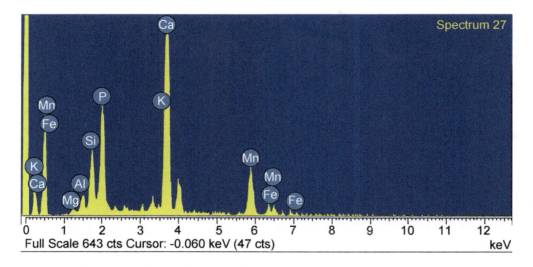

Element	Intensity Corn.	Weight%	Atomic%
Mg K	0.6456	0.52	0.80
Al K	0.8219	2.28	3.15
Si K	0.9893	8.14	10.78
P K	1.2417	25.23	30.31
K K	1.0348	1.36	1.29
Ca K	0.9204	45.41	42.16
Mn K	0.8067	13.83	9.36
Fe K	0.8435	3.23	2.15
Totals		100.00	

Fig. 2.6 Plot and table of chemical element composition of a hydroxylapatite mineral (bone = Ca, P), manganese mineralization (Mn) and sediment (silts [illite] = Mg, K, Al, Si, Fe) using an EDS detector fitted to the SEM. EDS chemical analysis at the SEM is not destructive, takes a few seconds and may analyze micrometric areas such as the infilling of Havers canal or modern unicellular diatom algae, made of silica

are obtained with the microprobe, which has another kind of detector that can also be attached to SEMs and which measures differences in wave length (WDS) to analyze chemical composition.

Confocal microscopes have great advantages as they are based on optical energy (laser) emissions, not electrons (Schmidt et al. 2012). The images obtained may penetrate several microns below the surface of the sample (not limited to the outer layer as in the scanning electron microscopy). Images are based on the fluorescence reflected and emitted, and depending on fluorescence properties of the specimen observed its use may be restricted, but, when appropriate, the advantage of this non-invasive technique is certainly high. A recent generation microscope, also named confocal because it provides focused images of irregular surfaces, is based on natural light. These microscopes provide good depth of field even using high magnifications (3,000×), and they have the advantage of maintaining natural color. Similarly, RAMAN spectroscopy is non-invasive, also based on laser energy and provides a fast chemical analysis (organic and inorganic). Its use is still limited in paleontology, as is catodoluminiscence (applied in electron microscopes), but see Chen et al. (2007) and Damas Mollá et al. (2006). These techniques are especially useful in research on diagenesis and fossilization processes.

Summary of Localities Sampled

The list of localities, samples, monitored and experimental collections for the images illustrated here are summarized below, including both fossil and recent sites. The taphonomic interpretations are based on our work at these sites, where we have either worked directly, been involved through students and associates, or have carried out taphonomic analyses on the collections. Thirty-four sites are represented here, 16 in Europe, 12 in Africa, 4 in the Middle East, and one each in China and South America; 19 sites are open air sites and 15 cave sites; and 21 are fossil sites from the Miocene to recent archaeological sites, and 13 are modern comparative sites, including 7 that are long-term monitoring sites where bones have been observed in naturalistic conditions over periods of time. Taphonomic modifications relate mainly to mammalian bone, both recent and fossil, large mammals and small mammals (less than 1000 g), with some descriptions of modifications to amphibian and bird bones where evidence is available.

Abric Romaní, Spain

Abric Romaní is a travertine rock shelter near Barcelona (Spain). Butchery processes, knapping, cooking and eating have been identified at this site in a complex space arrangement of 400 m^2 extent (Carbonell et al. 1996, 2008; Vaquero 1999; Carbonell and Vaquero 1998; Carbonell and Mosquera 2006). The site is characterized by intensive and recurrent Neanderthal occupations through an extended period of time (70.2 ± 2.6 to 39.1 ± 2.8 ka). Abric Romaní is especially interesting due to an excellent preservation of all types of remains (including wood), with occupations sealed by repeated collapses of the travertine ceiling, and where a pattern of activities organized around hearths has been distinguished (Cáceres 2002). The taphonomic study of macrofauna from Abric Romaní site was part of the doctoral thesis of Cáceres (2002) supervised by Y. Fernández-Jalvo.

Arrikutz, Spain

Cave site in northern Spain with assemblages of cave bear remains. The bears are all relatively large old adults, and the assemblage is attributed to hibernation deaths (Pinto Llona et al. 2005). Several carnivore species are associated with the bears, including hyena (*Crocuta crocuta*), wolf (*Lupus lupus*) and the large cat *Panthera spelaea*. A high proportion

of bones at the site (70%) have been modified by carnivore action, with many spiral breaks.

Arroyo Seco, Argentina

This is a complex of several archaeological sites, located in the eastern and central parts of the Argentinean Pampa region. The sites of Arroyo Seco have yielded human burials from the Late Pleistocene-Holocene associated with contemporaneous mega- and microfauna as well as manufactured tools. The taphonomic study of faunal remains was the subject of doctoral thesis by Gómez (2002) supervised by Y. Fernández-Jalvo.

Atapuerca, Spain

The Atapuerca fossil sites are located in the southern part of the Sierra de Atapuerca, 15 km from the town of Burgos (Spain). This region, located in the Duero Basin in north-central Spain, has an elevation of 1,079 m above sea level or less. The area is extremely rich in fossil sites of different ages, from over 1 Ma (Sima del Elefante site) to the Bronze Age (El Mirador site). We will discuss three sites where we have been directly involved in the taphonomic studies: Gran Dolina with 11 stratigraphic units identified, including TD6-Aurora, a stratum that provided cannibalized human remains of 750 ka (Díez et al. 1999; Fernández-Jalvo et al. 1999), Galería (a butchery site estimated to date to 400–200 ka; Cáceres 2002) and Sima de los Huesos at an age close to 400 ka (Arsuaga et al. 1997; Bischoff et al. 2003). The small mammal taphonomy of Gran Dolina, Galería and the site of Penal was the subject of Y. Fernández-Jalvo's doctoral thesis (1992), and the taphonomy of amphibians and reptiles of Galeria was studied by A. Pinto Llona, both investigations supervised by P. Andrews. The taphonomy of the first 28 human individuals from Sima de los Huesos was studied by P. Andrews and Y. Fernández-Jalvo (1997).

Azokh Cave, Nagorno Karabagh

Azokh cave is part of an extensive phreatic system in limestone rocks of the lesser Caucasus Mountains. The cave is a 200 m dark gallery with several entrances (King et al. 2003). Two of these connections are partially blocked by boulder chokes. The sediments are in part derived from within the cave and in part from avens extending down from the hill slopes above. The sediments consist of silts and breccias, some levels cemented, and they contain fossil animal remains and stone tools in most levels dating from the Middle Pleistocene up to the present (Fernández-Jalvo et al. 2004, 2010). There is extensive evidence of cave bear (*Ursus spelaeus*) and human occupation levels, as well as humanly induced damage, animal activity, bacterial and fungal attack, and diagenesis that produced remineralizations and mineral neoformation in some of these units due to the presence of bats in the cave. Human remains have been found, including *Homo heidelbergensis*, *Homo neanderthalensis* and *Homo sapiens*. Taphonomic, paleontological, geological and paleoecological studies of this site have been supervised by P. Andrews, T. King, L. Yepiskoposyan and Y. Fernández-Jalvo (Fernández-Jalvo et al. in press), with a taphonomic study by Marin-Monfort et al. (in press). The taphonomy of the macrofauna is the subject of the doctoral thesis by Marin-Monfort (2015) supervised by Y. Fernández-Jalvo. The small mammal taphonomy has been studied by P. Andrews.

Barranco de las Ovejas, Spain

The site at Barranco de las Ovejas is in a steep ravine where local shepherds abandon dead sheep in the Pyrenees (Huesca, Spain). The area is taphonomically interesting because of the presence of the Lammergeier or Bearded Vulture (*Gypaetus barbatus*). This specialized scavenger usually avoids rotting meat, and lives on a diet that is 90% bone marrow. It drops large bones from a height to crack them to get access to the marrow. The site provides information on gravitational dispersal of bones, weathering, and, of course, breakage by high falls (Y. Fernández-Jalvo unpublished data).

Çatalhöyük, Turkey

Çatalhöyük is a Neolithic town dating to approximately 9000 years BP occupying a low mound of approximately 15 ha on the Konya plateau. Both human and animal bones have been found in abundance, but whereas the animal bone is mostly found dumped in midden areas, the human bones come from burials beneath the floors of the houses. The taphonomy of the human bones was investigated to determine the nature of burial practice, in particular looking for evidence of secondary burial, the evidence for which was based on selection of skeletal elements in the burials or by the presence of cut or chop marks (Andrews et al. 2005). Many of the graves were reused, sometimes with many individuals in the same grave, the later inhumations disturbing the earlier ones so that the bones became mixed together. Most of the burials were deep enough so that disturbance from compaction of the surface of the floors above is not an issue, and they were further protected by the multiple layers of clay plaster by which the floors of the houses were sealed. The sediment infill of the burials is generally free of rocks and consists of fine silts. Taphonomic processes observed were both anthropogenic (selected and/or modified by humans), or natural, such as fungal staining of bones exposed during disturbance. The taphonomic study of the human burials at Çatalhöyük was carried out by P. Andrews and B. Boz (Andrews et al. 2005).

Concud, Spain

Concud is a reference site of the Neogene's EuroAsiatic Mammalian faunal stages (Late Miocene zone MN12,

7.5 Ma). The site has yielded an abundant fossil fauna bearing taphonomic modifications (Pesquero 2006; Pesquero et al. 2010, 2013; Pesquero and Fernández-Jalvo 2014). The paleoenvironment was an ancient lakeshore along which fossil accumulations were formed by attritional mortality of large mammal species consisting mainly of artiodactyls (giraffids, suids, bovids and cervids) and perisodactyls (mainly *Hipparion* and some rhinos). Carnivorous species (felids, canids and hyenids) are also present, and evidence of scavenging activity has been deduced from fossils of herbivores. The lake site was highly calcareous and provided the conditions that favored the preservation of delicate structures. The taphonomy of the macrofauna was the subject of the doctoral thesis by Pesquero (2006) supervised by Y. Fernández-Jalvo.

Cueva Ambrosio, Spain

This is a small cave shelter site located in the South-East of Spain in Almeria (Spain), and excavations have been lead by S. Ripoll (1988). Three levels of human occupation have been distinguished from the Solutrean period of the Late Paleolithic. Radiocarbon dates provided an age of 16.6 ± 1.4 ka, with abundant lithics but relatively few remains of vertebrates (mainly rabbit, and some wild goat, deer and horse). The environmental and climatic interpretation of the site suggests a temperate and humid period. The site has been considered to be a stone tool workshop with a high variety of raw material with different qualities and traits collected from the immediate vicinity to the site. The taphonomy of the macrofauna was studied by Y. Fernández-Jalvo.

Draycott, Somerset, UK

Draycott is the site of a single natural death of a cow. The cow fell down a limestone cliff onto a rocky substrate consisting of fallen limestone blocks and surrounded by dense low woodland on the middle slopes of the Mendip Hills (Andrews and Cook 1985). The carcass was left undisturbed except by scavengers. There is a cattle pathway at the bottom of the cliff along which cows pass regularly. The carcass was still partially articulated when first found. Trampling pressed the bones into the rocky substrate and caused multiple scratch marks as the bones were rubbed against the stones, with some bones being buried to shallow depths in the surrounding soil. The carcass was monitored annually from 1978 to 1982, when most of the bones were collected, and a second collection was made in 1984. Taphonomic processes observed were scavenging, trampling, breakage and weathering studied in detail over a period of seven years by P. Andrews and J. Cook.

Gorham's Cave, Gibraltar

Gorham's Cave is located on the Mediterranean side of the Rock of Gibraltar, at Governor's Beach, near the site known as Vanguard Cave (Stringer et al. 2000). The site, today at sea level, is a wide cave infilling of about 20 m thickness located 50 m inside in the cave, with a stratigraphic sequence of brown-reddish clays and silts. Sediments record successive Neanderthal occupations from 100 ka to the last representatives of this species, followed by *Homo sapiens* occupation levels. The top of the series has yielded fossils and artifacts of Phoenicians and Carthagenians (8th to 3rd centuries BC). Taphonomic analyses of the animal bone showed evidence of carnivore and human modifications (Fernández-Jalvo and Andrews 2000; Finlayson et al. 2006).

Gough's Cave, Somerset, UK
Gough's Cave is part of an extensive cave system in Cheddar Gorge, Mendip Hills, UK. Six human individuals from the Mesolithic Period were found with an assemblage of animal bones during excavations conducted by a team from the Natural History Museum (Stringer 2000). There were 269 fossils consisting of six human, ten equid and six cervid individuals found located between a large rock and the cave wall in sediment of fine gravel and silt. The bones were mixed together and extensively broken, and they included two partial human rib cages, all with a variety of percussion, chop and cut marks (Andrews and Fernández-Jalvo 2003). Human-induced damage was the main taphonomic process, and the similarities in modifications of human and animal bones suggests that they were all the products of human butchery and that cannibalism was being practiced. The evidence for cannibalism was studied by both authors of this book.

Grimstone, Dorset UK
Grimstone Down is an area of chalk downland in south Dorset (Andrews 1990). Predator accumulations include small mammals and remains from natural deaths of deer (*Capreolus capreolus*). Small mammals assemblages are the product of mammalian predators, particular fox (*Vulpes vulpes*), and the deer remains are probably scavenged by foxes also.

Ibex Cave, Gibraltar
Ibex Cave is a small rock shelter on the upper slopes in the Rock of Gibraltar. Taphonomic evidence suggested that humans were involved in the site as stone tools were recorded, but no evidence of butchery on the fauna has yet been found (Fernández-Jalvo and Andrews 2000).

Jebel Barakah, United Arab Emirates
Jebel Barakah is a low hill on the shores of the Arabian Gulf in western Abu Dhabi 24° N latitude. The hill is made up of late Miocene sediments, marine at the bottom trending into terrestrial deposits up the section. A single camel carcass was located soon after death in a remote valley, and the bones of the skeleton were monitored in five survey periods from 1984 to 1998 (Andrews and Whybrow 2005) to record bone dispersal, weathering and burial. After initial

scavenging by jackals the skeleton was undisturbed and parts of it were either dispersed down the valley, transported by occasional but usually violent rain storms, or buried by sediment washed down from the upper slopes. Taphonomic processes observed over the period were weathering, transport and burial (Andrews and Whybrow 2005).

Kajiado, Kenya
Sukuta hyena den is 4 km west of Kajiado. The den is situated half way up a steep rocky slope about 16 m high with very large boulders of in situ lava. There are two entrances close together and connecting beneath rocks into a low tunnel at least 10 m long. i.e., this is a true cave in rock (Sutcliffe archive, Natural History Museum). It was found to be accessible for about 3 m; but potentially it could be penetrated further by removing sediment from the floor. Numerous bones were visible inside the cave, mostly pushed against the sides of the tunnel. Human remains, with occasional giraffe and horse, were found, mostly fairly complete but with the ends sometimes gnawed. Many bone splinters were present. Bones were present as far as it was possible to see into the interior of the cave, including a human skull at about 8 m. Taphonomic modifications include hyena breakage and tooth marks studied by P. Andrews.

Langebaanweg, South Africa
Langebaanweg is located on the Cape west coast, where phosphate mining operations uncovered one of the richest fossil sites in the world, dated ca. 5 Ma (Matthews 2004). The site has yielded a wide range of fossils of animals now extinct. The past environment in the area included riverine forests, wooded savanna, the adjacent sea and offshore islands. Over the past 40 years, fossil bones of 200 species of animals, many of them new to science, have been recovered. Large carnivores inhabited the area together with the elephant-like gomphotheres (*Anancus*) and an okapi-like giraffid (*Palaeotragus*), antelopes such as *Miotragus*, three toed equid (*Hipparion*) and Sivatheres, a horned giraffe. The excavators have left fossils exposed as part of an in situ palaleontological museum. Fossils provided evidence of many taphonomic surface modifications and deformations related to wet environments. Taphonomic processes are currently under study by A.K. Behrensmeyer, previous to which, the Taphonomic Analytical Working Group of the RHOI (Revealing Hominid Origins Initiative) project (C. Denys, P. Andrews, Y. Fernández-Jalvo, A. Louchard, T. Matthews and D. Reed) was involved in a preliminary taphonomic study of the site.

Mumbwa Cave, Zambia
Mumbwa Cave has multiple entries in a limestone inselberg situated in a shallow basin. The cave is phreatic and extends about 60 m into the limestone, and sediment infill is

Table 2.1 Neuadd environments

	N	pH	Altitude (m)
Exposed environments			
Rough grazing/grass	11	3.5	380
Bracken moorland	12	3.6	460
Heather moorland	4	4.0	490
Stone scree	5	–	370
Closed environments			
Woodland with leaf litter	8	5.1	270
Scrub with ground vegetation	12	4.2	336
Wet environments			
Woodland-marsh	1	6.0	278
Moorland pool	3	5.1	460
Stream	4	5.0	270–380
Dry ditch	4	4.5	316

N gives the numbers of specimens in each environment

composed of outwash from the surrounding sandstone ridges. The major part of the cave infill is Middle Stone Age, with some Late Stone Age on top, and both animal bone and lithics are present in each level (Barham 2000). Worked bone is present in the Late Stone Age and may be present in the Middle Stone Age (Barham et al. 2000), and in both levels there is anthropogenic evidence of percussion marks on the fossils. Taphonomic features/processes observed were digestion of small mammal bone, percussion marks and rounding/polishing of ends of bones. The taphonomic study was made by E. Jenkins, supervised by P. Andrews.

Neuadd, Wales

A long term project was started in 1976 by P. Andrews at a near the village of Rhulen, Wales, latitude 52° N (Andrews and Armour-Chelu 1998). The purpose of the project was to investigate the rates of dispersal and decay of bones from undisturbed mammal carcasses in a temperate climate, and subsequently to record rates of weathering and burial. Over 100 specimens of domestic animal have been monitored over a period of 30 years. When the monitoring started, a surface collection of bones already present in the study area was made so that bones dispersed from monitored specimens could be more readily identified. The area consists of open moorland along the tops of the hills, with several moorland pools and areas of accumulating peat, and places where springs emerge from underground giving rise to locally wet areas of bog. The lower slopes of the hills are fenced for rough grazing but are rarely ploughed or managed in any way. Along the valley bottoms there are scattered trees, with woodland lower down. The main environments present are listed in Table 2.1, which shows the numbers of individuals monitored in each environmental type. The study area was approximately 900 ha, with 17 km of sampling lines, which were searched to a distance of 20 m on either side of the lines. Histological studies to investigate early diagenesis are currently in process (Fernández-Jalvo et al. 2010).

Monitored skeletons are numbered as Neuadd 1, Neuadd 2, etc. and samples collected from the skeletons are numbered ND1, ND2, etc. The long term monitoring over 30 years was carried out by P. Andrews (in preparation).

Olduvai Gorge, Tanzania

Olduvai Gorge has a long sequence of terrestrial sediments from Early to Late Pleistocene with abundant animal remains and human artifacts. The sediments have been described by Hay (1976) and the taphonomy by Fernández-Jalvo et al. (1998), Blumenschine et al. (2007, 2012), and Domínguez-Rodrigo et al. (2007), among others. There is abundant evidence of human and carnivore activities, small mammals accumulated by avian and mammalian predators, and many other taphonomic processes. The small mammal taphonomy of the FLKN and FLKNN excavations by Mary Leakey was studied by Y. Fernández-Jalvo, P. Andrews and C. Denys (Fernández-Jalvo et al. 1998). More recent excavations have yielded new small mammal fossils which are being studied by P. Andrews and Y. Fernández-Jalvo.

Olkarian Gorge, Tanzania

This is a narrow gorge with high cliffs inhabited by a large population of Ruppell's vultures (*Gyps rueppellii*). Evidence of incidental hyena incursions is indicated by regurgitations. These incursions, however, may not be prolonged as vultures defend their territory and nestlings, probably preventing hyenas from entering frequently. The main action observed along this gorge is vultures' and other predators' prey remains and barn owls nesting at the end of the gorge, and we made taphonomic investigations into the bone collecting and taphonomic modifications produced by vultures (P. Andrews and Y. Fernández-Jalvo unpublished data).

Overton Down, UK

Overton Down is an experimental earthwork built in 1960 (Bell 1996). Included within the earthwork when it was built were human artifacts and animal remains, and the intention was to investigate the effects of burial on these objects under experimental conditions. To this end, the earthwork has been sectioned on a progressive scale, i.e., in 1962, 1964, 1968, 1976 and 1992. The next section will not be made until 32 years after the last one, i.e., in 2024. Two animal bones were available from the 1992 excavation (Armour-Chelu and Andrews 1996), and these showed modifications from microorganism and insect attack and cracking and degradation from high humidity. More animal skeletons are still buried and will be available in the following years for taphonomic studies, lead by P. Andrews.

Paşalar, Turkey

The Paşalar deposits are Middle Miocene in age. They accumulated in a shallow basin as a flood deposit transported a short distance from the surrounding hills, where the bones had been accumulating for an unknown but probably brief period of time (Andrews and Ersoy 1990). Much breakage occurred before transport, and the weathering profile suggests that the bone accumulation was in equilibrium with the environment: that is all weathering stages were equally represented showing that the initial period of accumulation was sufficiently long for bones to enter the system and degrade and disappear as a result of weathering (Andrews 1995). Weathered bones were abraded during transport to the site, but fresh bones were not abraded, indicating that the transport duration/distance was not great. More complete bones transported to the site suffered cracking from variable humidity, and in some cases the bone disappeared completely after deposition, for example leaving tooth rows in anatomical position with no trace of bone. Finally, there is evidence of carnivore activity on both large and small mammals. The taphonomic study of the site was carried out by A. Ersoy, supervised by P. Andrews.

Riofrío, Spain

The Bosque de Riofrío is a natural reserve of about 700 ha located in the foothills of the Northern slope of Guadarrama Mountain, 41° N latitude, in Segovia (Spain). This is a natural reserve where hunting practices are forbidden. Since 2000 Y. Fernández-Jalvo in collaboration with I. Cáceres and M. Esteban (Universidad Rovira i Virgili, Tarragone, Spain) have been carrying out an experimental project in the forest (Cáceres et al. 2007, 2008, 2011). The main objectives are to study the natural dispersal of carcasses and to evaluate the activity of vultures and foxes. At present, the study focuses on 45 carcasses (fallow deer and red deer) to which these carnivores had access, making observations on the disarticulation and dispersal of bones of these two species of deer. The aim of the project is to find a consistent pattern that may differentiate between large avian and small terrestrial scavengers. Further observations have been made on other taphonomic processes, such as weathering, soil corrosion,

water abrasion and plant or micro-organism attack) This builds upon and expands experimental studies carried out at Neuadd (Andrews and Armour-Chelu 1998) and at Amboseli (Behrensmeyer 1978). The Riofrío natural reserve has different environments of forest, open grassland, seasonal streams, and in general there are low trees with almost complete absence of bushes due to cervid browsing.

Rudabánya, Hungary

Rudabánya is a Late Miocene site in north eastern Hungary with a large and diverse assemblage of fossil animals and plants (Kordos and Begun 2002). The sediments were laid down in two cyclical events with fossil soils forming during falls in lake level of the Pannonian Lake in what is now the Pannonian basin. Initial clays formed under wet conditions were succeeded by lignite formation under swamp conditions, and this was succeeded by transgression of the lake, then lake flats when the lake level fell again and muds and lignites formed above these (Andrews and Cameron 2010). Abundant plant remains occur throughout, many in growth position, but sediment compaction and limited transport have resulted in the breakage and abrasion of many of the fossils. Taphonomic modifications particularly include soft-sediment deformation of the fossil bone, studied by D. Cameron, supervised by P. Andrews.

Rusinga Island, Kenya

Early Miocene sites on Rusinga Island have sediments derived from alkaline carbonatite volcanoes and are preserved as flood plain over bank deposits with soil formation. Fossils are well preserved, occasionally retaining casts of soft parts of the body, and they occur as pockets of fossils dispersed through the sediments (Andrews and Van Couvering 1975). Large and small mammals are abundant, and some levels also contain plant remains (Collinson et al. 2010, Maxbauer et al. 2013). Both forest and woodland environments are indicated at the site, and both are associated with fossil ape remains (Collinson et al. 2010, Michel et al. 2014). The taphonomy of the site was studied by J. Van Couvering (now J. Harris) and P. Andrews.

Senèze (Haute Loire, France)

This site was formed in a Pliocene maar (volcano crater) filled by a lake during the Villafranchian (Nomade et al. 2014). The site represents the international biochronological reference of MNQ 18, with one of the most important Pliocene collections of fossil vertebrates of this period and several holotypes. It was the subject of international collaborative fieldwork from 2000 to 2006, whose taxonomic, geological and taphonomic results are currently in preparation (Delson, Faure & Guérin in prep.). It has yielded thousands of fossil vertebrates, several complete and partial skeletons, some in anatomical position, and taphonomic investigations show a high incidence of damage by falling blocks and trampling by animals that mimic carnivore chewing (Fernández-Jalvo et al. in prep.).

Songhor, Kenya

Early Miocene deposits with extensive soil development producing alkaline red paleosols. The stratigraphy is much interrupted by minor faulting (Pickford and Andrews 1981). Fossils are abundant and well preserved, with several intact skulls of small mammals, but all fossils are disarticulated. The taphonomy was studied by P. Andrews.

Spitalfields (UK)

The medieval cemetery site in the crypt of Spitalfields church is a large collection of human skeletons, many with known age and sex and some named individuals with documented relationships with others. The crypt was in use from 1729 until 1859. Several types of coffin were used, including wood, where soft parts have decayed except in some cases for the hair, and lead coffins where much organic matter has survived (Molleson and Cox 1993). Insect damage, damage from decaying hair, and decomposition products are common. The taphonomy was studied by T. Molleson, in the NHM collection.

Swildon's Hole and Wookey Hole, Somerset, UK

Swildon's Hole is a long phreatic cave in the Mendip Hills that exits via Wookey Hole several kilometers away. Accumulation of recent bones in the cave has been investigated to see how bones enter, are preserved and are dispersed in cave systems (Andrews 1990). Water flow from Swildon's Hole comes out at Wookey Hole, but the bones collected from the latter site come from the upper chamber, well above the present water table of the system. Some came from an old phreatic chamber that exits through an opening approximately 15 m above the present cave outlet. The cave was excavated and bones collected by E. Andrews and P. Andrews.

Tianyuandong, China

Tianyuandong is a Late Paleolithic site, 6 km from the core area of the Zhoukoudien Complex (site 27). Tianyuandong (or Tianyuan Cave) yielded one fossil skeleton of *Homo sapiens* (43–39 ka, among the oldest in China). Together with the human skeleton, abundant animal remains were found, but no stone tools were recovered. The animal fossil remains are extremely fragmented, in contrast to human skeleton elements that are, for the most part, complete (Fernández-Jalvo and Andrews 2010). A more comprehensive taphonomic study of the human and faunal remains was undertaken by P. Andrews and Y. Fernández-Jalvo and is now published (Fernández-Jalvo et al. 2015).

Troskaeta, Spain

Cave site in northern Spain with an assemblage of cave bears (*Ursus spelaeus*). Adults and many juveniles and infants are preserved, but no other carnivore species (Pinto Llona et al. 2005). It is likely that this cave was a denning site inhabited by adult bears with their young. It is inferred that tooth marks present on the cave bear bones were produced by the cave bears themselves, and 25% of the bones had tooth marks.

Tswalu Kalahari Reserve, South Africa

This Reserve is approximately 100,000 ha in size. It is situated in the Northern Cape Province, South Africa, about 100 km north-west of Kuruman. The grid references are between 27° 04′ S and 27° 33′ S latitude and 22° 10′ E and 22° 36′ E longitude. The terrain consists of plains, parallel dunes with lowlands between them and hills, and includes the Korannaberg mountains and hills in the north and east. There are sandy valleys between these hills. Extensive sandy plains with dunes and dune streets occur to the west and south of the Korannaberg mountains. A few pans and other depressions are found in the extreme west of the reserve. The altitude of the area varies from 1020 m near Rogela pan in the west, to 1586 m at Drogekloof on the Korannaberg in the east. Taphonomic studies on pollen taphonomy and brown hyenadens are in progress. Tswalu's dry environment preserves animal carcasses and scats, providing taphonomic patterns of preservation. We are making seasonal collections of scats from an area around a hyaena den. The study includes experiments in an environmental chamber in Madrid to determine what factors favour pollen preservation in coprolites (Y. Fernández-Jalvo unpublished data).

Vanguard Cave, Gibraltar

Vanguard Cave is one of the archaeological sites located at Governor's Beach on the eastern side of Gibraltar. It provides a rich record of Neandertal everyday life in combination with Gorham's Cave. Vanguard cave is a large shelter that contains a stratigraphic series of over 17 metres, consisting of coarse sand layers interstratified with sterile aeolian sand layers (Goldberg and MacPhail 2000) as well as brown silt and silty-clay lenses. The latter are frequently associated with periods of human occupation. The age of the Vanguard sediments is >41,800 radiocarbon years, based on a series of dates on charcoal from the top of the sequence. The taphonomic study provides evidence that humans fed on marine mammals (Stringer et al. 2008) and indications of human behavior and activity at the site (Stringer 2012; Cáceres and Fernández-Jalvo 2012). The taphonomy of large mammals was part of the Doctoral Thesis of Cáceres (2002) supervised by Y. Fernández-Jalvo.

Westbury Cave, Somerset, UK

Westbury Cave is a two-chambered cave with deposits dating back to the earliest middle Pleistocene (Andrews et al. 1999). The lowermost cave infills consist of water-transported sands and gravels older than 730 ka, common to both chambers, and succeeded by two separate sequences of cave breccias with varying amounts of silts and clasts. Two cave bear (*Ursus deningeri*) denning areas have been identified (Andrews and Turner 1992), at least two levels with water-transported animal remains, several small mammal assemblages accumulated by predators, and a major influx of surface material transported into the cave together with its biologic contents (Andrews et al. 1999). Included with the animal bones is a dispersed flake industry providing evidence of human presence in the cave, and cut marks have been found on the bear and cervid fossils (Andrews and Ghaleb 1999). Taphonomic modifications include breakage, trampling within a den, weathering, digestion by predators, and abrasion.Taphonomic investigations were carried out by B. Ghaleb, supervised by P. Andrews.

Wonderwerk Cave, South Africa

Wonderwerk is a large cave situated in the Kuruman hills (27.50° S; 23.33° E) on the western edge of the Ghaap plateau in the northern Cape Province of South Africa. It has a well-preserved fauna and flora in a continuous sequence from Early Pleistocene (2 Ma) to Holocene deposits (Chazan et al. 2008). Taphonomic studies on small mammals provide information about the predators involved in the site as well as possible evidence of fire (controlled/natural) and site formation (Fernández-Jalvo and Avery 2015).

Special Taphonomic Reference Collections

Behrensmeyer collection Several specimens from experimental work and monitoring studies from Behrensmeyer´s collection have been included to compare and describe some modifications, especially fluvial abrasion and insect damage, from sites including Amboseli Park, Kenya, South Plate, Lost Creek and Calamus rivers (USA). Amboseli occupies a flat basin covering approximately 600 km^2 to the north of Mount Kilimanjaro in Kenya. The climate is semi-arid, but there is an extensive swamp area fed by springs from Kilimanjaro. Over 20 different habitats have been recognized (Western 1973), ranging from woodland to open grassland, but it is a dynamic ecosystem subject to changes in temperature and rainfall patterns and fluctuations in the water table. Much of the woodland is disappearing, being replaced by edaphic grasslands as the water table rises and deeper-rooted trees die off. The naturally occurring vegetation beyond these limits is low *Acacia* woodland. The area has been the subject of a long-term taphonomic monitoring project since 1975 (Boaz and Behrensmeyer 1976, Behrensmeyer 1978, 1981, 1983, 1993, 2007; Western and Behrensmeyer 2009) to investigate aspects of bone dispersal, weathering and burial, and the modifications arising therefrom. Taphonomic modifications observed include a wide range from time of death to burial, and they cover a variety of environments and periods of time sufficiently long to show how processes such as weathering impact the animal bone. The actualistic taphonomic collection from Amboseli was available to Y. Fernández-Jalvo during a predoctoral grant in 1988 supervised by A.K. Behrensmeyer.

E.M. Geigl has provided specimens from sites in Europe, Asia and Near East that were histologically studied and analyzed for fossil DNA in the context of a wider project of research into the domestication of horses and cattle (mainly cows and sheep).

Gómez collection The small mammal actualistic taphonomic studies were performed by G. Gómez (2000). He conducted experiments with different raptors and mammalian carnivores fed with rodents at the Zoo of Olavarría (Argentina) which provides a reference collection of Argentinean predators.

Anthony Sutcliffe collection This is part of the taphonomic collection at the NHM and includes several actualistic collections from Holartic and African environments investigated by P. Andrews (Sutcliffe 1970).

Part I
Surface Modifications

Chapter 3
Linear Marks

Linear marks are defined as marks with lengths four times their breadths and longer. They are grooves penetrating the surface of the bone either by incision, where the bone tissue is cut into, or by chemical solution, where the bone tissue has been penetrated by chemical action.

Agents and Processes

The distinction is made throughout the book between inorganic processes, such as the effects of agents like stone or the weather, and organic processes entailing animal, plant or microorganism agents.

Inorganic processes:

Linear marks made by movement of stone against bone or bone against stone

Cuts and scrapes made by humans A.2 and A.89
Cuts and scrapes made by animal trampling (animals are the primary agent, but the process entails movement of stone or bone against bone) A.60 and A.61
Abrasion by wind or water A.118 and A.111

Organic processes:

Linear marks made by animals

Gouges by tooth, claw or beak against bone A.155 and A.202
Linear marks made by termites and other insects A.209

Solution marks made by plants

Linear root marks made by plants A.232
Lichen, moss, algae A.262, A.258 and A.265

Linear marks made by microorganisms: fungi, bacteria A.287 and A.295

Characteristics

The principal characteristic distinguishing different forms of linear marks is their cross-sectional morphology, and we base our classification on this. The categories of cross-sectional morphology of linear marks are based on the shape, width and degree of rounding of the mark:

Linear marks with V shaped cross-section (A.1);
Linear marks with U shaped cross-section (A.231).

The following descriptions are based on these divisions, and there are ten additional characteristics that add further information as to the nature of the linear marks and help to evaluate the process by which they were formed. These additional characteristics are as follows:

Cross-sectional widths of linear marks
Multiple or single linear marks
Presence/absence of internal micro-linear marks
Linear length of linear marks
Depth of linear marks
Location of linear marks
Direction of linear marks
Frequency of linear marks on single bones or bone assemblages
Branching of linear marks
Straight or curved

Description of Linear Marks

Combining morphology of linear marks with agents and processes forms the basis for the following descriptions:

© Springer Science+Business Media Dordrecht 2016
Yolanda Fernández-Jalvo and Peter Andrews, *Atlas of Taphonomic Identifications: 1001+ Images of Fossil and Recent Mammal Bone Modification*, Vertebrate Paleobiology and Paleoanthropology, DOI 10.1007/978-94-017-7432-1_3

Inorganic Linear Marks with V Shaped Cross-Section Made by Stone

Linear marks with V shaped cross-section on bone are produced by stone objects, which are harder than bone, and the agent can be either human action with a tool, movement of rock against bone during abrasion, or movement of the bone against a hard object, for instance during trampling or transport. These marks penetrate the outer surfaces of bones or teeth by physical action, and in the process they remove some tissue and displace other tissue. Stress patterns set up during the incision accumulate tissue in front of the cutting edge, and this tissue is periodically displaced to form Hertzian fracture cones (A.1) along both sides of the striation (Bromage and Boyde 1984). They are normally only evident along one side of the cut (Bromage et al. 1991), but they can occur on both sides (A.5). The cross-section of the linear mark is usually asymmetric, with one side steeper than the other, and in this case the Hertzian fracture cones are more prominent on the shallow side that is exposed to greater friction, and displaced bone may build up on this side to form a raised shoulder of bone running alongside the linear mark. In the absence of abrasion, this displaced bone may remain in place in fossils. The formation of Hertzian fracture cones indicates the direction of the cut (Fig. 3.1). Directionality of the cut may also be indicated by supination at the end of the linear mark, whereby the striation curves to one side as contact with the bone by the object making the mark ceases (Fig. 3.1). Directionality can also be determined for cut marks by the buildup of "pseudo-steps" formed in the interior of the cut in the direction from which the mark was made (Bromage and Boyde 1984). Pseudo-parallel thinner linear marks accompany the primary striation and were named by Shipman and Rose (1983) as shoulder effect, formed by irregularities of the cutting edge.

The size of the linear mark varies depending on the size and shape of the producing agent, being narrow with a sharp blade such as an obsidian flake, and broader with a core stone tool such as a hand axe or a retouched stone blade (Fig. 3.2). Depth and breadth of linear marks are related, with deeper cuts also being generally broader. Sharp single edges such as flakes, or metal tools also produce single striations without any additional marks outside the main striation but in general micro-striations are present within the main linear mark when marks are made by stone. Stone edges with uneven surfaces or retouched edges produce multiple striations both within a single major cut, because

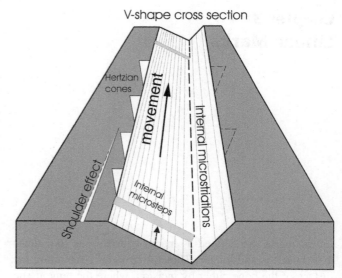

Fig. 3.1 Profile of a cut mark showing Hertzian cones and asymmetry of cuts; the arrow shows the direction of the cut. The profile of the cut is V-shaped, with shoulder effect and Hertzian cones formed on one or both sides of the cut. Shoulder effect is produced by irregularities of the stone tool edge arising from supination of the hand during cutting. Internal striations are normally formed along the length of the cut and internal microsteps may break up the smooth outline of the cut mark. Hertzian cones and microsteps provide criteria for directionality of the cut

several parts may contact the bone surface simultaneously, and outside the main cut, where protuberances from the stone edge come in contact with the bone. The former may cut deeply into the bone, but the latter produce shallow striations alongside the main linear mark. The widths of marks made by stone depend on the characteristics of the stone edge and the depth to which the linear mark penetrates, and width also depends on the narrowness of the stone edge. Narrow linear marks can be produced both by butchery and by trampling, as can broader, multi-striated marks (Andrews and Cook 1985; Behrensmeyer et al. 1986; Olsen and Shipman 1988; Shipman and Rose 1984). Different raw materials also produce different features of striations made by stone tools (Fernández-Jalvo and Cáceres 2010). Most cut marks are found on large mammal bone, and it is unusual to find them on small mammals, the bones of which would be likely to be destroyed in the process of cutting. One example we know of comes from Olduvai Gorge (Fernández-Jalvo et al. 1989).

Linear marks with V shape cross-section can be the result of natural agencies as well as through human activity (Domínguez-Rodrigo et al. 2009). Trampling of bones resting on a stony substrate may produce cut mark mimics as

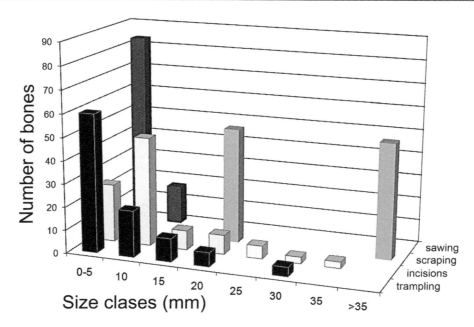

Fig. 3.2 Comparison of lengths of trampling marks versus human-caused cut marks, sawing marks and scrape marks. Number of bones is shown on the vertical axis, and size classes in mm is on the horizontal axis

the bones rub against sharp pointed stones (Andrews and Cook 1985; Behrensmeyer et al. 1986; Olsen and Shipman 1988). The marks are made by the movement against stone, not by the animals themselves, and they may closely resemble cut marks, including features such as long linear striations, Hertzian fracture cones developed (A.62) and internal striations within the main cut. They differ from cut marks, however, in being generally less deep, shorter, and very much more frequent (Fig. 3.3), with little or no proximity to areas of tendon or muscle attachment, and having preferred direction transverse to long axes of long bones (Fig. 3.4) (Shipman and Rose 1984).

Broader areas of linear marks are generally referred to as scrapes, and they can be caused either by movement of stone across a bone surface, or by movement of the bone against the sediment or a firmly fixed stone. The former could be a stone tool in the hands of a human (A.89), scraping bone surfaces to clean the bone or the stone edge, or to separate muscle fibers from the bone by moving stone tools transversely across bone. The latter entails movement of bones against a stony substrate by trampling or transport and may involve an isolated large rock or the side of a cave wall (A.93). In both cases, the morphology of the individual linear marks that make up the single scrape may vary

considerably, with some linear marks being typically V shaped, but some with more U shaped profiles. Several linear marks grouped together when a tool is used in a sawing action can also produce marks more like scrapes, but this is because of the close proximity of the individual marks. In general, however, scrape marks produced by different agents tend to be similar to each other, whatever the agent, but a small difference is that trampling marks tend to be randomly placed, while those made by humans may have some reference to muscle attachment (A.102) (Olsen and Shipman 1988). Trampling marks frequently run transversally to the long axis of the bone (A.60), while scrapings done by humans are often oblique or parallel to the long axis of the bone.

Linear marks may be simple or compound, grouped or independent, and they sometimes appear similar to decorative marks made by humans. Multiple marks may also be made by sawing action, where a bone is contacted several times in succession when tough tendons or muscle are being detached from the bone. This is seen as several cuts grouped together that may be continuous with each other. When this is done repeatedly along the length of a bone, the effect can appear similar to decoration of bone. Figure 3.5 shows a human radius with a row of compound marks along its

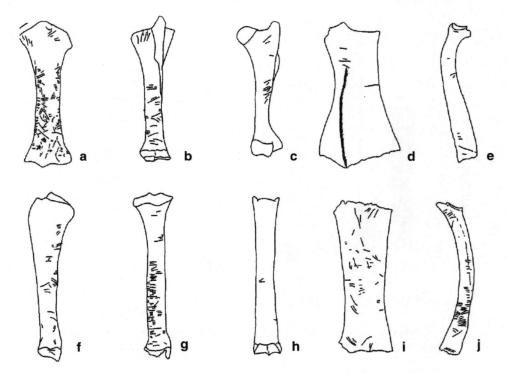

Fig. 3.3 Ten modern bones with trampling marks (short dark lines) identified during long-term monitoring of the specimens. The identifications of the bones and lengths of time they were exposed and monitored is as follows: **a** Draycott cow, 8 years; **b** Neuadd horse, 15 years; **c** Neuadd horse, 15 years; **d** Grimstone deer, 4 years; **e** Draycott cow, 8 years; **f** Neuadd horse, 15 years; **g** Neuadd sheep, 6 years; **h** Neuadd horse, 15 years; **i** Neuadd sheep, 8 years; **j** Draycott cow, 8 years

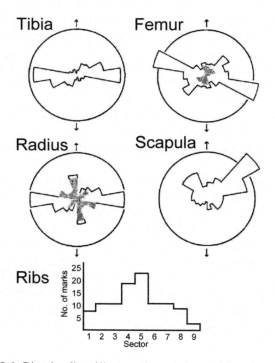

Fig. 3.4 Directionality of linear marks made by trampling of cow bone at Draycott, Somerset. The rose diagrams have arrows indicating the long axis of the bones, and the trampling mark directionality is mainly transverse to the long axis. The scapula marks are oblique on both blade and the neck of the scapula. The diagram for the ribs (bottom) has the long axis of the ribs signified by the numbers 1.9 on the x-axis, with 5 being transverse to the long axis

length that were first thought to be decorative marks made by humans (A.27). Closer examination shows that each of the marks is a compound mark made with a broad edge, with several striations made by a single movement converging to a point in each mark (Andrews and Fernández-Jalvo 2003). In some cases, particular types of tool can be identified with a particular form of compound mark, e.g., stone tools with retouched edges typically produce single striations with X shaped pattern (A.24) (Schick and Toth 1993).

Single linear cuts can also appear complex when the cuts are made by a retouched tool. In this case the irregularities of the edges contact the bone if the tool is not held precisely perpendicular to the bone surface. If it is held at an angle, one side of the tool contacts the bone, producing a shorter and shallower cut alongside the main cut, and as the movement of the tool across the bone progresses the angle at which the tool is held changes, so that the other side of the tool may contact the bone at the end of the stroke.

The length of linear marks with V shaped cross-section varies considerably, and there is a difference between intentional cut marks made by humans and trampling marks. The latter tend to be short on large mammal bone (Andrews and Cook 1975, with the great majority less than 5 mm in length, for example at Draycott and Neuadd; Fig. 3.3). In contrast, many cut marks are 10 mm or more. Sawing marks

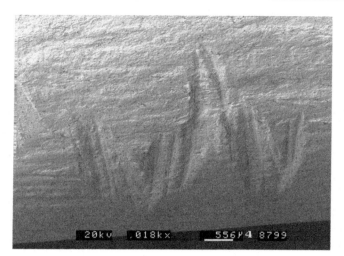

Fig. 3.5 SEM microphotograph of sawing cuts. These marks have been interpreted as engraving, but the 'groups' of incisions are actually compound marks made by a series of single strokes, with consistent directionality towards the superior aspect of the shaft. This is taphonomically interpreted as filleting of the arm muscles progressively along the shaft

have short lengths comparable to trampling marks, but the sawing action determines this and the morphology of the marks distinguishes these (A.14). Scraping marks are in general longer, and comparable to these is the special case of linear marks with V shaped cross-sections produced on bones modified into points by human action. These can be very long and straight (A.10) and are oriented along the length of the bones as they taper the end of the bone to a point (Barham et al. 2000). Linear marks made by abrasive agents such as wind or water tend to be short and have comet like shape (d'Errico et al. 1984).

There are few data on depths of V shaped linear marks. It is readily apparent that cut marks may penetrate more deeply into bone than trampling marks (Olsen and Shipman 1988), and although this is not always the case (Behrensmeyer et al. 1986), we have no quantitative data to demonstrate this.

The location on bones of V shaped linear marks shows a marked difference between cut marks and trampling marks. The former can usually be related to parts of bones where muscle or tendon insertions are present, and slicing or scraping marks may be located at concave protected areas that are not reachable by trampling unless the salient bone angles are damaged (A.12). The purpose of the cut is to detach muscle, tendon or skin from the bone during butchery

or defleshing. Trampling occurs randomly on exposed bone surfaces, especially on convex surfaces. It should be remembered that cut marks are often the result of accidental contact between the bone and the carefully shaped edge of the stone tool. The aim of the tool is to detach soft tissues, and scratching the edge against hard bone will reduce the sharpness of the edge (Olsen and Shipman 1988). Cut marks therefore tend to be few in number, whereas trampling marks may be numerous and their location has no relation to the insertion of muscles (Andrews and Cook 1975). Instead they tend to be concentrated along the shafts of long bones, and they form as the shaft rotates around its axis when pressure is applied by large animals trampling the bone (Fig. 3.3).

Orientations of V shaped linear marks relative to the long axis of the bone also differ between cut marks and trampling marks. The former can be in any direction, but since they are located at or close to epiphyses as noted above, they tend to cut across the line of the epiphysis and the articular surface so that they are oblique to the long axis of the shaft. Trampling marks on the other hand may have strong preferred orientation transverse to the long axis of the diaphysis of long bones (Fig. 3.4) (Andrews and Cook 1975). This is because the marks often are formed as the long bone rotates around its axis due to trampling, and contact with sharp-edged stones produce transverse marks on all surfaces of the diaphysis. Trampling marks have also been found to be mainly transverse on ribs, but on bones with more complex shapes they may form in any direction.

Frequency of linear marks is a contentious issue (Lyman 2008), being low on cut marked bones, but sometimes very high on trampled bones (Fig. 3.3). Nine ribs from the trampled skeleton from Draycott (Andrews and Cook 1985) had between them a total of 108 linear marks, ranging from 3 to 24 per rib, and 11 long bones had 510 linear marks, with over 100 each on the femur and tibia. A horse skeleton from Neuadd (Andrews and Armour-Chelu 1998) had five long bones with linear marks resulting from trampling ranging from 3 to 47 per bone, and a sample of five sheep bones also from Neuadd had an average of 15 linear marks per bone, the maximum being 68 marks on one tibia (Fig. 3.3). Our records from sites of large mammal skeletons that bear evidence of human induced damage (Gorham's and Vanguard Caves, Coldrum, Gough's Cave and TD6 Atapuerca) show numbers of cuts per long bone (limbs, metapodials and ribs) are usually less than 5 cuts per specimen and rarely as many as 20 (depending on the proximity of muscle or ligament attachment). Usually, long bones that have been butchered are broken during the final stage of butchery to extract the marrow, so our data are based mainly on bone fragments, but even when anatomical elements are almost complete, or when bone fragments are refitted, the number of cuts per specimen does not change, because not all refitted fragments bear cut marks and most of these cuts are isolated incisions. It is also interesting that these cuts are long and may reach more than 4 cm length, which is rarely seen for trampling marks.

Linear marks with V shaped cross-sections lack branches. Some cut marks appear to divide into two or more branches, but these are near the ends of strokes and result from supination of the cutting tool as has already been mentioned. A tool with rough edges may bring other parts of the edge into contact with bone as it rotates in the hand, resulting in the apparent division of the cut. Most V shaped linear marks are straight, but longer linear marks in particular may be curved (A.4). Cut marks may be obscured by weathering or abrasion, but detailed comparative work has not yet been done. On the other hand, weathering cracks covering the bone surface but not present on the surface of potential cut marks is enough evidence to dismiss them as cuts and consider them as trampling marks that were made after exposure to weathering.

There is also considerable variation in morphology of cut marks made using different raw materials. Shells may be used for cutting (A.44), and they usually produce narrow V shaped cuts (based on experimental cuts with shells broken to form a sharp edge).

- Characteristic traits of linear marks made by shells are: internal microstriations, few or no lateral hertzian cones, and frequent shoulder effects (A.45). The size of the cuts is bigger and longer than cuts made with quartzite or flint.
- Limestone tools produce a variety of marks ranging from V shaped to U shaped or even U shaped with flat bottoms. Characteristic traits include: internal microstriations, lateral hertzian cones, frequent shoulder effect due to the coarse grained raw material that produces more irregularities in the edge of the tool and a wide cut mark (A.42). The cuts are large and resemble trampling marks, which is not surprising when it is considered that the trampling marks depicted here were all made in a limestone area. Limestone stone tools used during butchery quickly become blunted so that the sides of the cutting edge touches the bone surface more frequently, marking the bone surface with multiple cuts. Limestone is not a good raw material for tool manufacture, but it is a readily available in limestone caves, and cut marks made by limestone tools suggests an immediate need for tools.
- Flint and obsidian, which can be worked to produce fine cutting edges, almost always produce thin V shaped cuts. These variations are illustrated in detail (A.36, A.37, A.38 and A.39).

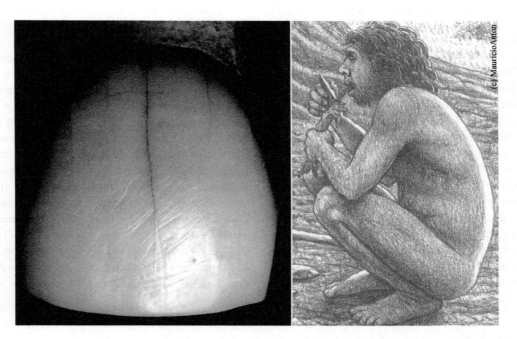

Fig. 3.6 Cut marks may also affect teeth. Striations shown on the buccal surface of this upper incisor are visible with the naked eye (left hand figure). Striations have an oblique distribution and a recurrent orientation. Striations are absent on the lingual surfaces and only affect anterior teeth (upper and lower incisors and canines). All these traits suggest that these striations were made while holding meat between the teeth and cutting pieces off before ingesting (reconstruction, right hand figure). The scratches run from top left to bottom right (with a characteristic rough lateral margin of the striation) indicating a right handed person. Specimen from Sima de los Huesos, Atapuerca

Linear Marks on Teeth

Cut marks may also affect teeth. Two sites (Abric Romani and Gough's Cave) have yielded horse mandibles with teeth cut obliquely on the buccal surface. The aim of such cuts is not clear (Cáceres 2002; Andrews and Fernández-Jalvo 2003) except for ligament extraction or skinning. Cut marks have also been observed on human teeth (Fig. 3.6). Striations appear on the buccal surface on the anterior teeth and are visible to the naked eye (A.105). Human teeth from several sites (Bermúdez de Castro et al. 1988) show oblique striations with a recurrent orientation. Striations are absent on the lingual surfaces and are only present on anterior teeth (upper and lower incisors and canines). All these traits suggest that these striations have been made while holding meat between the teeth and cutting pieces off before ingesting, scratching the teeth accidentally. When scratches were made, the teeth were in anatomical position and the hand also in relative anatomical position with the teeth. This situation has allowed us to characterize traits that indicate handedness (Bermúdez de Castro et al. 1988), and indirect evidence of brain lateralization (Bromage et al. 1991) (A.2, A.3, A.4 and A.5).

Similarly, criteria of handedness based on the orientation of cuts on butchered long bones have been proposed (A.3 and A.4) (Bromage et al. 1991; Noll 1995). In these cases, however, the anatomical position of the cut surface and the hands of the agent are not as constrained as is the case of these teeth. This is applicable in very limited situations, if and only if cuts are made before dismembering. In this case, the butcher gained access to the bone either at the bottom or at the top of the animal carcass, so that right handed cuts are orientated from top left to bottom right, and left handed cuts are oriented from top right to bottom (Noll 1995).

Organic Linear Marks with U Shaped Cross-Section Made by Animals

Linear marks with U shaped cross-section are commonly made by animal chewing. This entails physical cutting into the surface tissue of the bone in a manner analogous to that of cut marks and trampling marks. Unlike these, however, the marks left by teeth are often more abrasive, for their cutting edges are less sharp than those of stone tools. Here we must distinguish between the striations produced by incisors, which have a scraping action across the surface of bone, and striations produced by canines and premolars/molars, which penetrate more deeply into the bone.

Incisor Gnawing

This is typically seen with rodents (A.188), but it is also a characteristic of humans when using the incisors on the surface of bones (A.131). The width of the incisor marks depends on the size of the animal or human (A.369), although human incisor marks always seem to be narrower than the actual incisor width. Experimental incisor marks on animal bone produces a linear mark that starts off with superficial micro-striations and passes into a more abrasive stage, with displacement of bone in the direction of the bite and with greater displacement on one side than on the other, ending up with a deeper puncture which also has larger and more jagged micro-striations within it. Human incisors produce shallow linear marks (with internal microstriations) sometimes associated with shallow crescent pits on one end of the scratch, not easily distinguishable to the naked eye (A.139), and not as deep and distinct as tooth scores made by carnivores. A consistent pattern has been observed on an ethnological collection of bones chewed by humans (Koi people, Zoutrivier Village, Gobabeb; Brain 1969, 1981) and from experimental conditions with 18 different people eating meat off bones (Fernández-Jalvo and Andrews 2011). In the fossil record, striations resulting from human incisors has been recorded for fossils from Atapuerca TD6 and Gibraltar sites (Fernández-Jalvo and Andrews 2010).

Rodent incisor marks are common in the fossil record, and they are usually formed by the upper incisors, which generally have a flatter cross-sectional profile than the more pointed lower incisors. It is often the case that bones with flattened incisor marks on one side may have numerous punctures on the other side of the bone where the opposing lower incisors anchored the bone during the chewing process. Incisor widths vary according to the size of the rodent species (Tong et al. 2008), and even within one size group there are variations with some species having relatively broader incisors than others (A.189 and A.201). Some rodent species also have grooved upper incisors, and the grooves may sometimes be apparent on the chewed bones.

Rodent incisor marks are always unbranched, although the superimposition of marks may give the impression of branching or multiple marks. They vary in length from long to short, often on the same bone, and in general the shorter the mark the deeper it is. The location of marks has no relation to muscle insertions, although some rodents do eat meat (Rabinovich and Horwitz 1994; Klippel and Synstelien 2007). Other reasons for rodent gnawing are to gain minerals contained in the bone, and possibly to hone or wear their teeth (Brain 1981; Kibii 2009). Rodent gnawing is common on broken edges of bone. Transverse marks on long bone diaphyses are also common, as are inferior borders of mandibles, but gnawing of articular surfaces and epiphyses is less common. Bones gnawed by rodents generally have large numbers of marks that are consistent in size and morphology and are in parallel sets on the bone. These features are definitive diagnostic criteria by which rodent gnawing can be recognized.

Canines and Premolars/Molars

Linear marks made by the posterior teeth of carnivores are frequently produced by single cusped premolars. Linear marks produced by multi-cusped teeth are rare in our experience, although there are examples of punctures produced by them. We have developed a simple scheme for relating different types of carnivore chewing marks to location on bone, for we have found that their size and shape are directly affected by location (Andrews and Fernández-Jalvo 1997). The common type of linear mark is on diaphysis surfaces and broken ends of bones rather than on breaks or articular surfaces. Linear marks are indicated by the letter b in our classification, the full description of which is given in Chap. 4.

> b = linear marks on surface of bone diaphyses (A.155)
> b1 = linear marks on broken ends of bones (A.156)
> b2 = linear marks on articular bone surfaces (A.157)

The cross-sectional widths of linear marks caused by carnivore chewing are generally small when located on hard surfaces of diaphyseal bone. As for cut marks, when animals chew bone they run the risk of breaking or wearing down their teeth, and unless they have teeth adapted for bone penetration they may seek to avoid contact with bone. We have no data on the ranges of variation or how they relate to different sized predators. Most work of this nature has been done on pits and perforations (Domínguez-Rodrigo and Piqueras 2003), although based on average values that may not be as informative as ranges. Depending on the type of tooth making the marks, for example premolars, canines or incisors, large sized carnivores make both small and large chewing marks in the course of bone chewing. Smaller carnivores, on the other hand, can only make small chewing marks, and for this reason we suggest that size ranges of chewing marks, or even more simply the size of the five largest marks, give a better measure of the probable size of the carnivore species responsible. This issue will be discussed further in Chap. 4, but it is clear that additional data on sizes of linear marks made by different carnivore species are urgently needed (Pinto Llona et al. 2005).

Most carnivore linear marks have U shaped cross-sections lacking internal striations (except those made by human incisors). They may have smooth contours, but frequently they have uneven surfaces where fragments of bone have been stripped away by the pressure of the teeth (A.153). Linear tooth marks are single and unbranched, and they vary greatly in length. Their direction is partly determined by the shape of the bone, so that marks on mid-shafts of long bones tend to be transverse to the long axis of the bone, while marks on the ends of bones may be both transverse and parallel to the long axis, with all angles in between. Small carnivores such as viverrids and mustelids produce linear marks, usually on smaller bones, and they are similar in shape and morphology to those of large carnivores on large bones, but at a smaller scale. Even smaller are linear marks produced by small shrews. Soricids are mainly insectivorous, but they may scavenge carcasses of small mammals and leave marks that are again comparable to those of large carnivores but at a smaller scale (Andrews 1990).

Linear marks produced by human anterior teeth are shallower than tooth scores made by carnivores (White and Toth 2004). Also present is characteristic bending of the ends of thin bones, resulting from pushing the bone up or down with the hands while holding the ends between the upper and lower cheek teeth (A.1001) (see Chap. 9). The resulting bent shape of the ends results from the difficulties that flat–cusped molar teeth have in puncturing hard bone tissues. This shape has also been recognized on bones chewed by chimpanzees (Pobiner et al. 2007). A fossil example has been found of ribs in a modern human occupation level in Azokh Cave (Caucasus) (A.1004): the bones were exhumed from a hearth where they had been discarded.

Another type of linear marks are those made by herbivores, and they may be similar structurally to carnivore tooth marks (A.165) (Cáceres et al. 2007). Early stages of bone chewing by herbivores may be similar to linear marks left by carnivores. The main criteria distinguishing them is that herbivores mainly chew bones when they are dry, and the linear grooves resulting from herbivore chewing may sometimes be seen to penetrate through weathered surfaces of bone: i.e., weathering precedes the chewing. When herbivore chewing is at a more advanced stage, the ends of limb bones take on a characteristic forked shape that has been described by Sutcliffe (1973, 1977) with a warning of the possible misinterpretation of these shapes as human artifacts. This type of chewing may be produced by cervids, giraffes, camels, cows and sheep (Brothwell 1976). Herbivores have also been observed to predate smaller animals such as bird nestlings, which they kill and eat (Furness 1988). They bite off and eat their heads and legs in environments that are poor in nutrients such as *Calluna* moorland.

Linear Marks Made by Beaks of Raptors

Linear marks made by birds are usually very superficial as their beaks are not powerful enough to penetrate any but the thinnest bone, and beak marks are more frequently seen as punctures (see Chap. 4, where the marks made by eagles on monkey scapulae are described). The marks are broad and flat-bottomed (A.203 and A.205), may be long, and are either straight or curved. We have no data to show variation in this modification.

Linear Marks Made by Insects

Many species of insect can damage bone, creating both punctures and linear marks. Little systematic work has been done on this form of modification, but we have included what we have available. Coleoptera species are known to produce linear marks on bodies of corpses, for example in the Spitalfields cemetery in London (Molleson and Cox 1993). Termites (Fig. 3.7) produce linear marks where they penetrate into bone and dermestids may also produce linear marks observed on bones prepared by taxidermists (A.209). Moth larvae tunneling is common on bovid horns and antlers (Fish 1950). Numerous examples are known from the fossil record where insect damage has been inferred but for which comparative data are lacking. Forensic entomology identifies the presence of insect bodies found with human corpses, but their presence may also be detected by the modifications produced on the bone. We may still be far from this pseudo-taxonomic identification, but we can make preliminary identifications of differences in damage by insects based on different types of marks.

Organic Linear Marks with U Shaped Cross-Section Made by Plants

The most frequently seen linear marks on bones are the result of roots of vascular plants impinging on the bone surfaces. These marks are sometimes mistaken for carnivore chewing or cut marks, but they have a totally distinctive morphology. They are U shaped, as are all linear marks made by plants, and the contours of the marks are smoothly concave. In addition, root marks are rarely straight for more than a few millimeters; they may be single, but more commonly they are branched (A.227) and occur in some abundance, often so dense that the surface of the bone appears corroded and/or cracked and the surface texture of the bone is totally destroyed (A.249). Finally and most importantly,

Fig. 3.7 Termite hill near Maun (Botswana). To build up a mound this high (5 m), termites had to dig the equivalent amount of soil from the ground. To fix the sediment, termites use a sticky substance produced by them to give strength to the excavated soil. This substance is also recorded on bones and may also modify the bone surface

root marks commonly divide into branches. This is never seen in linear marks made by animal agents or by stone, whether by human action or during trampling, although root marks may rarely appear to mimic Hertzian cones (A.236). The branching of the marks is the product of the branching of the roots that make the marks.

Linear marks made by plants are the result of solution of the bone by chemicals produced directly or indirectly at the growing root tips of the plants. This contrasts with the action of stone against bone and by the activities of animals, which are all the product of mechanical action of one substrate against another. Roots of vascular plants do not themselves erode bones but rather it is the symbiotic association of fungi (mycorrhyza) or bacteria (rhizobia) with the plants that are responsible for what we identify as root marks. The cross-sectional width of root marks is wider than the size of the root, which makes it difficult to identify the plants responsible for the marks. The fungi or bacteria that envelope smaller or secondary roots and filaments (root hair or rhizoids) extend laterally into the bone and dissolve it to form the root mark (A.228 and A.229). Size difference may also be greater than expected in modern experimental conditions as a result of reduction in size of the root due to dehydration.

The difficulty in identifying the association between plant species and root marks may be illustrated with the example mentioned above of the Draycott cow skeleton (Andrews and Cook 1985). Some parts of this skeleton showed root marks on the bone surface when the specimens were collected. These bones had moved down slope away from the death site and were buried in accumulating debris of limestone boulders and soil, and the vegetation consisted of small trees and weak herbaceous growth, including nettles. Despite the monitoring and careful excavation of the bones, it could not be determined whether it was the herbs or the trees that produced the root marks (Andrews 1990). This is another type of taphonomic modification that needs careful investigation, and the best results will probably come from controlled laboratory situations rather than from naturalistic samples.

One other important factor to note is that root marks may be made before burial by subaerial plants. Vascular plants growing in extreme humid conditions, or plants forming an impervious mat covering the surface of the ground, or subaerial plants such as moss, may all produce root marks on bone. These subaerial plants, however, preferentially make perforations rather than linear marks (see Chap. 4), although there are some exceptions, and it cannot be assumed that root marks are necessarily evidence of burial. Root marks may also form on fossil bone, after it has been fossilized, and as it is emerging into the biologically active zone on outcrop surfaces. In this case the root impressions are usually a different color from the surface of the bone (see Chap. 5).

Striations may also be produced as a form of tunneling produced by algae and moss penetrating into the bone (lichen penetrate into the bone through fissures made by them). This has been observed on monitored modern bones under known conditions. Thin microtunnelling has also been described by Jans (2005) as "enlarged canaliculi" thinner than fungal bores. This microtunnelling (enlarged canaliculi) and loss of density (darker grey periosteal layer on the scanning electron microscope) seems to be histological evidence of what is traditionally known as "soil acidity corrosion" (A.275) which penetrates up to 100 microns into the cortical bone (Fernández-Jalvo et al. 2010).

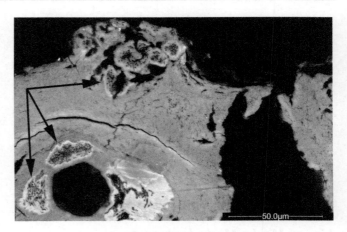

Fig. 3.8 Microscopical focal destruction (MFD) made by bacteria (black arrows)

Finally, it has been proposed (Fernández-Jalvo 1992) that diatoms may produce linear marks on bones immersed in water. The modifications by diatoms consist mainly of perforations, and these perforations may form in a line where water energy is relatively high (A.444). Diatoms link to each other in a chain to resist the water current, and where they anchor themselves to the bone perforations occur. In our experience, this is quite common in bones lying in water.

Solution marks on bone may also be caused by microorganisms (Fig. 3.8 black arrows). Sometimes, bones are heavily damaged by bacteria, and areas of bacterial attack can be distinguished on the surface (A.292 and A.908). Areas of damage may have any shape, sometimes with linear trajectories (A.294). There may also be exogenous bacteria that penetrate well into the bone, or endogenous bacteria derived from the animal's gut (Bell et al. 1996, 2012), and in both cases they gain access to the bone through the Haversian system as decay sets in and the bacterial activity is only evident when the bone is sectioned. Hackett (1981) described microscopic modifications found on bones which he named "microscopical focal destruction" (MFD) and which he considered were produced by bacteria producing striations that he described as "linear longitudinal", "budded" and "lamellae". Wider grooves (named by Hackett as Wedl tunnels) can be produced by fungi.

Atlas Figures

A.1–A.306

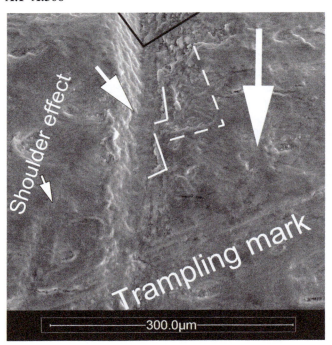

Fig. A.1 SEM microphotograph of a fossil bone, detail of individual cut mark showing V-shape cross section with inclination towards the lateral margin (towards the right for right-handed operator, as here, or towards the left side for left-handed operator) (Bromage 1987). Directionality (large arrow) is indicated by Hertzian fracture cones (Bromage and Boyde 1984) on the lateral margin (right, white lines). Hertzian cones are caused by the stress produced on the bone. Micro-steps on the cut mark floor are facing the movement (see Text Fig. 3.1) and oblique faults on the medial margin (left) are oriented obliquely forward into the cut mark (small arrow). These traits indicate right handedness according to (Bromage et al. 1991, p. 166)

Fig. A.2 Experimental cut marks on a modern rib of young pig. The cuts were made while the bone still had meat on it. Sawing marks (A.22) were made on the left side (double arrow) and cut marks on the right side of the bone (enlarged top right). Directionality is given by the lateral Hertzian fracture cones on the right side of the cuts (A.3). Taphonomic identifications may largely be done using binocular microscopes. Uncertain marks (A.62), however, need the use of SEM for correct identification (A.6)

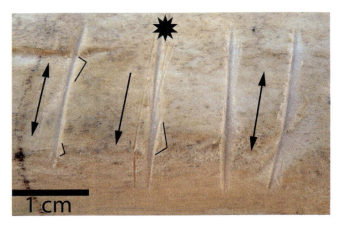

Fig. A.3 Experimental cut marks on recent bone by a right handed experimenter. Directionality was noted during the experiment and confirmed by lateral Hertzian fracture cones. The middle mark (*) was made incising repeatedly on the same cut, but keeping the movement from the top to the bottom. The other marks are the result of sawing motions (up and down repeatedly A.19). Hertzian fracture cones on the right side show opposing directions (A.22) or are abraded by friction of the stone tool against the bone surface

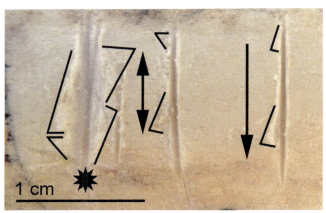

Fig. A.4 Experimental cuts made on recent bone by a left handed experimenter (A.5). The marks were made directly on the bone without meat. Hertzian fracture cones appear on the left lateral margin, rougher than the medial side, because of slight supination of the hand laterally during the cutting motion by a left-handed. Sawing motions produce opposite Hertzian cones, particularly when sawing cuts are dynamically made (*), so that both lateral and medial sides are rough and highly abraded

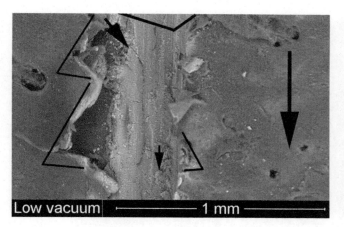

Fig. A.5 SEM microphotograph of a recent bone, detail of experimental cut mark made by left handed person directly on the bone without meat. The cut shows all the traits described as characteristic of cut marks (see A.1 and Text Fig. 3.1). Experimental studies show that Hertzian fracture cones may occur on both sides, with the largest cones on the lateral margin (here on the left). The smaller cones on the medial margin (shown here on the right) may eventually disappear by post-depositional abrasion or other taphonomic processes

Fig. A.6 SEM microphotograph of a fossil bone, detail of slicing marks made on a child's clavicle (A.98) from TD6, Atapuerca. Note that bone is smeared and exceptionally preserved on the medial margin (broken arrows). These deep and precise cut marks are in the region of the *deltoid* and *pectoralis major* muscles attachments. Hertzian fracture cones are not obvious on the lateral side so that directionality or handedness cannot be confirmed

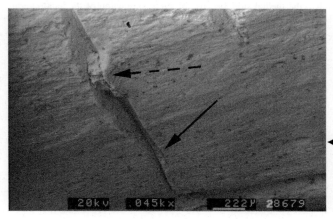

Fig. A.8 SEM microphotograph of a fossil bone, detail of a slicing mark on a bone from Gough's cave, leaving a single V-shaped incision. Lateral Hertzian cones occur on both sides (see A.5)

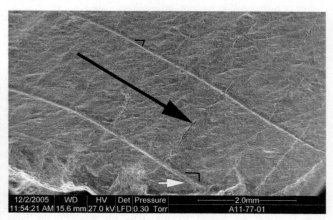

Fig. A.9 SEM microphotograph of a fossil bone, detail of thin slicing marks on a vertebra apophysis of a medium sized mammal from Gorham's Cave. Incisions are very thin with distinctive V-shaped cross section, with internal microstriations (white arrow) and lateral Hertzian fracture cones (A.1) indicating the directionality of the cutting motion (black arrow)

◄ **Fig. A.7** SEM microphotograph of a fossil bone, detail of single slicing mark (detail of central cut in A.6). The lateral incision marked by the solid arrow corresponds to a former cut rather than a sawing motion – otherwise the smeared bone on the side of the cut (broken arrow) should be abraded. The shape of this cut mark matches most closely marks made with a flint stone tool (see A.37)

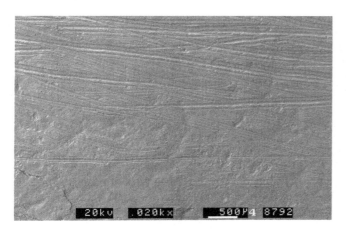

Fig. A.10 SEM microphotograph of a fossil bone, detail of scraping marks on a horse bone from Gough's Cave. These marks are characterized by holding the stone tool edge transversally to the direction of the motion. The purpose can be cleaning the bone from fat before breaking it to extract the marrow, or cleaning the tool edge when cutting. At Gough's Cave scraping and cut marks are present both on human fossils (see A.88 and A.100) and animal fossils, such as this hyoid of horse (*Equus ferus*)

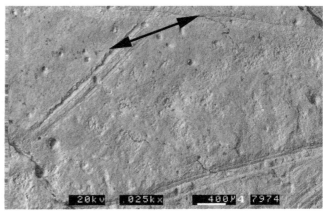

Fig. A.13 SEM microphotograph of a fossil bone, detail of linear marks on a bovid vertebra (ATA'98, TEi 31, n.1) from Sima del Elefante, Atapuerca. Sometimes the shape and morphology of marks made by a scraping may be difficult to distinguish from trampling marks. Location, orientation and similarities with other clear marks made by a stone tool are criteria to characterize scraping marks. Compare similar lateral traits pointed here by a double arrow with A.55, A.56 and A.82

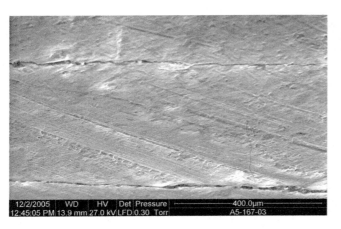

Fig. A.11 SEM microphotograph of a fossil bone, detail of scraping marks on a rabbit tibia from Gorham's Cave. These marks are made using the edge held transversally to the direction of motion. This produces multi-striations covering a relatively wide area along the bone. Scraping may affect large to medium sized mammals, and small game as well. Rabbit fossils from this unit also have human chewing marks (A.148) superimposed on the human induced tool marks (see A.370)

Fig. A.14 SEM microphotograph of a fossil bone, detail of a deep cut made by sawing motion on a rib of wild goat from Vanguard Cave near the articular end of the bone. Sawing is usually related to dismembering activities. Compare with U shape cross-section made by plant roots (A.231)

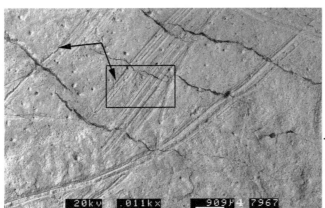

◀ **Fig. A.12** SEM microphotograph of a fossil bone, detail of scrape marks on a bovid vertebra (ATA 98, TEi31, n1, A.48) from Sima del Elefante (Atapuerca). These marks are found on recessed surfaces or protected areas, and there is no evidence of breakage or damage on the salient angles as would be expected if trampling produced these marks. The boxed area is shown in A.82, black arrows point to the lateral side (compare with A.56)

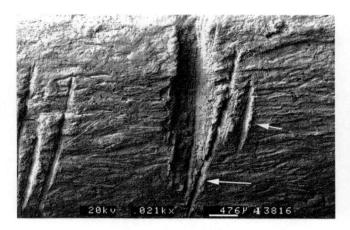

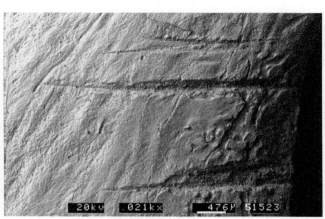

Fig. A.15 SEM microphotograph of a fossil bone, detail of linear marks on a right ulna shaft fragment from TD6 Atapuerca (ATD6 97). The bone is heavily altered by butchery and bears impact and percussion marks related to breakage. As obtained experimentally (A.3 and A.4), several cuts may occur with a wide diameter as a result of sawing. Sometimes individual cut marks can be distinguished (white arrows and striations on the left of the picture)

Fig. A.18 SEM microphotograph of a fossil bone, detail of cuts on a lumbar vertebra of a juvenile wild goat (*Capra ibex*) from Vanguard cave. Cuts are on the concave angle of the lateral apophysis, inaccessible to trampling without damaging the vertebra. In these positions, trampling is excluded because the area of bone affected is deeply recessed (as in A.49). The marks are also deeply incised, which is not usual in trampling marks

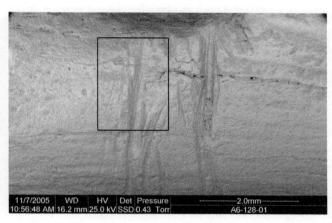

Fig. A.16 SEM microphotograph of a fossil bone, detail of cut marks on a human scapula from Gough's Cave (M54059, GC7). These cuts are located on the concave angle between the *acromion* and the scapular neck, cranial orientation. The scapula is intensively butchered, with abundant scraping marks (A.88 and A.91), percussion marks on the edge of the inferior border and peeling

Fig. A.19 SEM microphotograph of a fossil bone, detail of sawing cuts (see A.3*) on a rib fragment of immature *Cervus elaphus* from Gorham's Cave. Cuts are located at the caudal end, near the vertebral articulation. The *outlined square* is shown in close up view in A.20

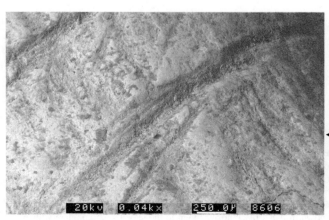

◄ **Fig. A.17** SEM microphotograph of a fossil bone, detail of cut marks, detail of the human scapula (A.16). The cuts are protected by the curvature of the bone, and they pass over and around curvatures in the bone without interruption, occurring on areas that are inaccessible to trampling (see also A.18). Trampling marks are also present (not shown in this view) with a different pattern and superimposed on the butchery marks, indicating that they were formed later

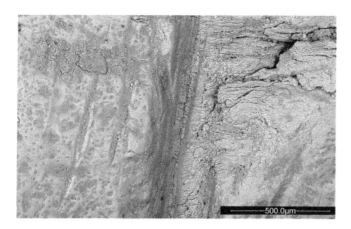

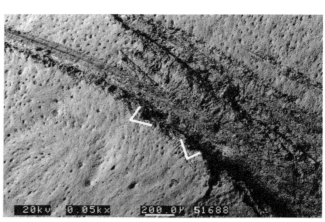

Fig. A.20 SEM microphotograph of a fossil bone, close up view of A.19 sawing cuts found on a rib fragment of an immature *Cervus elaphus* from Gorham's Cave. Cuts are located on the rib metaphysis near the vertebral articulation. Sawing cuts are deep and located near a strong muscle attachment. The location is congruent with dismembering activities (as in A.98, A.99 and A.100)

Fig. A.22 SEM microphotograph of a fossil bone, detail of cuts from A.21. Some Hertzian fracture cones can be distinguished on the medial side that indicates reiterative opposite movements (sawing motion). The lateral sides of the cut are heavily abraded by reiterative sawing movements in contrast to A.23

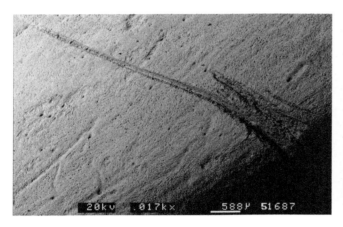

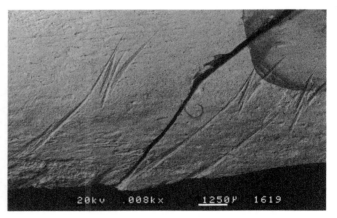

Fig. A.21 SEM microphotograph of a fossil bone, detail of a cut mark on a middle sized mammal rib fragment (unidentified species) from Vanguard Cave. Cuts are located on the anatomical border of the caudal epiphysis. The general aspect of these cuts is similar to A.23, but a close up view of the cuts seen in A.22 indicates sawing motion. Location and type of cuts (sawing), suggest dismembering processes

Fig. A.23 SEM microphotograph of a fossil bone, detail of cut marks on a rib fragment shaft from TD6 Atapuerca. Marks are slightly different from individual incisions and sawing marks. The analysis of these cuts (A.24 and A.25) suggests that the unusual shape of these cuts is a consequence of irregularities of the stone tool edge. They were made by three different movements, each producing identical cuts. Compare with GC87-74, A.27. Note top right the fossil surface has a glue drop exuded from the break slightly covering the cut marks

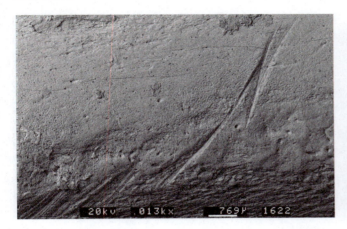

Fig. A.24 SEM microphotograph of a fossil bone, detail of another mark on the same fossil shown in A.23, a rib fragment from TD6 Atapuerca. Probably the tool moved laterally while cutting due to hand supination. A close up view of the cut (A.25) shows that the cut was made in a single movement and the resulting shape was produced by irregularities of the tool edge. Compare with A.27

Fig. A.26 SEM microphotograph of a fossil bone, detail of A.23 on a rib fragment from TD6 Atapuerca. The atypical shape of these cuts is a consequence of irregularities of the stone tool edge, possibly by a retouched flake, made by single incisions along the bone shaft (sawing motion is discarded). Compare with GC87-74, A.27 and A.31

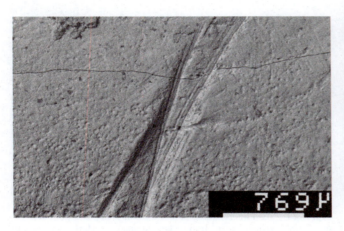

Fig. A.25 SEM microphotograph of a fossil bone, close up view of A.24, a rib fragment from TD6 Atapuerca, showing the cut has been made in a single movement (A.65). The irregularity of the edge produces an X shape in a single movement as the angle of the tool changes during the cutting motion. The characteristic X shape of these cuts is produced by retouched flakes on both sides (see Schick and Toth 1993, Domínguez-Rodrigo et al. 2009)

Fig. A.27 Fossil bone. Right human radius GC87-74 from Gough's Cave. The diaphysis has extensive cut marks on the side of the shaft. These marks were formerly interpreted as decorative engraving. 'Groups' of incisions, however, are made by single strokes. The cut marks are shown enlarged in the lower figure and in A.28 and A.29 and Text Fig. 3.5

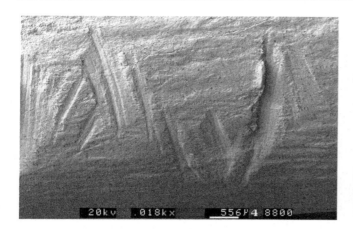

Fig. A.28 SEM microphotograph of a fossil bone, detail of two sets of marks of A.27 on a right human radius (GC87-74 from Gough's Cave, UK). Cuts can be superimposed upon each other when cutting. Compare with A.14 and A.21, which are common sawing marks related to dismembering, and with A.23 and A.26 (single cuts with multi striations) more related to single movements

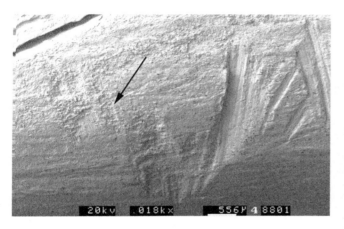

Fig. A.29 SEM microphotograph of a fossil bone, detail of cut marks on a right human radius GC87-74 from Gough's Cave. Marks on the left of the image have less marked incision on left side of the cut (black arrow). This may be due to a lateral movement or slight hand supination while cutting, producing a shallower penetration on the side of the mark. In contrast, marks on the right of the image are deeply marked and also show indication of lateral supination of the hand. Compare with A.28 right side mark

Fig. A.30 SEM microphotograph of a fossil bone, detail of cut marks on a human radius from Gough's Cave (A.27). The 'groups' of incisions are compound marks made by single strokes, with consistent directionality towards the superior aspect of the shaft. The marks are made by the same stone tool edge and with individual movements that follow the same direction for all cuts along the bone shaft. The accompanying striations at each movement may be the result of a retouched edge at both sides (see Schick and Toth 1993). Field width equals 7.3 mm

Fig. A.31 SEM microphotograph of a fossil bone, detail of cut marks on a human radius, detail of A.30 from Gough's Cave. The fossil was formerly interpreted as an engraved bone. The taphonomic study interpreted it as result of filleting, cutting the arm muscles progressively along the shaft, not a decoration (see Andrews and Fernández-Jalvo 2003 and Text Fig. 3.5). Field width equals 3 mm

Figures A.32, A.33, A.34 and A.35: experimental cut marks using quartzite stone tools.

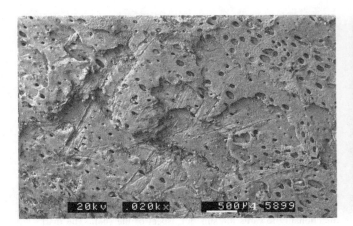

Fig. A.32   Magnification: 20x, A.36, A.40 and A.44

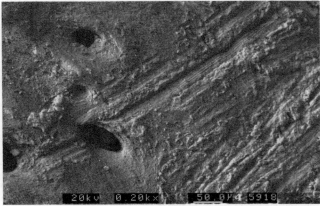

Fig. A.35   Magnification: 200x, A.39 and A.43

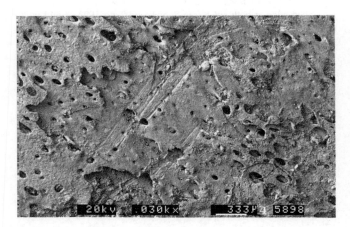

Fig. A.33   Magnification: 30x, A.37 and A.41

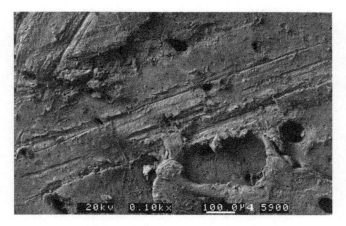

Fig. A.34   Magnification: 100x, A.38, A.42 and A.45

SEM microphotographs of modern long bones, detail of experimental cut marks using quartzite stone tools. The bones are from a young sheep individual and the surface appears irregular. Striations have a well defined V-shape cross section and internal microstriations along the cut. Shoulder effect and lateral accompanying striations are abundant and more frequent than cuts made with flint stone tools. The cut edge is slightly irregular, although lateral Hertzian fracture cones can be easily distinguished.

Our experiment was performed by non-skilled butchers and as a result the bone surfaces have many cuts, with some traits that are indicative of the cut capacity of each type of raw material used, but not essentially characteristic. We observed that cutting is most effective when inclining the tool edge for meat filleting, but more effective if incising perpendicularly to the bone surface.

We also observed that penetration of cuts using quartzite stone tools may not be deep, resulting in narrower and finer striations. More skilled and precise butchers performing similar experiments have produced wider striations with quartzite stone tools than with flint stone tools (Fernández-Jalvo and Cáceres 2010).

Figures A.36, A.37, A.38 and A.39: experimental cut marks using flint stone tools.

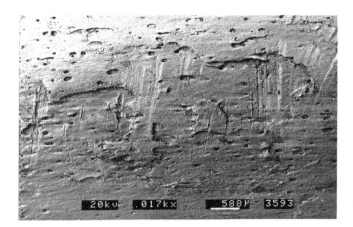

Fig. A.36 Magnification: 17x, A.32, A.40 and A.44

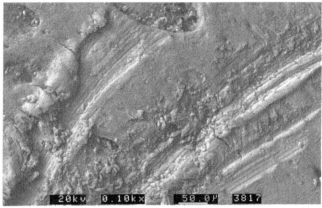

Fig. A.39 Magnification: 200x, A.35 and A.43

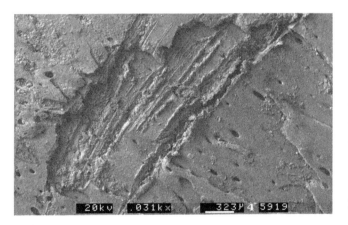

Fig. A.37 Magnification: 30x, A.33 and A.41

SEM microphotographs of modern long bones, detail of experimental cut marks using flint stone tools. The bones are from a young sheep individual and the surface appears irregular. Striations have a well defined V-shape cross section along the striation and internal microstriations. Shoulder effect and lateral accompanying striations are relatively rare or less abundant than cuts made with quartzite. The cut edge is well defined and lateral Hertzian fracture cones are well defined on the lateral margin.

Our experiment was performed by non-skilled butchers and has a resulting highly cut bone surface. The experience provided information of the cut capacity of each type of raw material used, but cuts are not totally characteristic of the raw material used. In contrast to quartzite tools, we noticed that cutting for meat filleting using flint tools is effective even when the tool edge is inclined relative to the bone surface, and that the edge could penetrate deeper when inclined. The resulting cuts using flint edges were wider than those produced by quartzite stone tools (deeper penetration makes wider striations). This also depends on the sharpness, shape and thickness of the tool edge. More skilled and precise butchers performing similar experiments have produced thinner striations with flint stone tools than with quartzite stone tools (Fernández-Jalvo and Cáceres 2010).

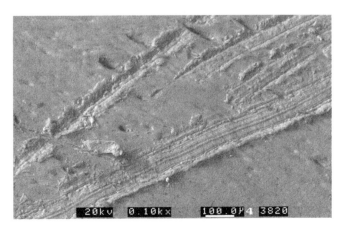

Fig. A.38 Magnification: 100x, A.34, A.42 and A.45

Figures A.40, A.41, A.42 and A.43: experimental cut marks using limestone stone tools.

Fig. A.40 Magnification: 16x, A.32, A.36 and A.44

Fig. A.43 Magnification: 200x, A.35 and A.39

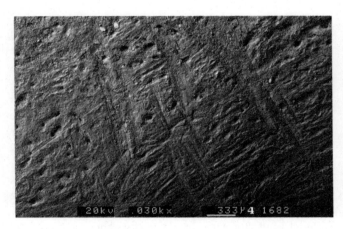

Fig. A.41 Magnification: 30x, A.33 and A.37

SEM microphotographs of modern long bones, detail of experimental cut marks using limestone stone tools. The bones are from a young sheep individual and the surface appears irregular. Striations have a V-shape cross section, though it may become U-shaped along the striation. Lateral Hertzian cones can be distinguished. Abundant internal microstriation and frequent shoulder effect may be caused by the grainy texture of the limestone.

Our experiment was performed by non-skilled butchers and as a result there are many cuts on the bone surface. Traits indicative of the cut capacity of limestone tools are as follows: the raw material is more porous than any other material; the stone tool edge soon becomes blunted by meat and needs to be frequently cleaned, usually against the bone surface, sometimes detaching the meat from the stone tool edge using a cloth or the hands. The cuts are wider and shallower than any other cut made with quartzite, flint or shell. More skilled and precise butchers performing similar experiments using limestone stone tools have produced similar striations, with better defined edges and deeper cuts (Fernández-Jalvo and Cáceres 2010). The appearance of these cuts is very similar to trampling marks where bones were resting on limestone substrate (Andrews and Cook 1985). Location of marks on the anatomical element related to muscle or ligament attachment or filleting are the main criteria distinguishing butchering and the use of stone tools from trampling marks, locations of which are not related to anatomical traits.

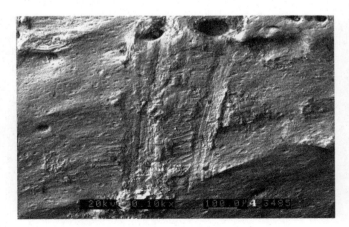

Fig. A.42 Magnification: 100x, A.34 and A.38

Figures A.44, A.45, A.46 and A.47: experimental cut marks using mollusk shell.

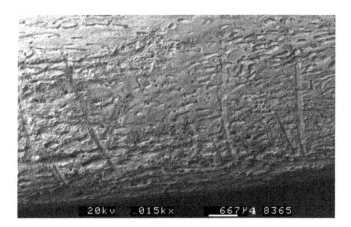

Fig. A.44 Magnification: 15x, A.32, A.36 and A.40

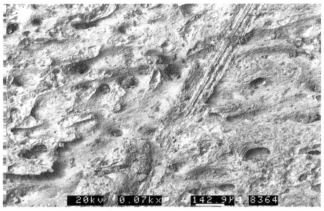

Fig. A.47 Magnification: 70x, A.42

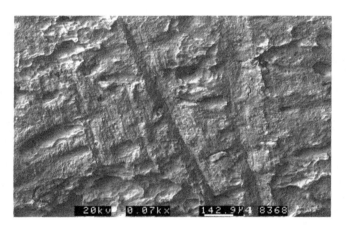

Fig. A.45 Magnification: 70x, A.38

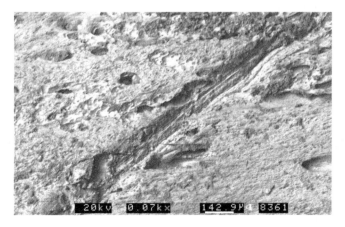

Fig. A.46 Magnification: 70x, A.34

SEM microphotographs of a modern rib, detail of experimental cut marks using mollusk shell to make the cuts. The bones are from a young pig individual and the surface appears irregular. Striations have a V-shape cross section and abundant internal microstriations, but lateral Hertzian cones are smaller (or absent) than cuts made with flint, quartzite or limestone stone tools. The size of the cuts is wider than cuts made with flint or quartzite, but thinner than cuts made with limestone stone tools.

Our experiment, performed by non-skilled butchers produced highly cut bone surfaces, with some traits that are indicative of the cut capacity of each type of raw material used. We observed that mollusk shell has an ergonomic shape and high cutting capacity. A shell was broken to make a sharper edge, but results were similar to using unbroken shells. More skilled and precise butchers cutting with mollusk shells commented on the unexpected capacity of shell to make cuts and they produced extraordinarily long striations, especially on flat surfaces (deer scapula, Fernández-Jalvo and Cáceres 2010).

Figures A.48, A.49, A.50 and A.51: linear marks on a fossil.

Fig. A.48 Magnification: 22x

Fig. A.51 Magnification: 50x

Fig. A.49 Magnification: 19x

SEM microphotographs of a fossil bone, detail of linear marks on a vertebra of bovid (ATA'98, TEi 31, n.1) found at the base of Sima del Elefante sequence, from Atapuerca Cave system (~1 Ma). The vertebra was found accompanied by limestone flakes which were questioned if they were made by humans. The vertebra showed a type of striation which by their shape could be attributed to trampling. These striations were, however, located in cavities and recessed surfaces (A.18) that were inaccessible to stones or sediment grains due to trampling. Most of the striations on this vertebra have V-shape cross-section (A.54 and A.55) and they were located on ligament insertions of the bone vertebra. All these traits suggest that the marks were the result of human action a using stone tools made of limestone (compare A.51 with A.43 and A.57). The vertebra came from a partial bovid carcass that was probably butchered by humans, and this is one of the oldest sources of evidence of human butchery in Europe (~1 million years BP). The Sima del Elefante site recently yielded human remains (Carbonell et al. 2008) supporting the taphonomic indications of human involvement at the site.

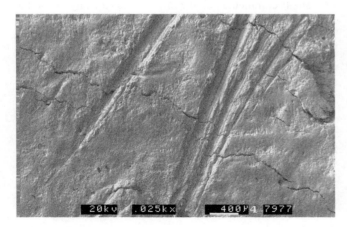

Fig. A.50 Magnification: 25x

Fig. A.52 Fossil bone. Second human phalanx of the hand from Atapuerca, ATD6-46, with both peeling at the proximal end and percussion marks on the palmar side as result of dismembering. Striations (A.53) observed on this human phalanx show traits that suggest the use of limestone stone tools. Photo M. Bautista

Fig. A.54 SEM microphotograph of a fossil bone fragment (ATD6-97) from Gran Dolina, Atapuerca. This fossil is a right ulna shaft fragment intensively butchered, bearing cuts and percussion marks on the bone surface and chop marks possibly to dismember and impact marks to break the bone to extract the marrow. Cut marks show traits of stone tools made of limestone. Compare with A.48 and A.50 both at similar magnifications ∼30x)

Fig. A.53 SEM microphotograph of a fossil bone, detail of scratches on the human phalanx (A.52) from Atapuerca, ATD6-46 showing transverse striations affecting the *flexor digitorium* tendon attachment, apparently related to dismembering tasks (see Fernández-Jalvo et al. 1999) Compare with experimental cuts A.41

Fig. A.55 SEM microphotograph of a fossil bone, showing details of the cut marks on ATD6-97. Two different types of cuts can be distinguished. A perpendicular group of wide and deep cuts (white arrows), similar to the vertebra from Sima del Elefante (ATA'98, TEi 31, n.1, see A.50) and an oblique cut (black arrow) that has similarities with experimental cuts made with another stone tool (A.56). Unfortunately, none of these cuts are superimposed to know the sequence of cuts

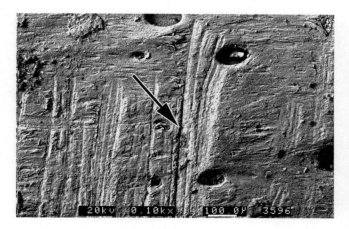

Fig. A.56 SEM microphotograph of a modern bone, detail of experimental cut marks using flint stone tools. The cuts show strong similarities with the cut indicated by a black arrow in A.13 and A.55

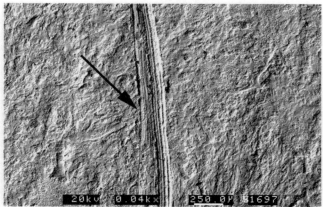

Fig. A.58 SEM microphotograph of a fossil bone, detail of a cut mark on a fossil from Vanguard Cave. The traits of the cut marks (see A.1) such as internal microstriations, shoulder effect (black arrow) or V-shape cross section may vary according to the raw material used to make the tool. Sometimes, however, cut marks are mimicked by trampling marks (see A.59). Location, distribution and organization of the striations help to distinguish between cuts and trampling

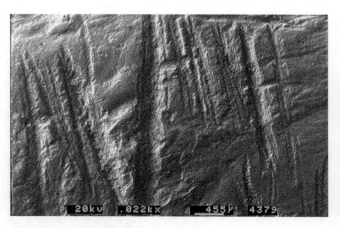

Fig. A.57 SEM microphotograph of a fossil bone, detail of cut marks found on a fossil bone fragment from Gough's Cave. The morphology of these striations and the context in which the fossil was found (a limestone cave) suggests that cuts could be made by a limestone tool. Compare with experimental limestone cuts A.42 and fossil A.51

Fig. A.59 SEM microphotograph of a modern bone from Draycott monitored carcass (Andrews and Cook 1985). Trampling marks may bear many of the characteristic features of a cut mark (see Text Fig. 3.1 and A.1), including deep penetration, linear grooves internally in the cut mark, lateral Hertzian cones, and internal micro-steps (one indicated by a white arrow). This particular striation, however, is not V-shaped but has a flat cross section. In general, trampling marks are shallower than cuts. Field width equals 400 microns

Fig. A.60 Linear marks on a modern cow rib due to trampling. The cow carcass was monitored for 8 years exposed on a rocky limestone pathway (see text, and site figure Draycott). Bones were intensively trampled by other cows passing daily along the path (Andrews and Cook 1985). Sediment friction or trampling marks refer to the movement of bone against a rocky/sandy substrate causing multiple scratches dispersed along the length of the bones and usually transversally to the main axis of long bones. See Text Figs. 3.3 and 3.4

Fig. A.62 Fossil bone. Rhinoceros bone fragment from Senèze with a deep striation that mimics human made cut marks. The site is Villafranchian. No humans were present at this age (2 Ma) in the region. Fossils from this site were frequently damaged by falling blocks and trampling. This mark has a clear V-shape cross section, Hertzian cones and "shoulder effect" (black arrow). Compare with A.1

Fig. A.61 Fossil bone. Elongated thin linear marks strongly mimic human made cut marks using stone tools (A.8 and A.9). The fossil comes from Concud. The absence of humans at this age (7 Ma) excludes any possibility of the marks being cut marks. The site, a paleolakeshore, was frequently visited by animals to drink. Bones exposed on the ground, free of soft tissues, would have been pressed against the substrate producing abundant trampling marks. Courtesy of D. Pesquero

Fig. A.63 Fossil bone. Trampling and falling blocks may mimic tooth marks. The fossil comes from Galeria, Atapuerca, the sediment of which is rich in gravels. The fossil does not show any evidence of carnivore action (e.g., chewing, licking or breakage). Scratches here are probably made by trampling when animals stepped on the bone and rubbing it against the gravel-sandy substrate. See Text Figs. 3.3 and 3.4

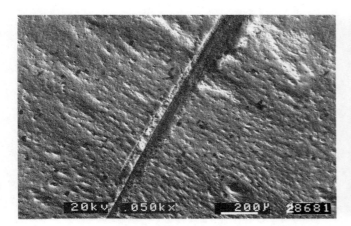

Fig. A.64 SEM microphotograph of a fossil bone, detail of an individual cut mark from TD6, Atapuerca. The stone tool in the hands of a human is moving (active object) against the bone (passive object). The physical result is similar to trampling (see text and A.68). The purpose of the cut is to detach muscle, tendon or skin during butchery (Spennemann 1990). The shape of the mark, together with its location and the distribution of marks on the skeletal elements are the main criteria to distinguish the use of stone tools versus trampling action

Fig. A.66 SEM microphotograph of a fossil bone, detail of cut marks on a bovid vertebra from the lower units of Sima del Elefante Atapuerca (ATA'98, TEi 31, n.1). Cuts were probably made with a limestone tool to extract meat and dismembering a carcass. The lateral side of the cut (squared box) has similarities with the lateral sides of trampling marks (see A.70)

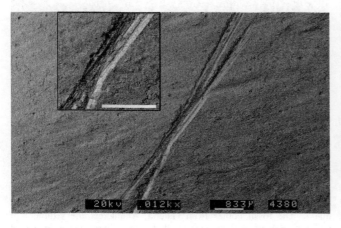

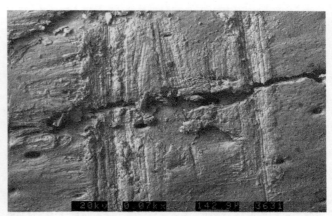

Fig. A.65 SEM microphotograph of a fossil bone, detail of a cut mark from Gorham's Cave, showing the characteristic X shape striation when cutting is produced by a stone tool with retouched edge (see Schick and Toth 1993, Domínguez-Rodrigo et al. 2009, see text). The irregularity of the edge produces an X shape as the angle of the tool changes as the tool passes along the surface of the bone. Compare with A.69

Fig. A.67 SEM microphotograph of a modern bone, detail of cuts from experimental butchery of a sheep humerus using a limestone tool. The width of the cut is greater than that produced by most other raw materials. Similarities with trampling marks on a limestone substrate are remarkable (compare with A.71) but they can be distinguished when the cuts are located in areas that are congruent with butchery purposes or in areas that are inaccessible to trampling (e.g., recessed surfaces)

Fig. A.68 SEM microphotograph of a modern bone, detail of a single trampling mark from Draycott in monitored experimental conditions (Andrews and Cook 1985). The bone being trampled is moving (active object) against the limestone substrate (passive object). The physical result is similar to scraping or cutting (see text and A.64). The shape and distribution of marks, together with their location on the anatomical element, are the main criteria distinguishing trampling from the use of stone tools. Field width equals 2 mm

Fig. A.70 SEM microphotograph of a modern bone, detail of a trampling mark from Draycott in monitored experimental conditions. The shape of trampling scratches may be similar to cut marks. Location of marks on the anatomical element (congruent vs. inconsistent with butchery purposes), their number and orientation (Andrews and Cook 1985) are the main criteria to distinguish the action of trampling from the use of stone tools (compare with A.66). Field width equals 800 microns

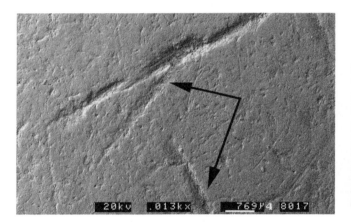

Fig. A.69 SEM microphotograph of a fossil bone, detail of trampling marks from TD4, Atapuerca. Trampling marks may produce by accident the effect of X shape striation (compare with A.65). They, however, have been formed in two different motions or due to two different grains of sediments. Notice shape similarities between the two lowermost scratches in the middle half of the photograph (black arrows)

Fig. A.71 SEM microphotograph of a modern bone, detail of deep scratches produced by trampling on a monitored cow carcass from Draycott. The carcass rested for a long time on the way of a cow track with limestone substrate. Depending on the gravel-rock sharpness of the substrate, trampling marks may have V-shaped cross section (compare with A.67). It is, however, more frequent to have U-shaped cross sections. Field width equals 4 mm

Fig. A.72 SEM microphotograph of a fossil bone, detail of a cut mark on a fossil bone surface from Gorham's Cave, detail of A.65. Lateral Hertzian fracture cones, internal microstriation and accompanying-shoulder lateral striation are all present, but note that they may also occur in trampling marks (see A.76)

Fig. A.74 SEM microphotograph of a fossil bone, detail of long and thin slicing marks made by stone tool on a fossil from Vanguard Cave. The cuts are not accompanied by shallower scratches as seen in A.78, and they run over curved surfaces. The area indicated by the square box is shown in A.75

Fig. A.73 SEM microphotograph of a fossil bone, detail of single cut marks, from Gough's Cave. The cuts are a combination of slicing and probably sawing motions. All of them have a deep V-shape cross section and the group of cuts is located near muscle attachments. Cut marks are often the result of accidental contact between the bone and the carefully shaped edge of the stone tool (see A.77). Cut marks therefore tend to be few in number. Field width equals 7 mm

Fig. A.75 SEM microphotograph of a fossil bone, detail of individual and thin slicing marks shown in A.74 a fossil from Vanguard Cave. These cut marks have internal microstriations and well defined V-shape cross section with later Hertzian cones. These traits can only be distinguished using the SEM. Compare with A.79

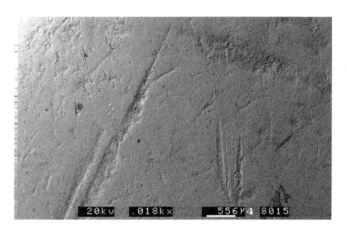

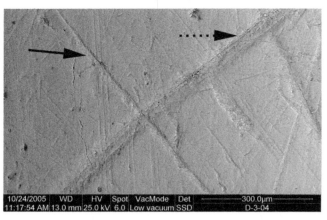

Fig. A.76 SEM microphotograph of a fossil bone, a rib fragment from TD4, Atapuerca. Similarities with cut marks are strong (see A.72). These marks, however, are too flat in cross-section and too shallow, they run transversally to the length of the shaft and they are accompanied by other sets of parallel shallow scratches (not shown in this view), making them more consistent with trampling

Fig. A.78 SEM microphotograph of a fossil bone from Azokh Cave showing the characteristic aspect of trampling marks. There are several sets of multi-striations with different random directions superimposed. A deep and narrow mark strongly mimics a cut mark (black arrow). It covers previously made marks with U-shape cross section (dotted arrow) and the accompanying set of shallower marks, with identical directions, exclude them as cut marks Compare with A.74

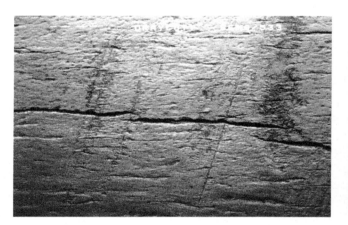

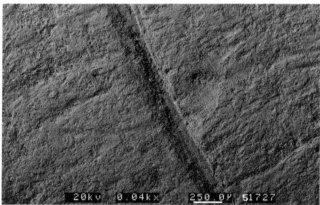

Fig. A.77 SEM microphotograph of a modern bone. Trampling marks on a monitored cow bone from Draycott. Individual scratches can be distinguished, but they are usually accompanied by other more superficial and shallower scratches having identical direction. The trampling scratches are generally less deep than cut marks (A.73). Field width equals 4 mm

Fig. A.79 SEM microphotograph on a fossil from Abric Romani. This is an isolated trampling mark similar to cut marks (compare with A.75). Traits to distinguish trampling marks from stone tool cut marks are: shallow marks with irregular and flat cross section, multiple marks with identical directions, and located without respect to muscle insertions. These traits are clear on the scanning electron microscope

Fig. A.80 SEM microphotograph of a fossil bone, detail of a linear mark on a bovid vertebra from the lower units of Sima del Elefante, Atapuerca (ATA'98, TEi 31, n.1). The striation shape itself looks similar to trampling (see A.84), however, the striation is located on a recessed surface, difficult to be reached by trampling. This together with the distinctive V-shaped cross-section and the general context of the site provide strong indications that the striation may be a cut mark

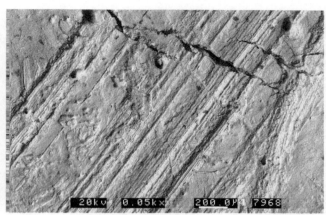

Fig. A.82 SEM microphotograph of a fossil bone, detail of a scraping mark on a bovid vertebra (ATA 98, TEi 31, n.1) from Sima del Elefante. Scraping marks are characterized by moving the stone tool edge transversally to the direction of the motion, running oblique or longitudinally to the length of the bone. Compare with A.86

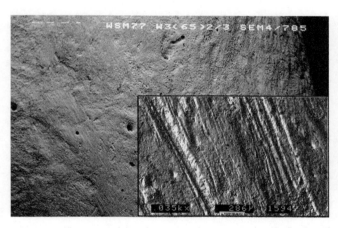

Fig. A.81 SEM microphotograph of a fossil bone, detail of scraping marks. Scraping marks may be similar to trampling marks (compare with A.85), but scraping marks made by humans are often oblique or parallel to the long axis of the bone, while trampling marks tend to be perpendicular to the long axis of bone. Field width of the background picture equals 7 mm

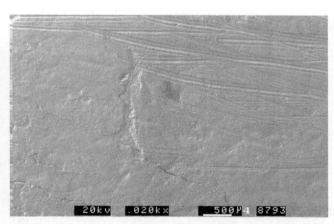

Fig. A.83 SEM microphotograph of the fossil hyoid of a horse from Gough's Cave. This shows the inner surface of the symphysis, deeply recessed and inaccessible to trampling (compare with A.87). This picture shows scraping marks passing over a transverse irregularity (black arrow), possibly made when cleaning the stone tool edge

Fig. A.84 SEM microphotograph of trampling marks on a modern cow vertebra from Draycott. Trampling marks are usually formed transversely to the long axis of long bones and ribs but may be oblique on axial bones. The location and distribution of marks on exposed or flat areas on the anatomical element is the main criterion to distinguish between the use of stone tools or trampling action (compare with A.80). Field width equals 1 mm

Fig. A.86 SEM microphotograph of trampling marks on a modern bone from Draycott monitored cow carcass mimicking scraping marks made by stone tools or vice versa. The presence of shallow accompanying marks is often present with trampling marks (see A.82). Field width equals 858 microns

Fig. A.85 SEM microphotograph of trampling marks from Draycott modern monitored cow carcass. Scratches by trampling may be similar to scraping marks made by humans. Trampling marks occur dispersed on the bone surface and transversally to the long axis of long bones as this is the most stable position of bones when rubbed against the sediment. Long bones tend to rotate about their long axis when trampled, and this produces the transverse marks (see A.81). Field width equals 5 mm

Fig. A.87 SEM microphotograph of trampling marks on modern bone from Draycott monitored cow carcass. Wide scratched areas and isolated and deeper scratches may interfere each other due to the characteristic random distribution of trampling marks that mimic scraping marks (A.83). Trampling is always produced when the bone is free of flesh and meat, not linked to muscle or ligament attachments but randomly dispersed on the bone surface according to the most stable position of the bone shape as it is rubbed against the substrate. Field width equals 1 mm

Fig. A.88 SEM microphotograph of a fossil bone, detail of stone tool marks on a human scapula from Gough's Cave. Cuts and scrapes show different directions related to the strong muscle attachments. Scrapes occur on the deeply recessed angle between the spine and the *infraspinatous fossa*, over areas protected by the curvature of the bone. The shape of the marks is similar to trampling marks (see A.92)

Fig. A.90 SEM microphotograph of a fossil bone, detail of incision marks from Gough's Cave, showing long curved striations. Cuts over and around curvatures of the bone, on uneven bone surfaces or across strong muscle/ligament attachments may end in a curved cut shape (compare with trampling A.94). Field width equals 5 mm

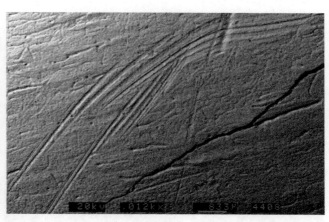

Fig. A.89 SEM microphotograph of a fossil bone, detail of long bone fragment of a medium-sized animal from TD6 Atapuerca. The bone surface has many scraping marks. Scraping marks may be organized in groups along the surface of the bone. They are in general longer than cuts, sawing or chop marks, and they may be oriented along the length of the bones (compare with A.93)

Fig. A.91 SEM microphotograph of scraping marks on a fossil from Gorham's cave. Scraping marks may be the result of cleaning the edge of the stone tool when cutting, and this action produces interrupted scraping marks. We have experimentally observed that when the edge becomes blocked up, the easiest way to clear it is to clean the edge on the bone surface. Compare with A.95

Fig. A.92 SEM microphotograph of trampling marks on modern monitored cow from Draycott. The internal shape of individual trampling scratches may mimic stone tool marks made by humans, as these marks look very similar to those of A.88. The Draycott specimen was monitored from shortly after time of death and no human agency was involved at any time after death (Andrews and Cook 1985). Field width equals 6 mm

Fig. A.94 Trampling marks from Draycott on a monitored modern cow bone. Linear marks made by trampling tend to be straight. However, the foot/hoof of the animal may produce a circular-like movements while trampling (compare with A.90). Pressure of sediment grains under the animal's weight may also change the direction of the striation. Field width equals 6 mm

Fig. A.93 Trampling mark on a modern bone from a monitored cow carcass from Draycott. Scratches may by themselves be similar to scraping marks. Location of marks on the anatomical element is one of the main criteria to distinguish the use of stone tools or trampling action (compare with A.89). Field width equals 1 mm

Fig. A.95 SEM microphotograph of trampling marks on a modern monitored cow bone from Draycott. Trampling by large animals causes friction of bone surfaces against the limestone substrate. Other agencies related to transport (water and sediment) or environmental conditions (wind and sediment) may also produce linear marks, but they can be only distinguished at a high level of magnification. Compare with A.91. Field width equals 230 microns

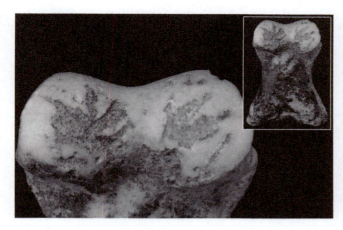

Fig. A.96 Fossil first phalanx of horse from Gough's Cave. Deep grooves are present on the ventral surface of the distal articulation seen as two rosette-shaped multiple marks with a single additional mark laterally. On either side of the proximal articular surface there are two extensive areas of percussion damage (compare with A.307 and A.308) ventrally and extending round on to this surface (not shown here). The location of these marks on articular surfaces suggests dismembering activities by humans

Fig. A.98 Fossil bone. Cuts along the edge of a child clavicle (ATD6-55) from TD6 Atapuerca. These deep and precise cut marks are at attachments of *deltoid* and *pectoralis major* muscles from the chest. The *trapezius* attachment (the neck muscle) from this clavicle is also heavily affected (Photo M. Bautista). Trampling marks tend to be randomly placed, while those made by humans have some reference to muscle attachments (see Chap. 3, A.6 and A.7, compare with A.60)

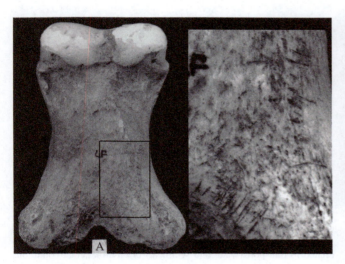

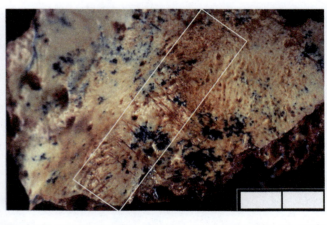

Fig. A.97 Fossil bone. Slicing marks along the lateral side of a first phalanx of horse from Gough's Cave related to the lateral ligament. The location of cuts, sawing or chop marks close to muscle, ligament and tendon attachments confirms the human origin of the marks. Examination of linear marks on fossil bones may indicate different butchering purposes (see also A.99, A.102 and A.103), in this case the disarticulation of the foot

Fig. A.99 Fossil bone. Fragment of human temporal bone from TD6, Atapuerca, bearing cut marks located transversally along the *mastoid* crest where the *sternocleidomastoid* muscle is attached. Location and distribution of cut marks suggest dismembering (detachment of the head, see A.103 and A.100) and defleshing activities. Scale bar 2 cm

Fig. A.100 Fossil bone. Human calvaria broken and cut marked from ▶ Gough's cave (detail in A.101). The human calottes from this site show similar patterns to each other regarding cut marks, percussion marks and breakage. Extensive and long cut marks are present on the parietal and occipital bones, along the sutures covering the insertion area of the *trapezius* and *sternocleidomastoid* muscles. They suggest removal of skin or scalping

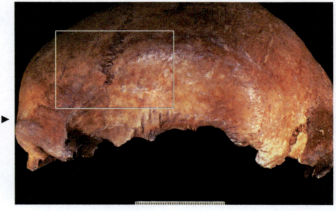

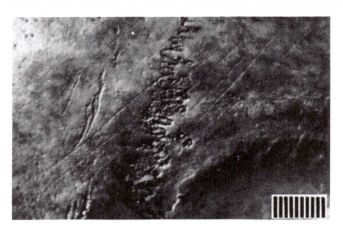

Fig. A.101 Fossil bone. Detail of the parietal (right) and occipital bones of the same calotte A.100 from Gough's Cave showing the cut marks in this region. Location and distribution of cut marks suggest removal of skin or scalping. Scale bar 1 cm

Fig. A.103 Fossil bone. Slicing marks oblique to the vertebral body of this adult human individual from Çatalhöyük. The cut marks are on the atlas vertebra of skeleton 4593 from which the skull had been removed after death (see A.99). The skeleton was complete except for the removal of the skull

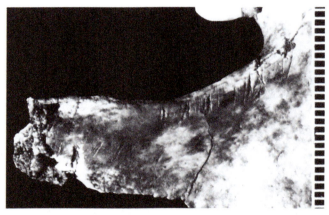

Fig. A.102 Fossil bone. Cut marks on the ascending ramus of a human mandible from Gough's Cave. The location of these cut marks suggest disarticulation of the mandible as a result of *masseter* muscle removal. Cuts are also present on the lingual side of the symphysis (*fossa digastrica* and *spina mentalis*). It is remarkable that one horse symphysis from Gough's Cave also shows cuts on the inner margin of the symphysis, similar to those seen on humans (A.10 and A.83). Scale bar in mm

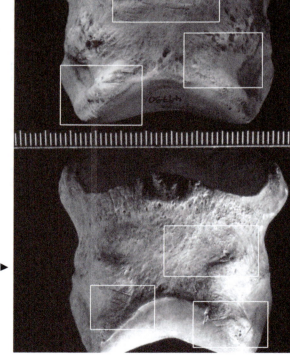

Fig. A.104 Fossil bone. Oblique slicing marks near the articular edge ▶ of opposing second phalanges of a horse from Gough's Cave. The association between muscle attachments and cut marks when considered on the same anatomical element may provide a recurrent pattern of cut marks in an animal assemblage (A.96 and A.97). The *digital flexor* tendons attach to the second phalanx in the place where the cuts occur. It is a long tendon running along most of the limb

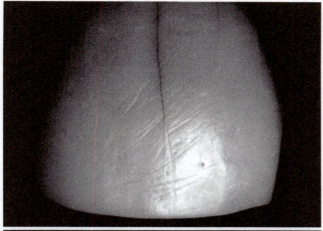

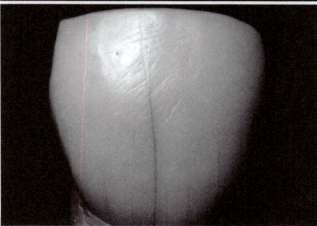

Fig. A.105 Fossil teeth. Stone tool striations may also affect teeth (Sima de los Huesos, Atapuerca). Striations shown on the buccal surface are visible to the naked eye and SEM examination showed indications of handedness. Striations only affect the anterior teeth, are absent on the lingual surfaces, have an oblique distribution and recurrent orientation. All these traits suggest they are the result of *antemortem* processes. Hominins could have cut materials held between the anterior teeth with a stone knife, inadvertently scratching the enamel at the same time. This is an exceptional case in which handedness criteria may be tested, because the anatomical relation of both the passive object (teeth) and the active object (hand and flint) are known. These results provide indirect evidence for lateralization of the brain. Hemispheric dominance could have been present in early hominids, tested in Middle and early Late Pleistocene hominids (Bermudez de Castro et al. 1988). Photo M. Bautista (see Text Fig. 3.6)

Fig. A.106 Modern bone. Scratches produced experimentally after breaking bones with heavy stone blocks to extract the marrow. Bone breakage can only be managed when placing a stone underneath the bones (see Text Fig. 9.1), and several attempts were needed before breaking them (see A.969 and A.972). Detail of scratches above right

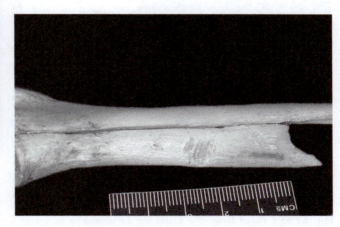

Fig. A.107 Modern bone. Strong similarities to experimental scratches (A.106) were recognized on sheep bones eaten by Koi people (Zoutrivier Village, Gobabeb, on the 7th December, 1965; conditions of the experience are described in Brain 1981). This sheep's radius was laterally broken, as experimentally obtained (see A.972 broken metapodial), and marrow had been extracted. Scratches and breakage are related on this sheep radius

Fig. A.108 SEM micrograph of a fossil bone, detail of linear marks ▶ on a fossil from Vanguard Cave resulting from an impact notch or percussion mark that also produced fine scratches underneath (see A.310). These scratches associated with notch marks are named hammerstone-anvil striations (Turner 1983), as this may be the cause of the striations. These scratches suggest that the bone was already clean of meat (to break the bone already defleshed for marrow extraction) in contrast to chop marks to dismember a bone still covered by meat

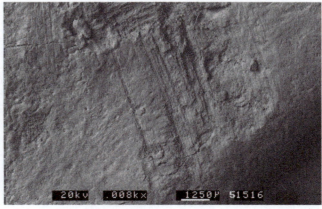

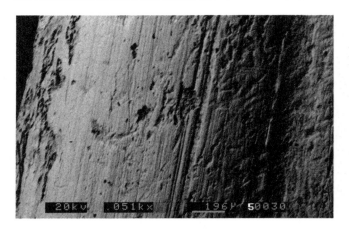

Fig. A.109 Fossil bone. SEM microphotograph of a bone-tool from Mumbwa Cave worked by humans to produce a pointed shape (A.587). The edges of the bone surface have abundant linear and vertical striations. The pattern of longitudinal linear striations suggests human manufacture (see Barham et al. 2000)

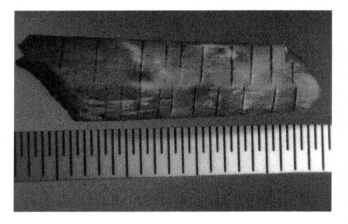

Fig. A.110 Fossil bone. Small bone fragment from a Solutrean-Late Paleolithic site (Cueva de Ambrosio, Spain). The distribution of marks and similar length precisely made cuts do not fit with any butchery purpose. (Specimen provided by S. Ripoll, Photo M. Bautista.). Compare with A.27

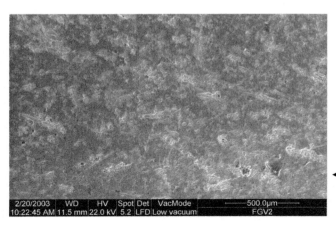

Fig. A.112 SEM microphotograph detail of a grain of sand (white arrow) intruding into the fossil surface and leaving a comet tail groove extending diagonally across the centre of the image. This sample is from an experiment of abrasion where the fossil bone surface was abraded by coarse sands and water for 360 h in a stone polishing tumbler. Compare with A.347 sand and wind abrasion

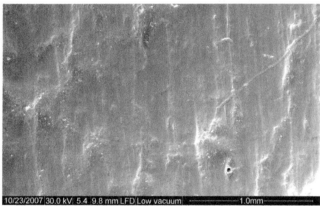

Fig. A.113 SEM microphotograph of a fossil bone, detail of the surface of a horse first phalanx found in Torralba (Spain). The appearance of the phalanx is anomalously shiny compared to other fossils from the site, and the surface is very smooth. Linear striations along the length of the phalanx on the surface of the bone are regular. Compare with A.114

◄ **Fig. A.111** SEM microphotograph of a experimentally abraded fossil bone with coarse sand sediment in water. Abrasion may produce scratches on bone surfaces, but pitting is the main damage. Compare with A.350 and A.351

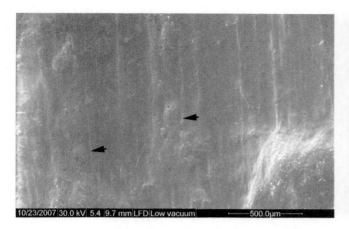

Fig. A.114 SEM microphotograph of a fossil bone, detail of the longitudinal striations found on the horse phalanx from Torralba (Spain). In detail, these striations have the comet tail shape as indicated by d'Errico et al. (1984) as in A.112 and A.115

Fig. A.117 SEM microphotograph of a fossil bone, close up view of the previous striation (A.116) that suggests rolling rather than sliding on the surface. This indicates that the bone has been modified by natural abrasion rather than a human made abrasion

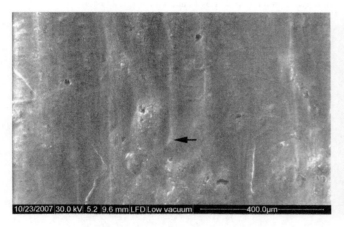

Fig. A.115 SEM microphotograph of a fossil bone, close up view at the centre of A.114 (horse phalanx from Torralba, Spain)

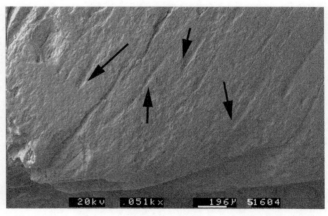

Fig. A.118 SEM microphotograph of a fossil bone from Vanguard Cave abraded by wind. Notice the superficial pitting and the longitudinal and "comet like striation" (as in A.114) on the bone surface (black arrows)

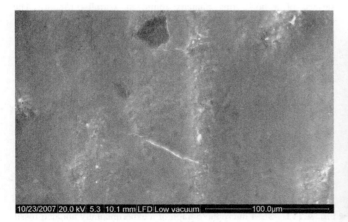

Fig. A.116 SEM microphotograph of a fossil bone, close up view of one of the striations in A.114 and A.115 (horse phalanx from Torralba, Spain), showing an irregular texture at the interior of the scratch, suggesting the sand grains rolled on the bottom of the scratches. Compare with A.337

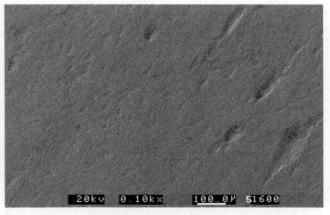

Fig. A.119 SEM microphotograph of a fossil bone, detail of previous fossil from Vanguard Cave showing the superficial pitting and the "comet like striation" on the bone surface (as in A.114)

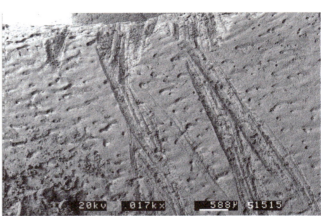

Fig. A.120 Fossil bone. Fragment of rib from Galeria, Atapuerca. Superposition may be observed between two stone tool cut marks. An incision mark (from top left to bottom right) is superimposed on scraping marks (from top right to bottom left). This may support experimental butchery observations i.e., cleaning the edge of the tool against the bone, and then continue cutting remains of meat or ligaments still attached (A.83 and A.91). Photo M. Bautista

Fig. A.123 SEM microphotograph of a fossil bone from Vanguard Cave. Cut marks are interrupted by rodent chewing (top of the image, see A.187 and A.199). This case is interesting because rodents do not usually chew green bone. They tend to chew dry bone that has been exposed for several years on the ground. These chewing marks may, therefore, indicate a prolonged period of humans' abandonment of the site

Fig. A.121 Fossil bone. Fragment of bone from Abric Romani. Conchoidal scars are superimposed on previous cuts (black arrow) indicating the butchery sequence, i.e., dismembering and filleting preceded the breaking of the bone, probably to extract the marrow. In addition, tooth marks (white arrow) cover previously made cut marks (see also A.123), indicating scavenging activities by carnivores after the human activity (A.122). Courtesy of I. Cáceres

Fig. A.124 SEM microphotograph of a fossil bone. Superposition of trampling marks on stone tool cut marks (as in A.1) on a fossil fragment from Abric Romaní. The cut mark runs from bottom middle to top right, and the trampling marks run almost vertically down the bone

◄ **Fig. A.122** Fossil bone. Superposition of tooth marks over incision marks on a fossil bone fragment from Galeria, Atapuerca (as in A.121). Courtesy of I. Cáceres

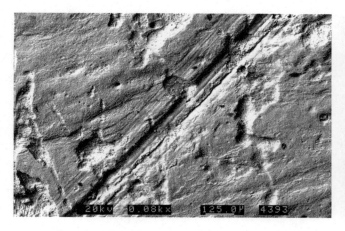

Fig. A.125 SEM microphotograph of a fossil bone. Superposition of root marks over incision marks on a fossil fragment from Vanguard Cave (as in Text Fig. 1.2 and A.172)

Fig. A.127 SEM microphotograph of the hedgehog mandible. Looking at a close up view of the fossil hedgehog mandible from Olduvai Gorge, Bed I, FLKN5, one of the oblique (cut mark) striations (black arrow), with identical inclination to those shown in A.126 was covering one of the longitudinal and wider scratches (considered to be caused by trampling). This evidence would indicate that all of these striations were trampling marks

◀ **Fig. A.126** SEM microphotograph of a hedgehog (*Erinaceus broomi*) fossil mandible from Olduvai Gorge, Bed I, FLKN5 published by Fernández-Jalvo et al. (1989). Based on the traits of these linear marks, the interpretation was that striations obliquely arranged on the mandible were very fine and delicate cut marks. Longitudinal and wider scratches (here indicated by white arrows), were interpreted to be caused by trampling and show similar internal microstructure to cut marks, see A.128). The boxed area is shown further enlarged with an SEM microphotograph showing two of the cut marks characterized by a V-shaped section. Internal microstriations running linearly along the length of the cut and lateral Hertzian cones, indicated by white half-triangles, at the right side of the striations, indicate direction of the cut (shown by the black arrow). The location of these fine cut marks on the anterior end of the mandibular body relates the cuts to skinning activities rather than defleshing

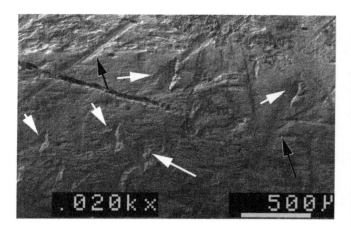

Fig. A.128 SEM microphotograph of the fossil hedgehog mandible at this second inspection showed another type of damage. This consists of a triangular regular pitting (see A.409), all of them similar in size and covering the mandible surface (white arrows). Notice the identical directions between the superimposed mark (bottom right – black arrow) and the striations shown in A.126 (top left margin at this picture)

Fig. A.129 SEM microphotograph of the fossil hedgehog mandible. This triangular pitting has been observed to occur at the end of oblique striations (white arrow, see A.409) as well as some of those vertical. The average measurements (108.72 μm) fit with hedgehog tip spine size

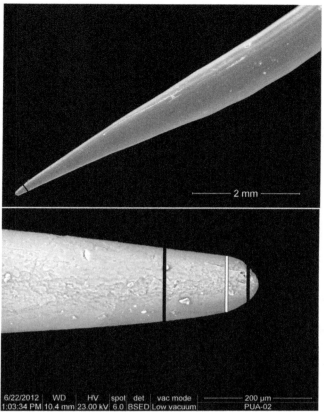

Fig. A.130 SEM microphotograph of modern hedgehog spines. Triangular shaped pits (A.128 and A.409) measured in the hedgehog mandible provide an average (over 14 measurements) of 108.72 μ (shown in the bottom figure as white line), ranging between 140.21 and 65.42 μ (shown at the top figure the maximum measurement taken (140.21 μ, and at the bottom figure as black lines). These measurements, therefore, fit with hedgehog spines dimensions of the tip. Striations and triangular shaped pitting observed on the hedgehog mandible were then likely made by humans 1.7 Ma. *Homo habilis* from Olduvai Bed I might have not used stone tools to skin hedgehogs, but detached the meat using an immediate tool, the spines. Similar behavior has been observed when humans used limestone stone tools (at cave sites) or shells (at seashore sites) to cut. Therefore, these linear marks were not finally made by an inorganic subject, but most likely by the organic pointed hedgehog spine. This links with the next chapter

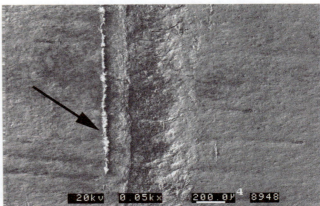

Fig. A.133 SEM microphotograph of a modern bone, experimentally chewed. Detail of the (incisor) tooth mark A.131. "Shoulder effects" (black arrow) as observed on cut marks are produced by irregularities of the tooth incising the bone surface, either because of slightly twisted teeth or because the edge can be unevenly flaked

Fig. A.131 Experimental human chewing marks on a modern pig rib. Top, a light microscope image and bottom, the same image on the SEM. One of the most characteristic traits made by human chewing is a series of shallow linear marks or grooves on the bone surfaces which have been shown experimentally to be made by human incisors. Shallow linear marks have distinctive microstriations restricted to the interior of the mark produced by irregularities of the incisor's edges (see A.134). Bottom, SEM microphotograph of the same marks as above. Linear marks made by humans are not as deep and distinct as tooth scores made by carnivores. These shallow linear marks are too superficial for their morphology to be distinguished by the naked eye, needing high magnification and SEM to examine traits. Left box enlarged in A.132, right box enlarged in A.133

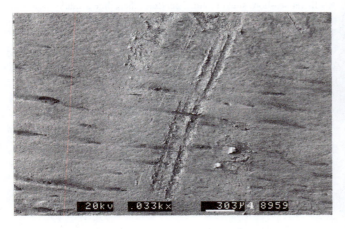

Fig. A.132 SEM microphotograph of a modern bone, experimentally chewed. Detail of the mark made by an incisor (left hand mark of A.131). As seen in cut marks, lateral Hertzian fracture cones are present along the right side of the scratch. These Hertzian fracture cones are caused by physical stress and resistance of the bone to the incisive-friction action. The Hertzian cones are so superficial that they would be likely to be destroyed by subsequent taphonomic processes

Fig. A.134 Modern long bone of a young sheep individual. Experimental chewing by humans. Top, a picture taken with a light microscope. Below, SEM microphotograph of the left linear mark. The tooth mark has formed a shallow crescent pit at one end of the groove and internal striations along the bottom of the groove. Internal striations result from irregularities of the incisive edge of the tooth. The association of a pit and scratches shares similarities with percussion marks described by Blumenschine and Selvagio (1988) or Pickering and Egeland (2006) and the accompanying hammerstone/anvil scratches (Turner 1983). Internal microstriations from human chewing are restricted to the interior of the shallow groove (see here and A.145, A.151), while scratches associated with percussion marks can extend beyond the percussion puncture mark (A.310). The former can only be seen on the SEM

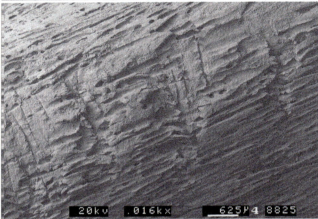

Fig. A.135 Modern rib of a young pig. Top, linear striations caused by human experimental chewing. Bottom, SEM microphotograph of the same area photographed at higher magnification. Linear shallow marks show varying dimensions, width and depth. Fernández-Jalvo and Andrews (2010) provided the dimensions of these chewing marks, ranging between 500 microns and 1.8 mm, but more recently we found scratches that are still narrower (see A.138) suggesting that these experiments have to be extended to obtain a more consistent and statistically significant sample. The range of tooth mark sizes made by humans is clearly much greater than we found in our experiment

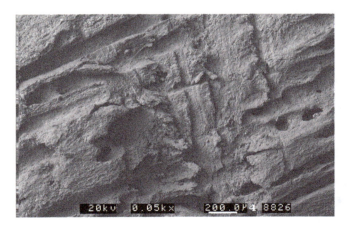

Fig. A.136 SEM microphotograph of a modern pig rib, detail of the central area of A.135, showing linear striations caused by human experimental chewing. These marks could also be confused with shallow linear marks made by trampling, although they differ for human chewing since the striations are restricted to the interior of the groove rather than shallow scratches randomly dispersed all over the bone surface, as present for trampling

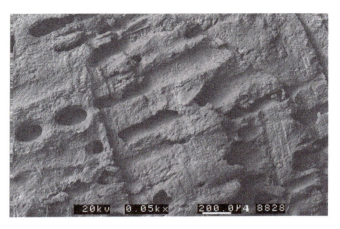

Fig. A.137 SEM microphotograph of a modern pig rib, detail of the left area of A.135, showing linear striations caused by human experimental chewing. Shallow linear marks may appear transverse or oblique to the long axis of the ribs (see A.139), and they also occur on fragile vertebral apophyses, and anatomically salient borders that are not affected by trampling

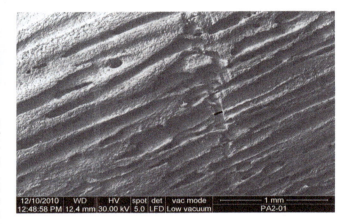

Fig. A.138 SEM microphotograph of a modern pig rib, detail of an isolated linear mark experimentally made by humans chewing rib bones. The width of the scratch is around 80 microns. These tiny and isolated marks are rare (see A.136), but they have to be included in the statistical measurements of human shallow linear marks

Fig. A.139 Modern pig rib. Shallow linear marks on bones chewed by humans are short (A.131). The different texture and, sometimes, color that can be observed in the interior of the modern marks are characteristic in both modern and fossil bones

Fig. A.140 SEM microphotograph of experimental shallow linear mark made by human chewing on a modern young pig rib. The cross-section of the mark outlined at the top right side of the picture has been seen sometimes (A.136) but not so frequently as to be considered a characteristic trait of human chewing

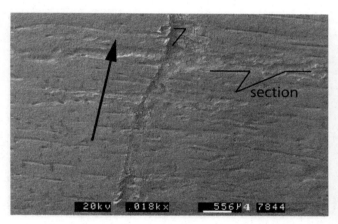

Fig. A.141 SEM microphotograph of experimental human gnawing made by the same experimenter as A.140 and A.142. The groove has a Hertzian fracture cone in the most abraded side of the scratch indicating directionality (following criteria described by Bromage and Boyde 1984). The cross-section of the mark varies along its length

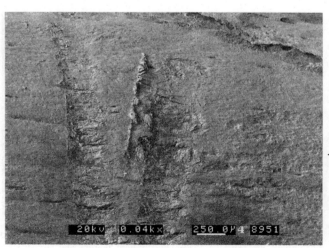

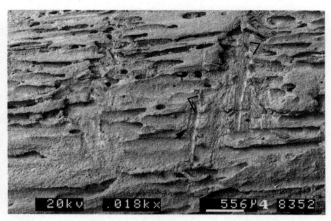

Fig. A.143 SEM microphotograph of experimental human chewing marks on a young pig rib. Apart from Hertzian fracture cones (A.141, directionality of the movement from the bottom to the top), internal multi-striations are distinctive (A.137 and A.147). Differences from trampling are that microstriations are not randomly all over the bone surface as for trampling but located in the interior of the grooves, and the groove is well defined. The cross-section is shallow and flattened

Fig. A.144 SEM microphotograph of linear marks on a fossil bone from Gorham's Cave. Similarities can be observed between these marks and those of A.143. Multi-striations are present at the interior (top right white arrow) of these probable human chewing marks. Scale bar 500 microns

◀ **Fig. A.142** SEM microphotograph of a young sheep long bone experimentally chewed by humans. The shape of this groove is similar to A.134, with a shallow crescent pit. Light internal multi-striations along the bottom of the groove can be distinguished. Lateral Hertzian fracture cones can be distinguished on the sides of the scratch, indicating the directionality of the movement (from top to bottom). A "shoulder effect" as seen in A.133 is distinguished on the left side of the grooves

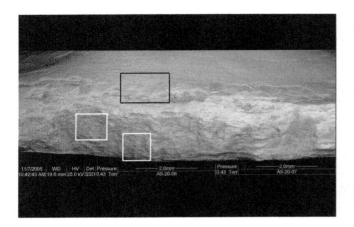

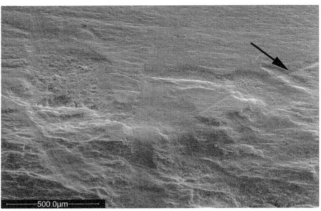

Fig. A.145 SEM composite microphotograph of the edge of a fossil rabbit pelvis found at a Neanderthal occupation level from Gorham's Cave. The fossil edge has a series of grooves, both on surface (black square, A.147) and on broken edge (white squares), similar to experimental marks seen in A.134

Fig. A.148 SEM microphotograph of shallow linear marks on a fossil rabbit pelvis recovered from Gorham's Cave. Microstriations are superimposed on each other, indicating several tries of chewing. Note that these chewing marks interrupt previous incision marks made by a stone tool (black arrow) see also A.370

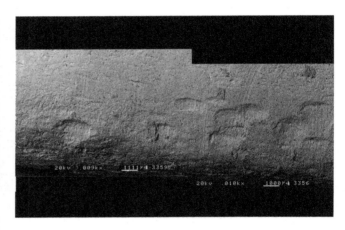

Fig. A.146 Detail of SEM microphotograph (A.145, left side) of a fossil rabbit pelvis from Gorham's Cave. Grooves at the broken edge of the bone (white squares) have both size and shape similar to those on the bone surface, with a crescent pit shape on one end as seen at A.134

Fig. A.149 SEM composite microphotograph of tooth marks along the edge of a fossil rib fragment from TD6, Atapuerca. The arrangement of marks is similar to that of A.145. The grooves have the characteristic shallow crescent pit shape described in A.134 and A.142 at one end of the groove and internal micro-striations. The marks are superimposed on each other

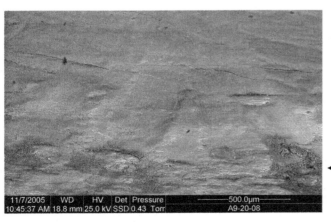

◄ **Fig. A.147** SEM microphotograph of the edge of a fossil rabbit pelvis from Gorham's Cave. Micro-striations are restricted to the interior of the shallow linear marks described in experimental chewing marks A.134, A.136 and A.137. The marks are superimposed on each other

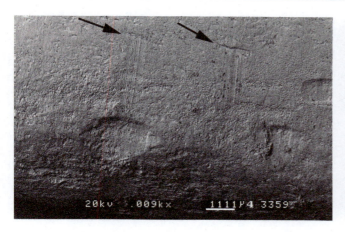

Fig. A.150 SEM microphotograph of a fossil bone, showing a detail of the left side of A.149 (TD6, Atapuerca). Chewing marks are superimposed on each other. The more superficial marks (black arrows) are interrupted by deeper marks at the bottom of the image. Microstriations along the bottom of each groove result from irregularities of the incisor incisive surface

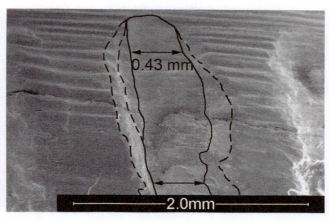

Fig. A.153 SEM microphotograph of a fossil bone. When measuring the width of a linear mark, the narrowest width along the bottom of the groove (solid line) is taken as the representative measurement (see also A.387 and Text Fig. 4.7). The reason for is to avoid over-estimating the width due both to deeper penetration of the conic tooth cuspid into the bone and to flaking of bone along the edges of the groove (broken line). Measurements are usually taken with the light microscope and a caliper

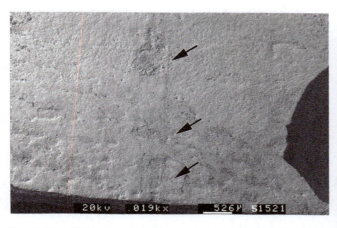

Fig. A.151 SEM microphotograph of a fossil bone fragment from a medium sized mammal from Vanguard Cave. Microstriations are restricted to the interior of the shallow linear marks described in experimentally chewed marks A.134, A.136 and A.137. These shallow marks could be observed using a magnification glass. Description and confirmation, however, may need the use of higher resolution equipment

Fig. A.154 Top, linear marks made by carnivore chewing of a fossil bone from Concud (Spain, 7 Ma). Bottom, SEM microphotograph of the carnivore marks (category b, see text). The carnivore tooth marks on this fossil bone are covered by later root marks (black arrow). These carnivore tooth marks may appear be similar to chop marks (see A.317), but microscopic traits (A.153) and the date of the site excludes hominin involvement. Courtesy of D. Pesquero

◄ **Fig. A.152** Modern cow bone diaphysis chewed by a domestic dog (the bone was originally cut transversally by a butcher). Grooves run parallel to the long axis of the shaft, up to and into the transverse broken edge (black arrows). The shape of these grooves is linked to transverse breakage and grooves are deeper (see A.156) than on the shaft (see A.155 and A.158)

Fig. A.155 Linear marks made by carnivore chewing of a fossil bone from Concud. Marks are shallower and narrower but no shorter than those from A.154. Identifying carnivore species making the tooth marks is based on the size of the marks, the anatomical element that has been chewed and their location on the bone (see text). The range of tooth mark sizes on this fossil is 0.8–2.1 mm width, skeletal element is unknown and location is on the diaphysis. Courtesy of D. Pesquero

Fig. A.158 Isolated linear striations (category b) on a fossil from Concud. These may be difficult to identify as carnivore tooth marks (see A.155 and A.159). The association of grooves with other chewing traits (i.e., punctures on surface or on broken edges), the site and bone assemblage contexts (e.g., bone fragments with smooth surfaces produced by digestion or licking, i.e., heterogeneously rounded bone ends), as well as, traits of bone breakage will help in the interpretation. Courtesy of D. Pesquero

Fig. A.156 Linear marks made by carnivore chewing of a fossil bone from Concud. Tooth marks on the edge of a fossil bone fragment may sometimes be present as grooves as in this example. It is, however, more frequent to find puncture marks associated with broken edges (A.152). Grooves in this case are sharp. Category b1 (category b, see text). Courtesy of D. Pesquero

Fig. A.159 Linear striations produced by carnivore teeth on the diaphysis (category b) of a fossil long bone from Concud. When these marks are analyzed, the minimum width of each of these grooves is measured 1.4–2.6 mm, category b. Courtesy of D. Pesquero

◄ **Fig. A.157** Modern monitored bone. Head of humerus of zebra intensively chewed from Amboseli. Articular bones have thin compact cortical bone that covers a thick and softer layer of cancellous tissue. This contrasts with the thick compact bone on long bone diaphyses. The difference in bone anatomical traits produces differences in size and shape of chewing marks even when made by the same carnivore (A.163). Category b2 (see text). Photo: T. Jorstad, Collection of A. K. Behrensmeyer

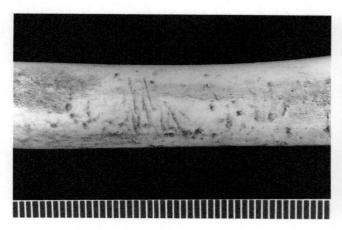

Fig. A.160 Linear striations (category b) and puncture marks (category a) on a recent long bone shaft scavenged by a fox (A.163 and A.363) that chewed a sheep carcass monitored in Neuadd. Reference collections may help in identifying traits of carnivore chewing. The size distribution of these category b marks has been described by Andrews and Armour-Chelu (1998)

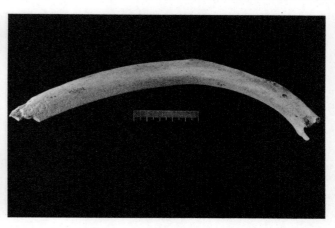

Fig. A.163 Linear striations on a recent rib of horse monitored in Neuadd. Chewing marks made by a small sized carnivore (red fox) are present on the shaft surface (category b) as well as the ends (category b1). Specimen from ND25, collected on the 30/9/90 (label VT54)

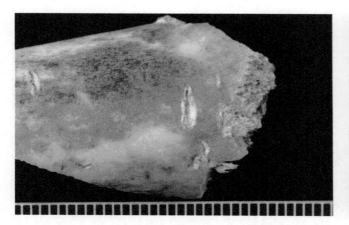

Fig. A.161 Linear striations on a modern bone predated by a lion from a modern monitored carcass (see also A.1022). Collection courtesy of D. Western

Fig. A.164 Linear striations on a modern deer bone of a monitored carcass from Riofrío. The proximal end of this ulna of fallow deer has been chewed by foxes. Abundant grooves as well as punctures can be distinguished at the edge of the proximal end (category b1) as seen in A.163

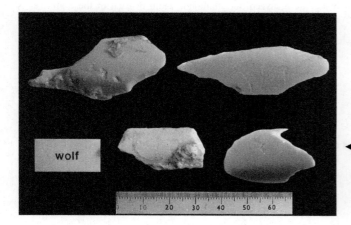

◄ **Fig. A.162** Linear striations (Category b) and surface pits (category a) on modern bones predated by wolves (see also A.1020). Wolves, like hyaenas and dogs, intensively break bone into fragments, and they may then either digest or lick the surface of the bone, producing rounding of the broken edges. To differentiate between abrasion, soil corrosion and digestion, go to Chaps. 6 and 8

Fig. A.165 Linear striations on a modern deer bone of a monitored carcass from Riofrío (A.1036). Herbivores may also chew bones to obtain minerals deficient in their diet (osteophagia). The grooves left on bones may be confused with carnivore chewing marks, but they may be distinguished by evidence that the bones chewed by herbivores were old, such as presence of cracks produced by weathering (Cáceres et al. 2011). Compare with A.166 and A.1034

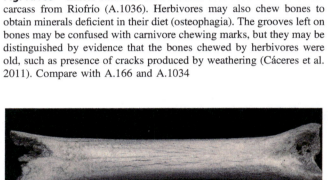

Fig. A.166 Deer, fallow deer, sheep, camel, cow are some examples of herbivores that have been seen to chew bones (Sutcliffe 1973, 1977). This cow metapodial collected from a South African farm (Kwazulu Natal) is weathered and chewed by living cows in the farm. A forked shape at the end of bones is characteristic of bones chewed by herbivores. Cracks are interrupted by incipient chewing (A.165) and the ends have early stages of the distinctive forked shapes (A.1035 and A.1040)

Fig. A.168 Linear striations on a modern bone from Riofrío. Although any bone is susceptible to be chewed by herbivores, it is more frequent to observe signs of osteophagia on long bones (A.1035). This is an incipient case of chewing by herbivores on a proximal end of a fallow-deer metatarsal

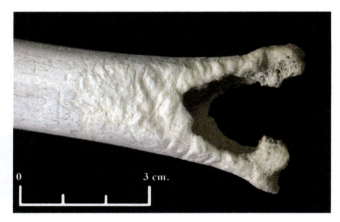

Fig. A.169 Linear striations on a modern bone from Riofrío. The ends of bones intensively chewed by herbivores acquire a characteristic forked shape. Herbivores chew dry and old bones often associated with cracks produced by weathering desiccation (A.165 and A.1034). These two features distinguish herbivore from carnivore chewing. Courtesy of I. Cáceres

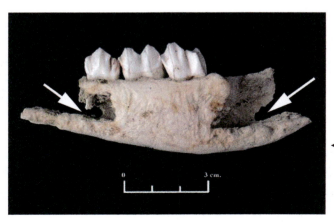

◀ **Fig. A.167** Linear marks and forked ends of a modern mandible chewed by deer at Riofrío. Carnivores (avian or mammalian) may consume mandibles in similar ways, but they chew the inferior border as it has a high content of blood (A.405). The ends of the mandible chewed by herbivores have a forked shape (white arrows see A.1040), similar to long bones. Courtesy of I. Cáceres

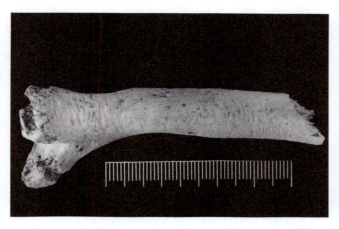

Fig. A.170 Linear marks on a modern rib from monitored specimens in Neuadd (ND 25, collected on 30/9/90 (VT54). Extensive grooves on the rib shaft have been made by chewing by a small sized carnivore (red fox). Note the marks on the left end of the bone have wider grooves and more rounded shape indicating herbivore (sheep) chewing. The left end is also rounded by the licking action of herbivores (A.1037)

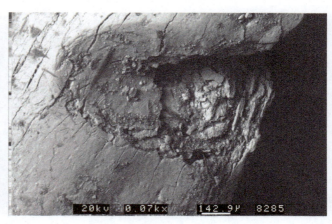

Fig. A.173 SEM microphotograph of a linear mark on a modern small mammal femur chewed by white tailed mongoose, *Ichneumia albicauda* from Meswa Bridge, Kenya (see also A.394). The bone comes from a modern latrine that was being monitored in 1979 (Andrews and Evans 1983)

Fig. A.171 Linear marks on a recent weathered bone fragment from the Abu Dhabi desert bearing tooth marks on the bone surface. Superimposed modifications of subsequent taphonomic agents, such as weathering in this case, may distort features and dimensions (such as width of the grooves A.153)

Fig. A.174 SEM microphotograph of linear marks on the distal end of a recent monitored vole humerus from Neuadd. Chewing marks are also produced by small mammals such as soricids. Marks on the lateral sides of this distal end of a rodent humerus were made by a species of soricid, probably *Sorex araneus*. The rodent had been trapped, and while the carcass was still in the trap it was scavenged by shrews. Field width equals 9.5 mm. Compare with A.182

◄ **Fig. A.172** Linear marks on a fossil bone from Concud. Superimposed root marks (A.125) or soil corrosion may alter or obscure distinguishing traits of the earlier marks, making it difficult to obtain good measurements of tooth marks recorded on the bone surfaces. Courtesy of D. Pesquero

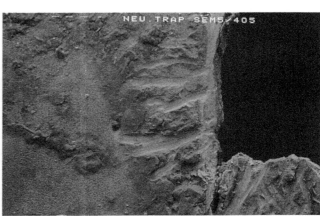

Fig. A.175 SEM microphotograph of linear marks on the distal end of a recent monitored vole humerus from Neuadd. Small mammal postcrania were heavily chewed by shrews on the outer surfaces. Linear marks made by soricids are similar to those of other carnivores except for difference of scale and striated flattened grooves. Field width equals 3 mm. Compare with A.183

Fig. A.178 SEM microphotograph of the proximal end of a monitored vole humerus from Neuadd. Some tooth marks are present as deep sinuous groves more flattened than usual in carnivore chewing, which usually have a more U-shaped cross section. Enlargement of A.177. Field width equals 400 microns

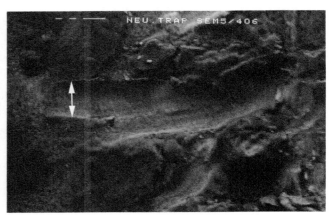

Fig. A.176 SEM microphotograph of the proximal end of a recent monitored vole humerus scavenged by shrews from Neuadd. The specimens were trapped inside a confined roof space in a house. Remains were eaten while still in the trap, totally protected from any agent except scavenging by other small mammals. Articular ends of limb bones were partially or totally eaten away, resulting in local damage, very different from the corrosive effects of digestion. Field width equals 2.6 mm. Compare with A.182

Fig. A.179 SEM microphotograph of the proximal end of a monitored vole humerus from Neuadd. Enlarged view of soricid chewing mark. Some chewing grooves present a multiple striated flatenned grooves. The width of the groove shown by the arrow is 17 microns. Field width equals 140 microns. Compare with A.180

◄ **Fig. A.177** SEM microphotograph of linear marks on the interior surface of a monitored vole humerus from Neuadd. Enlargement of the individual gnaw marks of A.176. Some vole postcranial bones were intensively chewed and broken by shrews, with linear chewing marks visible on the inner surfaces of broken bones. Field width equals 1.3 mm

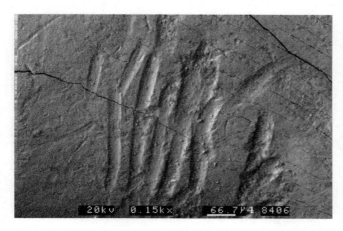

Fig. A.180 SEM microphotograph of fossil rodent incisor from Arroyo Seco site, detail of linear marks at a similar magnification to A.178. These marks are slightly larger than those of the Neuadd specimen A.179 (approximately 30 microns in breadth). The shape and traits are similar to those found in Neuadd experiment as made by soricids. Courtesy of G. Gómez

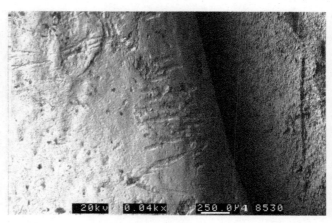

Fig. A.183 SEM microphotograph of linear marks on fossil a rodent incisor from Arroyo Seco. The bone surface shows straight sets of scratches along the edge following a particular pattern similar to A.181 and A.186. The likely agent was soricid, but the actual agent is unknown. Courtesy of G. Gómez

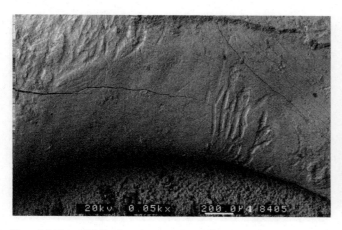

Fig. A.181 SEM microphotograph of linear marks A.180 from Arroyo Seco (similar magnification to A.177 and A.184). Scratches are present on both dentine and enamel of this fossil rodent tooth. Courtesy of G. Gómez

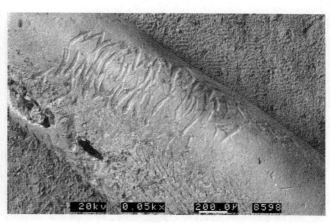

Fig. A.184 SEM microphotograph of linear marks on a fossil rodent incisor from Arroyo Seco. Scratches are showing apparent gnawing using upper and lower incisors (same magnification to A.181). The likely agent was soricid, but the actual agent is unknown. Courtesy of G. Gómez

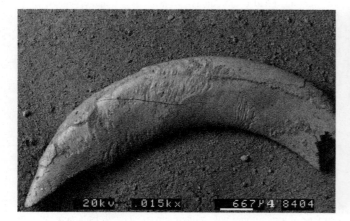

Fig. A.182 SEM microphotograph of linear marks on a fossil rodent incisor than from Arroyo Seco showing location and distribution of scratches on a rodent incisor at a similar magnification to A.185. Courtesy of G. Gómez

Fig. A.185 SEM microphotograph of linear marks on a fossil rodent ulna from Atapuerca. Scratches are showing apparent gnawing using upper and lower incisors (similar magnification to A.182). The likely agent was soricid, but the actual agent is unknown

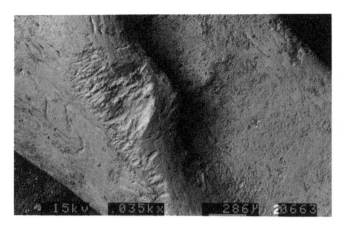

Fig. A.186 SEM microphotograph of linear marks on a fossil rodent ulna from Atapuerca, with close up view of straight scratches following a similar pattern to those observed in the vole humerus of Neuadd (A.176) and in fossil specimens (e.g.: A.183)

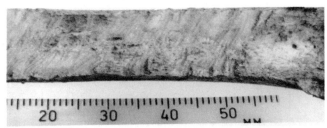

Fig. A.189 Linear marks on a recent monitored bone from Neuadd. These marks are located at the edge of the bone, and groove width and extension of the marks depend on the size of the rodent (see A.195 and A.200). The flat bottomed marks are made by the upper incisors, and on the reverse side of the bone may often be seen puncture marks made by the lower incisors

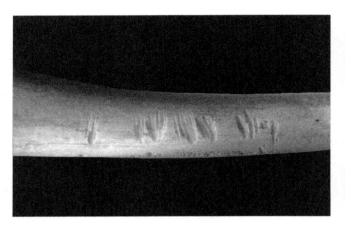

Fig. A.187 Linear marks on a recent monitored bone from Neuadd showing gnawing marks made by rodents (A.123 and A.196). Rodents may chew bones to make up for mineral deficiency in their diet or to wear down their continually growing teeth. Most rodents and squirrels need to gnaw old and dry bones to wear out their teeth and they do not usually gnaw greasy bones which would occlude their incisors. Rats may chew bones for feeding purposes. The diameter of single marks is 1.0 mm

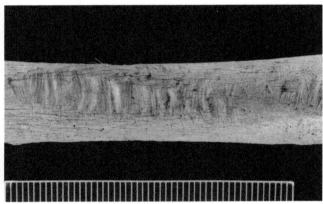

Fig. A.190 Linear marks on a recent monitored bone from Neuadd. The intensity of rodent chewing is variable and depends on a number of environmental factors: physiological (e.g.: lack of phosphates or need of wearing out their teeth), or the state of the bone (dry vs. green bones). They usually form sets of marks (A.123) with flat or slightly rounded bottoms (A.198)

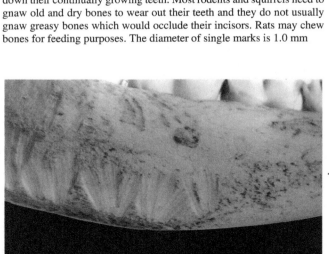

◄ **Fig. A.188** Linear marks left by rodents on a recent mandible monitored in the Neuadd fields. Gnawing by rodents is very characteristic with elongated and parallel sided double grooves (A.193). Along the inferior border of the deer mandible may be seen the roseate shaped groups of grooves that occurs when the rodent anchors its lower jaw on one side of the mandible and pivots against this with its upper jaw to produce a series of grooves converging downwards. The diameter of single marks is 0.9 mm. Field width approximately 10 cm

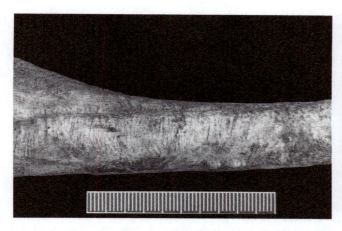

Fig. A.191 Linear marks on a recent monitored bone from Neuadd. Intense gnawing on bones probably depends on the state of the bone. The drier the bones the more intensive the chewing (A.189 and A.1033). The width of grooves of Neuadd specimens varies between 0.57 and 1 mm which is the size of rat gnawing

Fig. A.194 Linear marks on fossil phalanges of *Hipparion* from Concud chewed by rodents. Picture Courtesy of D. Pesquero (see also A.1033)

Fig. A.192 Linear marks on a recent skull of a young fallow deer from Riofrío. The presence of bones gnawed by herbivores (A.165 and A.1035) and by rodents at the same site has the potential to provide information on the environment such as lack of minerals in the deer's diet

Fig. A.195 Linear marks on fossil bone from Concud. The shape and size of these gnawing marks suggest the incidence of a tiny rodent (notice the scale bar is only 0.5 cm). Courtesy of D. Pesquero

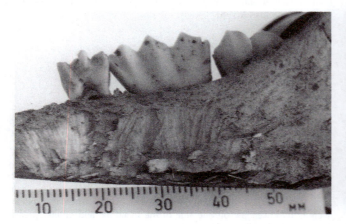

Fig. A.193 Linear marks on a recent monitored bone from Neuadd made by rodent gnawing. No apparent selection is made by rodents regarding the anatomical element selected to gnaw. Mandibles, flat and long bones or skulls are all gnawed (A.188). Long bones and long bone fragments, however, have been found to be slightly more frequently gnawed (A.190 and A.195)

Fig. A.196 Linear marks on a fossil from Concud. The site at Concud is a lake shore environment, therefore, a wet environment. The frequency of gnawed bones by rodents is limited to 4 fragments (0.08%): two bones and two tortoise carapaces. This picture is one of the tortoise carapace fragments gnawed by rodents (see A.197). Courtesy of D. Pesquero

Fig. A.197 Linear marks on fossil from Concud. Incidence of bone gnawing by small rodents appears restricted to small areas of the site, like this tortoise carapace (and A.194, A.195 and A.196). The low incidence of rodent gnawing (0.08%) fits with the lakeshore wet environment of Concud that would have had few dry bones. Courtesy of D. Pesquero

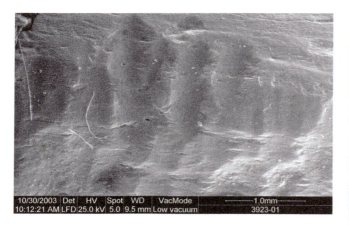

Fig. A.198 SEM microphotograph of linear marks on fossil bone from Concud (A.196). The width of these grooves varies between 0.16 and 0.38 mm. (Pesquero 2006)

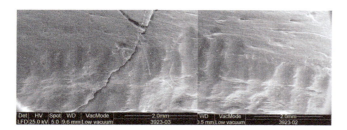

Fig. A.199 SEM composite microphotograph of linear marks on A.196 fossil bone from Concud. The bone has been gnawed by a small mammal along the broken edge of the fossil bone

Fig. A.202 SEM microphotograph of linear marks on a modern long ▶ bone fragment collected from Olkarien Gorge where an extensive community of Ruppell's vultures (*Gyps rueppellii*) nests along the gorge. At the bottom of the gorge, flat bone fragments were frequently punctured, especially skulls (see A.403 and A.658), and grooves were observed on large bones. These linear marks are assumed to be vultures' beak marks

Fig. A.200 Linear marks on a recent bone made by a porcupine (*Hystrix*). This species is a frequent gnawer of bones in Africa, and it also brings them to its den. The activity of porcupines has been seen in many fossil sites (A.201). A special reference is given to Swartkrans (South Africa) where C. K. Brain made a complete study of anatomical elements brought to the den (Brain 1981). Collection of A. J. Sutcliffe

Fig. A.201 Linear marks on fossil bone made by a large rodent at the Galeria site of Atapuerca. Grooves vary between 0.78 and 1.6 mm fitting with porcupines (A.200). The site is a karstic cavity, relatively humid, but directly connected to the exterior through a vertical shaft. Bones chewed by rodents suggest long periods of bone exposure on the ground

Fig. A.203 SEM microphotograph of a close up view of linear marks on a modern long bone fragment collected from Olkarien Gorge shown in A.403 showing a set of superimposed scratches on the bone surface apparently made by vultures' beaks

Fig. A.204 SEM microphotograph of a close up view of linear marks on a modern long bone fragment collected from Olkarien Gorge shown in A.202 and A.203. A characteristic trait of these linear marks is the smooth bottom, absence of internal microstriations and flat cross section. The width of the marks is 200–300 microns

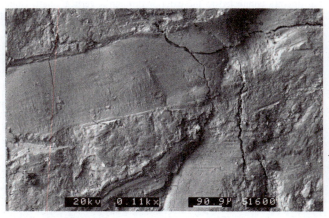

Fig. A.206 Carcass of a monitored male fallow deer from Riofrío scavenged by vultures and foxes. At this stage (3 months after death) only some tissues remained (mainly skin and ligaments), and dermestids fed on the carcass (A.1061). These beetle larvae are frequently used by natural history museums and collectors to clean and prepare skeletons for study and exhibition. Their population has to be carefully controlled as they may become a pest. Photo: Z. San Pedro

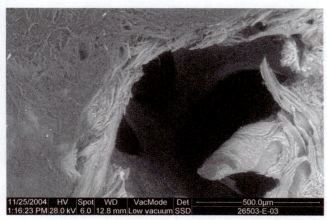

Fig. A.207 SEM microphotograph of a modern vertebra of *Milvus*, from taxidermic MNCN collection, prepared using dermestids (A.410). The larvae of these beetles have strong jaws and are able to cut thin bone as seen here. Forensic entomology may identify the presence of insect's bodies found with human corpses, but their presence may also be detected by the effects they produced on the bone. We are still far from this proxy-taxonomic identification, but we can make preliminary identifications based on differences in damage caused by insects

◄ **Fig. A.205** SEM microphotograph of a close up view of linear marks on a modern long bone fragment collected from Olkarien Gorge A.204, one of the grooves observed on the modern bone fragment from Olkarien Gorge (Tanzania). Notice the impact or puncture mark (top right of the picture)

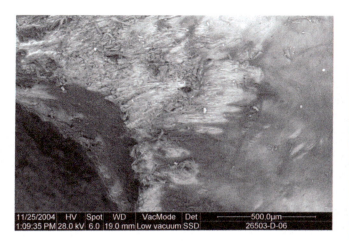

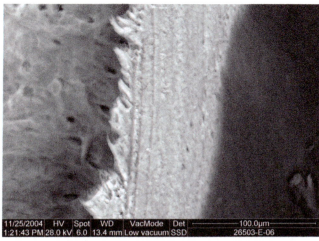

Fig. A.208 SEM microphotograph of a modern bone (from taxidermic MNCN collection). Detail of linear marks made by dermestid larvae when eating soft tissues (A.410). Modern bones prepared using dermestids are not fully suitable for SEM analyses, because most of the bone still keeps a greasy layer that covers the bone surface and prevents surface studies

Fig. A.211 SEM microphotograph of a modern bone (from taxidermic MNCN collection). Detail of the edge of another chewing mark made by dermestids showing the 'sawed' edge (see also A.207 and A.410)

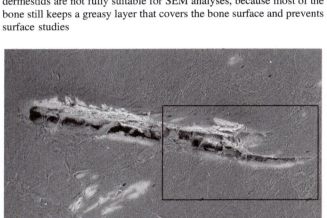

Fig. A.209 SEM microphotograph of a modern bone (from taxidermic MNCN collection). Detail of an isolated linear mark on a recent bone surface made by dermestids. This deep and straight scratch shows the strength of the jaws of these beetle larvae. The box on the right side shows a detail of the incision (A.210) made by dermestids on the bone surface. Note the depth and precise incision of the linear mark

Fig. A.212 Vertebrae or ribs, or any bone are incorporated into the structure of termitaria. This modern rib fragment was transversally cut by a butcher. This picture is a detail of the termite mountain near Maun (Botswana) shown in Text Fig. 3.7 located near a long term established camp site

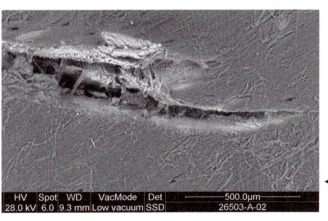

◀ **Fig. A.210** SEM microphotograph of a modern bone (from taxidermic MNCN collection). Detail of A.209 showing the edge of the chewing mark made by dermestids. It has a 'sawed' appearance

Fig. A.213 Modern vertebrae (see also A.414) collected from a termite nest in Kenya. The interior of the vertebrae is filled up by a sticky mud that termites use to build their tunnels and termitaria Photo: T. Jorstad Collection of A. K. Behrensmeyer

Fig. A.214 SEM microphotograph detail of one of the modern vertebrae in A.213 and A.414 showing linear marks on the edge of a hole which penetrates this recent bone. The hole was made by termites. The grooves are seen to be sinuous in shape, probably due to the capacity of cutting by termite jaws. Specimen from A. K. Behrensmeyer's collection

Fig. A.215 SEM microphotograph of sinuous linear marks on a modern vertebra (A.213 and A.414) penetrated by termites. Specimen from A. K. Behrensmeyer's collection

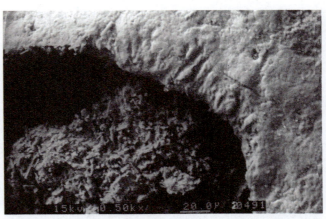

Fig. A.216 SEM microphotograph of a modern human skull fragment from Spitalfield Crypt showing linear marks around a hole similar to previous A.215 and A.214. These marks, however, are straight and more similar to those made by dermestids (A.208), except for the size which is smaller than linear marks made by dermestids

Fig. A.217 SEM microphotograph of a modern bone, detail of grooves and linear striations possibly made by Coleoptera larvae on human burials in the Spitalfield Crypt (A.419). Compare with A.421

Fig. A.218 SEM microphotograph of a modern bone, detail of the interior linear striations possibly made by Coleoptera larvae on human burials in the Spitalfield Crypt. Detail of A.217. The insect that made these grooves produced vertical and straight scratches (white arrow) on the walls of the grooves, differing in this respect from dermestids and termites (A.208 and A.216)

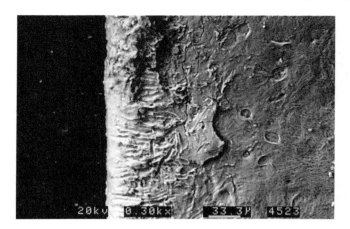

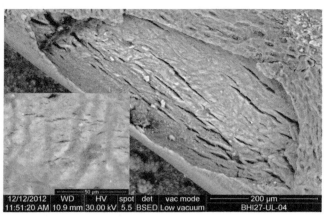

Fig. A.219 Linear striations on a recent bone from a rodent carcass buried by earthworms. The scapula which is shown here has scratches along the lateral edge. The rodent carcass was intact initially, and no taphonomic agent other than soil micro-organisms and earthworms were present under the experimental conditions laid down. These linear marks are similar in size and shape to A.216 (see details of the experiment in Armour-Chelu and Andrews (1994))

Fig. A.221 SEM microphotograph of linear marks on a modern bone from experimental burials. The grooves follow a characteristic pattern of linear marks at the interior of a hole apparently also made by insects. The pattern differs from A.218 and A.219

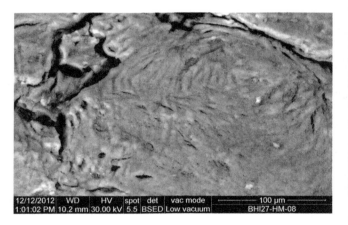

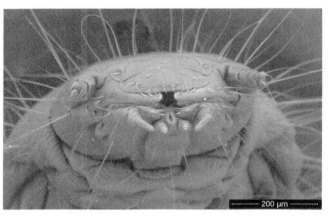

Fig. A.220 SEM microphotograph of linear marks on a modern bone from experimental burials. The bone surface shows sets of scratches dispersed along anatomical edges that appear to form a pattern. The likely agent is a species of insect, but the identity of the species is unknown (A.222)

Fig. A.222 SEM microphotograph of a possible maker of linear marks on bones (A.220 and A.221). It was recovered from burial experiments of rodent carcasses in different soil qualities, more specifically from horizon A (silty and organic soil). It is the larval stage of an insect so far unidentified. Identification of insects from traces left on bones is a promising field that needs highly controlled laboratory experiments. We are still far from this pseudo-taxonomic identification, but it may certainly have a high taphonomic and forensic value

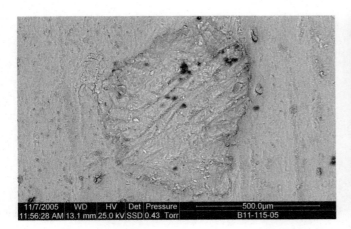

Fig. A.223 SEM microphotograph of a fossil bone fragment from Gorham's Cave (A.422). The shape of this modification suggests a combination of chemical damage, that produced the circular scar, and physical damage by which the interior was physically scratched. Such a combination could be the result of arachnids feeding. Compare with A.221

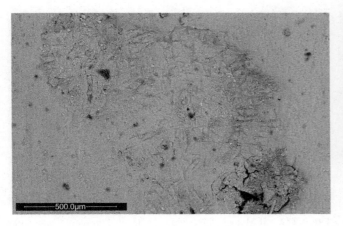

Fig. A.224 SEM microphotograph of a fossil bone, detail of linear marks with modifications similar to A.223 from Gorham's Cave. The resulting physico-chemical combination may probably be made by spiders. Arachnids exude digestive juices produced in their stomachs for external pre-digestion. When the bone is softened, they may then eat the surface of the bone, scratching the bone surface with their pedipalps and chelicerae

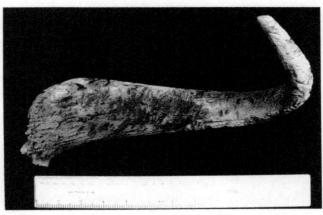

Fig. A.225 Linear striations on a recent horn formed by moth larvae tunneling through wildebeest horn from Kenya (A. J. Sutcliffe's collection, A.226)

Fig. A.226 Detail of linear striations on a recent horn (A.225) by moth larvae tunneling through wildebeest horn from Kenya. Detail of grooves made by moth larvae on wildebeest horn, showing deep and sinuous large-scale tunneling. A. J. Sutcliffe's collection

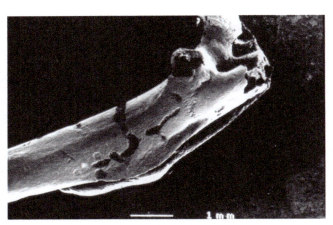

Fig. A.227 Linear striations on a fossil bone bearing root marks. Root marks are rarely straight for more than a few millimeters and commonly divided into branches. They may be single, or more commonly they occur in some abundance, as in this picture, often covering the surface of the bone to the extent that the surface texture of the bone appears chemically corroded (compare with A.228). Photo: M. Bautista

Fig. A.230 SEM microphotograph of a rodent femur from Monte di Tuda (Corsica, Holocene) a rock shelter (Sánchez et al. 1997). The presence of root marks characterizes both the depositional environment, if a bone is resting in or on top of a biologically active soil, and the diagenetic environment of fossilization. Linear marks made by root marks may mimic tooth mark grooves by carnivores (A.153, also compare with stone tool marks A.14) .

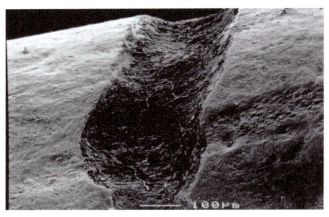

Fig. A.228 SEM microphotograph of a modern monitored bone damaged by plant roots. The diameter of the root mark is wider than the size of the root. Fungi or bacteria envelope smaller or secondary roots and filaments (root hair or rhizoids) and extend out into the bone corroding it (A.880). The benefit of this association is that the cell membrane of fungi/bacteria excrete organic acids to obtain nutrients from the soil, and some of these are taken up by the roots of vascular plants

Fig. A.231 SEM microphotograph of a characteristic U-shaped section of a root mark on bone (from Monte di Tuda, Corsica, Holocene). The interior of the mark has a corroded aspect produced by the symbiotic association of fungi or bacteria with roots of vascular plants (compare with A.14). Description and confirmation of isolated uncertain root marks may need the use of higher resolution equipment (SEM) and high magnification images

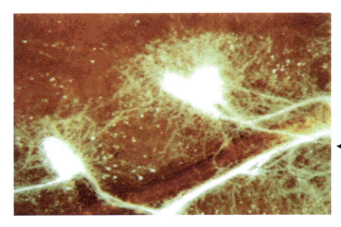

◀ **Fig. A.229** Mycrorhiza, symbiotic association of fungi with plant roots. The "fluffy" white hyphae of the mycorrhizal fungus *Rhizopogon rubescens* have enveloped the smaller roots of a Virginia pine seedling. Note that some of the mycelium extends out into the surrounding environment. http://tolweb.org/tree?group=Fungi&contgroup=Eukaryotes. Copyright © J. B. Anderson 1996. (see A.228)

Fig. A.232 SEM microphotograph of a U-shaped section linear mark on the bone surface of a recent monitored bone buried for 8 years. This is a specimen from the cow carcass monitored at Draycott (see also A.252 and A.253). Like insect damage, traits of root marks may indicate the plant type. Further experimental studies are needed and are currently in progress to find criteria to distinguish groups of plants, which would provide invaluable environmental information

Fig. A.235 Fossil bone. Linear marks identified as root marks are rarely as elongated as incisions (white arrow), but where they may mimic cut marks. Typical traits of cut marks as described in A.1 or trampling (grey arrow) are the result of physical action, while root marks (black arrows) are a product of chemical damage

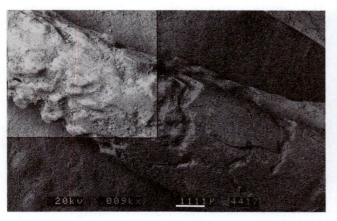

Fig. A.233 SEM microphotograph of root marks on a small mammal fossil bone from Vanguard Cave. Root marks may mimic digestion effects when root marks are densely concentrated on the bone surface, such as seen at the highlighted part of this SEM image (see A.869 and A.870)

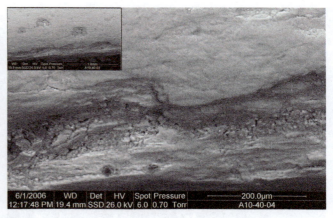

Fig. A.236 SEM microphotograph of a linear striation on fossil bone from Gorhams Cave. This mark is identified as a root mark, but its shape slightly mimics Hertzian fracture cones. This is exceptional, however, and is rarely seen on root marks. Where it does occur, the mark might be mistaken for a cut mark if the fossil is analyzed at inappropriate magnification. A close up view of the groove shows chemical damage at the interior of the mark which confirms it is a root mark. Compare with A.240

◀ **Fig. A.234** Photograph of linear marks on a fossil bone from Cueva Ambrosio where the shape of isolated root marks may mimic tooth marks, cut marks and/or trampling marks. To identify this modification as the result of root marks, the internal surfaces must show evidence of chemical corrosion in the interior of the groove (as explained in A.228). This trait is in contrast to grooves made by teeth, cuts or trampling that are result of physical action. Specimen provided by S. Ripoll

Fig. A.237 SEM microphotograph of a fossil from Abric Romani showing an isolated long root mark, describing a slightly sinuous groove. This sinuous shape is more frequent for root marks, than long straight marks as seen in A.235

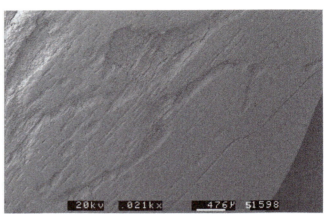

Fig. A.240 SEM microphotograph of root marks on a fossil from Vanguard Cave. The area affected by the root has internal cracks as in A.239, A.242 and A.425

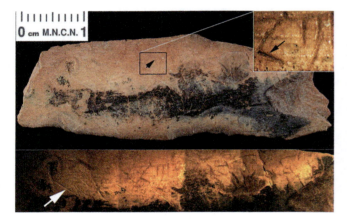

Fig. A.238 Fossil bone. Linear striations from Cueva Ambrosio. Some root marks may imitate engraved bones. This case is still more exceptional than A.236. Schematic figures could be distinguished and be interpreted as a product of human action, especially when the fossil also has cut marks on the surface (one indicated by a white arrow). These "figures", however, are root marks, surrounded by a fringe that confirms their chemical origin (see the small square on the top figure)

Fig. A.241 SEM microphotograph of root marks on a fossil from Vanguard Cave showing a detail of the linear striation on A.240, and cracks inside the root mark

◄ **Fig. A.239** Linear striations on a fossil bone from Cueva Ambrosio. Some root marks have cracks in the interior of the groove. These cracks are not always present, but when this happens, it is a feature that distinguishes root marks, because it is a response of the bone to chemical damage produced by the root. Compare A.240 having cracks in the interior of the groove to A.245 having no internal cracking. Specimen provided by S. Ripoll

Fig. A.242 SEM microphotograph of root marks on a small mammal fossil long bone from Atapuerca. The grooves made by roots have internal cracking that distinguishes their origin as root marks. The field width equals 7 mm

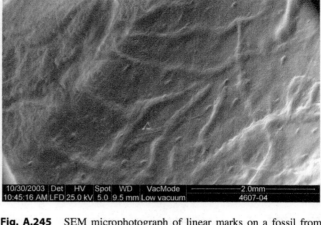

Fig. A.245 SEM microphotograph of linear marks on a fossil from Concud. The typical sinuous shape pattern of grooves and branching show these marks to have been made by roots (compare with A.425)

Fig. A.243 SEM microphotograph of linear root marks on a fossil bone fragment from Atapuerca showing cracking in the interior of the root mark (see also A.425)

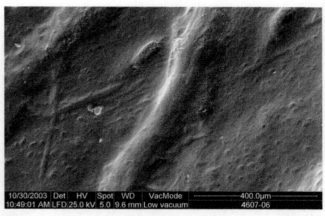

Fig. A.246 SEM microphotograph of linear marks on a fossil from Concud. This root mark is a detail of A.245 and it does not exhibit cracks inside the groove. This may be due to different type of plant. Traits to distinguish root marks from tooth or trampling marks consist of the presence of branching and the sinuous shapes of the marks. Compare A.239 having cracks in the interior of the groove versus A.245 with no internal cracking in the groove

Fig. A.244 SEM microphotograph of a sinuous root mark on a fossil bone fragment from Atapuerca showing cracking in the interior of the root mark (as in A.429)

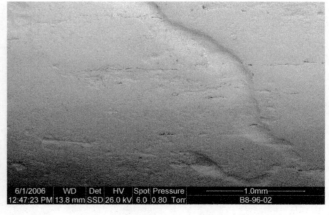

Fig. A.247 SEM microphotograph of a fossil bone, detail of a linear striation from Gorham's Cave. The sinuous shape identifies it as a root mark bearing no internal cracks

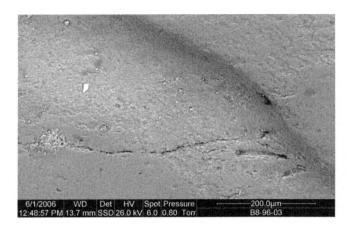

Fig. A.248 SEM microphotograph of a fossil bone, detail of a linear striation from Gorham's Cave, detail of A.247, showing no cracks in the interior of the root mark

Fig. A.251 SEM microphotograph of linear marks on a fossil from Abric Romaini (as in A.252). Many fossils from this site have concentrations of root marks

Fig. A.249 Intense root marked modern human bone found in a modern burial at Paşalar, Turkey. The entire surface of the bone appears corroded because of the dense concentration of root marks, so dense that little of the surface bone remains (Scale in millimeters). Compare with A.227

Fig. A.252 SEM microphotograph of linear marks made by roots on a modern bone from Draycott. The marks vary greatly in width, and they are narrower than those on A.232. The vegetation at this site was a mixture of small trees and herbaceous vegetation, and it is possible that the two sizes of root marks were made by separate types of plant. Further experimental studies may help to find criteria to distinguish groups of plants. The field width equals 5 mm

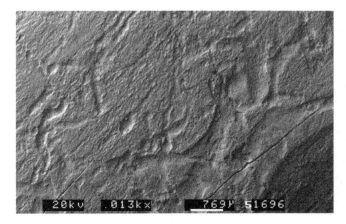

Fig. A.250 SEM microphotograph of linear marks on a fossil bone from Abric Romaini showing several grooves made by roots. The site is a travertine rock shelter that had abundant vegetation in the past

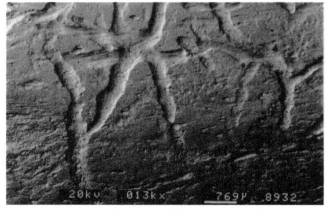

Fig. A.253 SEM microphotograph of linear marks made by roots on the bone surface of a large modern bird bone from Links of Noltland. The interior of the marks is smooth as in A.245

Fig. A.254 Linear striations on a fossil bone from Cueva Ambrosio that are branched and straight. They overlie a previous cut mark (black arrow). The fossil comes from a cave shelter with open air contact, showing the presence of plants that damaged the fossil surface (see also A.430). Specimen provided by S. Ripoll

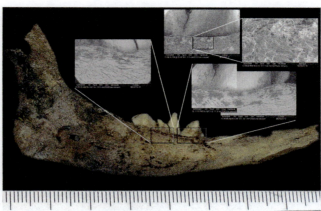

Fig. A.257 Mandible of the modern young fallow deer from A.256 photographed after moss was removed. Some areas seen with the SEM show that the root marking is not linear as is the case with root marks from vascular plants (see also A.258)

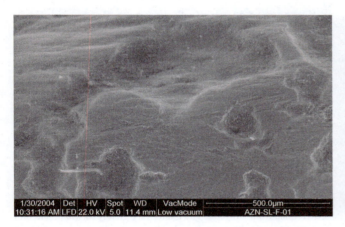

Fig. A.255 SEM microphotograph of a fossil bone, detail of linear striations from Azokh North Cave. The root marks have a darker color in the interior of the grooves as seen on the SEM using BSE mode, which shows differences in density. The darker color suggests the presence of organic matter, so that these root marks could feasibly be recently made (see A.228 and A.880)

Fig. A.258 Linear striations of a recent monitored bone from Neuadd showing marks made by moss. These long bones were densely covered by moss when collected from the surface, and these images show the bones after the moss was removed (A.257)

Fig. A.256 Mandible of a modern monitored young fallow deer from Riofrío collected from the surface and covered with moss, photographed before moss removal (see A.257 and A.784)

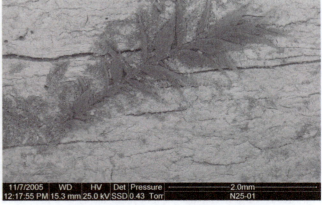

Fig. A.259 SEM microphotograph of a moss growing on a modern bone surface from Neuadd. This is superficial at an incipient stage. Around and underneath the moss, an area of apparently damaged bone (grayish color) can be distinguished (A.260 and A.261)

Fig. A.260 SEM microphotograph of a superficial linear mark on a recent monitored bone from Neuadd. The bone was covered by moss that was removed after recovery. The bone surface appears corroded and the linear marks made by the moss can hardly be distinguished (A.259)

Fig. A.263 Linear marks found on the bone surface of modern bones buried deep underground for 32 years from the Experimental Earthwork Project at Overton Down. It was initially thought to have been made by fungi (Armour-Chelu and Andrews 1996), but the origin remains uncertain

Fig. A.261 SEM microphotograph of a shallow groove made by moss on a recent monitored bone from Neuadd. The interior of the groove appears irregular, and the surface aspect of the bone appears corroded (A.260)

Fig. A.264 Specimen N25 from Neuadd, a recent monitored horse carcass that rested in calm water for 10 years at a measured acid pH 5.1. The surface of the scapula (ND21) is covered by a dense layer of algae (A.266), and the bone surface is heavily corroded (see Chap. 8, A.890)

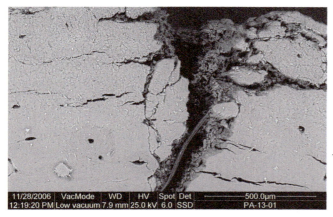

◄ **Fig. A.262** SEM microphotograph of a modern monitored bone, detail of sectioned radio-ulna from Neuadd 41/1 exposed on the surface for 25 years. The bone was covered by moss (A.784) and lichen (A.439), and after removal the surface is seen to be etched and damaged, and the section shows the penetration of lichen into the bone

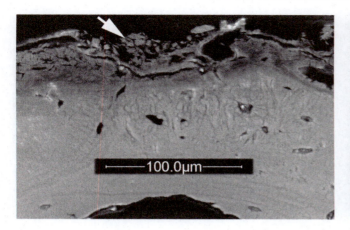

Fig. A.265 SEM microphotograph of a modern monitored bone, detail of A.264 from Neuadd (specimen N25) prepared and sectioned for histological analysis. Algae covering the bone surface has heavily corroded the periosteal cortical layer (see the top edge of the bone white arrow), and it has penetrated into the bone and enlarged the canaliculi similar to that made by moss (see A.274), although in a lighter and shallower manner. Acid fluids may penetrate bone tissues and enlarge the original canaliculi

Fig. A.267 Linear marks on a fossil from Concud, covered by short and shallow grooves. The type of plant that made this modification remains unknown. The agent that could act in the environment of Concud would be algae or subaerial-subaquatic plants. These marks appear smooth and rounded (A.268 and A.269), but the specimen is not rounded by abrasion

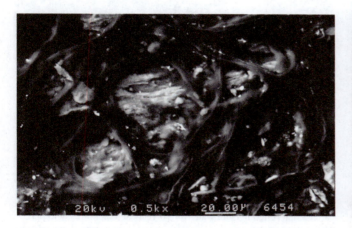

Fig. A.266 SEM microphotographs of the modern monitored bone surface of the horse scapula from Neuadd 25 covered by algae. Algae cover the whole surface, and pseudo-roots can be distinguished. Damage observed on the bone surface does not show linear marks due to algae, but a generalized surface corrosion (see A.265 and Chap. 8)

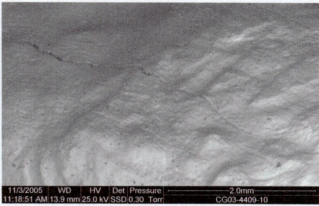

Fig. A.268 SEM microphotograph of a fossil bone, detail of linear marks on a specimen from Concud (not the same specimen seen in A.267, which could not be observed with the SEM because it was covered by consolidant). The surface seen in A.267 shows similar short and shallow marks and edges rounded and smooth

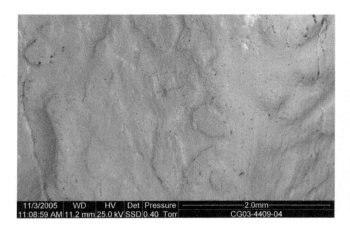

Fig. A.269 SEM microphotograph of a fossil bone, detail of linear marks on A.268 another area of the specimen from Concud. The marks are short and smooth with slightly rounded edges. Marks are here slightly better defined than in A.268

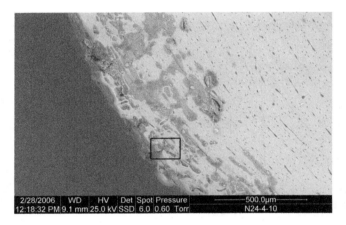

Fig. A.270 SEM microphotograph of a sectioned bone of a modern monitored femur (ND15) of an adult sheep from Neuadd 24. The carcass was monitored and the femur collected after 12 years in bog vegetation (pH 6). In contrast to the scapula covered by algae (A.264, A.265 and A.266), bog vegetation penetrates into the bone across histological traits, forming wide tunnels, more in the shape of roots, but at a smaller scale

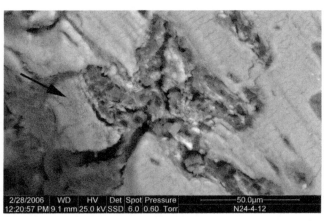

Fig. A.271 SEM microphotograph of a sectioned bone of a modern monitored femur (ND15) of an adult sheep from Neuadd 24 after 12 years in bog vegetation, pH 6. Detail of specimen A.270 (the square drawn in A.270). This environment is seen to produce thin micro-tunneling (black arrow) or enlarged canaliculi by acid fluids from elements of the bog vegetation penetrating several microns into the bone. Note the rounded and smooth texture of the bone on the edge of the penetrating vegetation (bottom and left side of the picture)

Fig. A.272 SEM microphotograph of a sectioned bone of a modern monitored femur (ND15) of an adult sheep from Neuadd 24. The area was permanently wet with moss and herbaceous vegetation. Some local areas on the bone have been penetrated by roots and algae leaving a hollowed surface, but in general the bone surface is macroscopically undamaged. Thin microtunnelling (enlarged canaliculi by acid fluids) can be related to the vegetation, mainly moss and sphagnum (A.270)

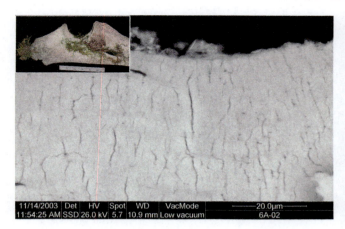

Fig. A.273 SEM microphotograph of a sectioned modern monitored horse radio-ulna (ND6) from Neuadd 41/1. The bone was exposed for 25 years on the surface of the ground showing light weathering (stage 1) and it was covered by moss and lichen. The periosteal cortical layer (PCL, see Text Fig. 2.2) is affected by randomly dispersed and some enlarged canaliculi

Fig. A.275 SEM microphotograph of a sectioned modern monitored bone of an adult sheep tibia (ND25) from Neuadd 25/2 (A.532). It was half buried for 6 years in soil (pH 5.1) and also covered by moss. The bone section shows microtunnelling (enlarged canaliculi by acid fluids) which have a zig-zag shape (black arrow) near the cortical edge

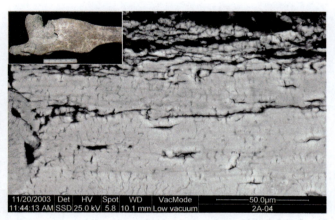

Fig. A.274 SEM microphotograph of a modern monitored sheep humerus (ND3, A.897) from Neuadd 19 exposed on the ground for 17 years. It had a thick covering of moss. The sectioned bone shows intense thin microtunnelling as result of acid fluids enlarging some of the canaliculi. Similar microtunnelling was described by Jans (2004), mainly concentrated on the cortical margin, as also seen here

Fig. A.276 SEM microphotograph of a sectioned bone of a modern monitored femur (ND10) of a juvenile horse from Neuadd 2. The surface was covered by lichen and heavily corroded after being exposed on the ground for 13 years on a damp soil (pH 4.0). Microtunnelling is not as intense as in A.274 and A.275, but the periosteal cortical layer appears heavily cracked by lichen

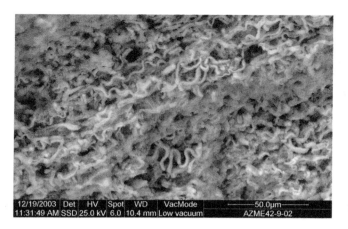

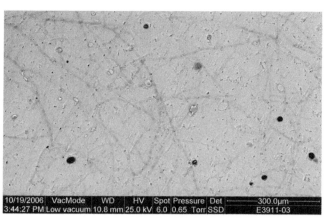

Fig. A.277 SEM microphotographs of a fossil from Azokh Cave. Most of the fossil bone has disappeared and casts of fossilized microorganisms remain on the fossil surface. No identifiable histological traits of the bone have been preserved, for the action of the microorganisms is destructive. In this case, the type of microorganism affecting these fossils is unknown

Fig. A.279 SEM microphotograph of modern fungal hyphae on the cut section of a fossil bone, from Azokh Cave (A.280). This SEM micrograph is taken at BSE mode, showing modern fungal hyphae that have penetrated deep into the fossil bone cortical

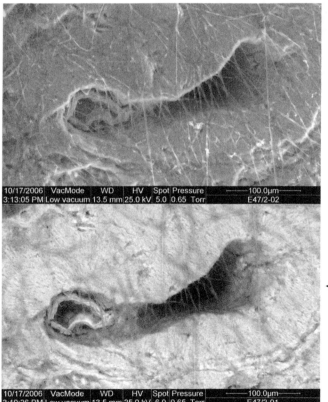

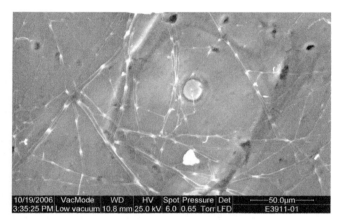

Fig. A.280 SEM microphotograph of modern fungal hyphae on the cut section of a fossil bone, from Azokh Cave (A.278). This SEM micrograph is taken at SE emission mode to better distinguish the topography. The SE mode confirms that hyphae are here superimposed on the fossil surface and apparently do not penetrate into the fossil bone histology

◀ **Fig. A.278** SEM microphotographs of modern fungal hyphae covering a fossil from Azokh Cave (A.280). Observations made in SE emission (top picture) show better topography. SEM images at BSE mode (bottom picture) show changes in density, i.e.: grayish less dense organic matter, (modern hyphae) and whiter mineralized matter (fossil bone). These pictures show fungi hyphae entering the interior of a Volkam's canal of a sectioned fossil bone cut transversely. The modern fungal hyphae have penetrated into the interior of the fossil cortical tissue. Azokh Cave is currently inhabited by a large population of bats dwelling in a nearby cave hall. Bat guano is rich in fungi and bacteria

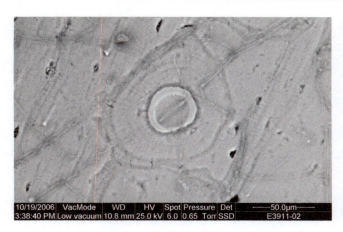

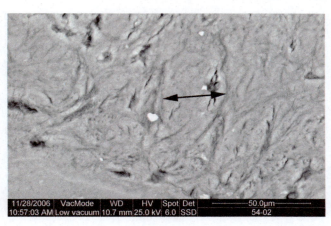

Fig. A.281   SEM microphotograph of hyphae taken at BSE mode in order to distinguish changes in density (A.278). Although hyphae are only covering the fossils, they cross the osteon from the Canal of Havers and through interstitial lamellae, running independently of histological traits (lacunae, lamellae or canaliculi) and with radial distribution

Fig. A.284   SEM microphotograph of a sectioned modern sheep vertebra ND18, from Neuadd 1 after 6 years exposure on the ground (pH 3.5). Fungal tunneling can be distinguished in this specimen (double headed arrow): see A.283. Fungal attack in this case is old damage, not just superimposed hyphae (see A.280)

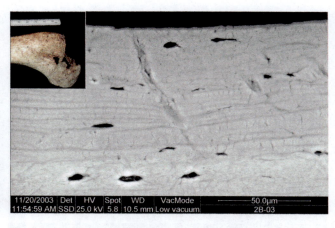

Fig. A.282   SEM microphotograph of experimental *Chaetomium* fungi cultured for 5 years on modern bone. General aspect of a bone section showing thin tunnels, also known as Wedl tunneling According to Hackett (1981) that affects the PCL. Fungal action is still difficult to identify. Like insects or plant roots, fungal attack is more diverse than formerly classified. Further experimental and laboratory cultures are needed (see A.928 and A.941)

Fig. A.285   SEM microphotograph of a linear striation on a sectioned modern humerus (ND9) of a juvenile horse carcass from Neuadd 2. This is Wedl tunnelling caused by fungi. The surface of the bone was heavily corroded, mimicking carnivore chewing (see A.902 and Text Fig. 8.1) shown in the small box top left, after being exposed on the ground for 11 years on a damp soil (pH 4.0)

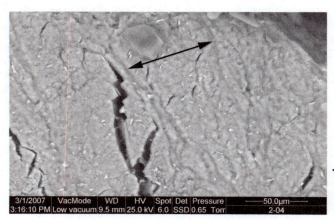

◀ **Fig. A.283**   SEM microphotograph of experimental *Chaetomium* fungi cultured for 5 years on modern bone (A.928). This is a transverse cut section of a bone showing fungal tunneling (indicated by a double headed black arrow) into the bone

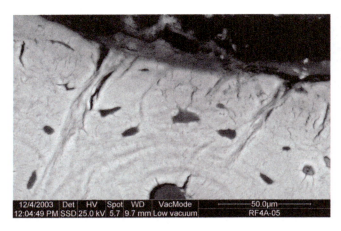

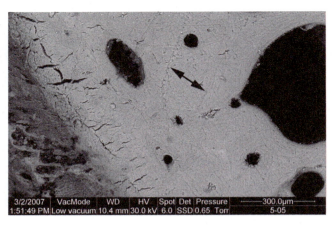

Fig. A.286 SEM micrograph of a modern distal tibia of fallow deer from Riofrío which was half buried and partially covered by moss. The bone surface, apart from being affected by moss, shows Wedl tunneling by fungi in the cross section of the bone (see also A.928 and A.287)

Fig. A.289 SEM microphotograph of experimental *Mucor* fungi cultured on modern bone in laboratory conditions for 5 years. This is a sectioned bone (cut transversely). The double headed arrow shows two fungi tunnelling (grayish) and forming a characteristic zig-zag trace (A.290). This species was isolated and described by Marchiafava, Bonuci and Ascenzi (1974) as osteofagus fungi

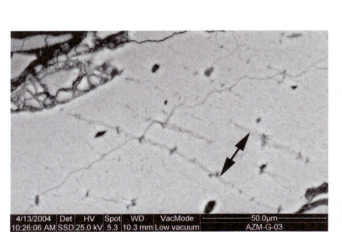

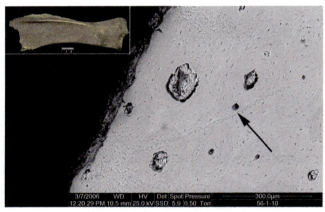

Fig. A.287 SEM microphotograph of a sectioned fossil from Azokh Cave cut transversely. This shows fungi or Wedl microtunnelling and photographed using BSE mode in order to distinguish bone densities (see A.288 in SE mode)

Fig. A.290 Modern scapula (ND22) of a monitored adult horse from Neuadd 25 exposed for 12 years in swampy vegetation (soil pH 5.1). The SEM microphotograph of the cut section of this scapula shows Wedl tunnelling, forming a characteristic zig-zag trace (black arrow) similar to *Mucor* cultured bone damage (A.289)

◄**Fig. A.288** SEM microphotograph of a sectioned fossil from Azokh Cave cut transversely. This picture has been taken in SE mode to see the topography. This picture confirms that hyphae penetrated into the bone (A.287) and have formed microtunnelling

Fig. A.291 Monitored horse, metapodial ND13 from Neuadd 2 buried soon after death in acid subsoil and excavated 18 years later. The area was protected by dense vegetation (heather) that kept the ground at a constant and high humidity (damp soil) and pH 4. The metapodial has a rotten surface, (dusty when touched, and exfoliated) with intensive thin microtunnelling and pitting in section (see A.275), shown in detail below on the SEM

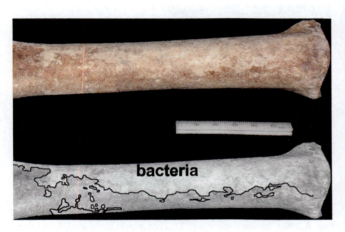

Fig. A.292 Horse monitored metapodial ND5 from Neuadd 41/1 exposed for 15 years in relatively dry conditions. The surface was heavily attacked by bacteria with a rotten bone surface and distinctive focal corrosion. Similar intensive bacterial attack has been observed on both periosteal (PCL) and endosteal (ECL) surfaces, with little bacterial modification in the medial layers (MCL). The extent of the bacteria is shown below A.298, A.465 and A.915

Fig. A.295 SEM microphotograph of linear striations on the surface ▶ of human bone from Apigliano, a late Medieval burial, Southern Italy. These linear striations have been made by incipient bacterial attack (black arrows). Similarly to A.294, some bones show linear trajectories of bacterial attack (see A.908). Specimen provided by C. Smith

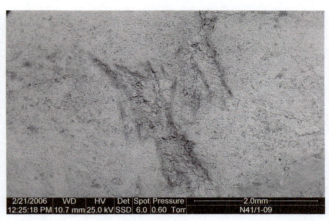

Fig. A.293 SEM microphotograph of linear striations on monitored horse metapodial ND5 from Neuadd 41/1 exposed for 15 years in relatively dry conditions (A.292), showing linear scratches apparently made by insects on the bone surface. Bacteria on bone may be eaten by insects. The bone surface has been intensively attacked by bacteria (A.463), and this may have attracted the attention of insects which marked the surface

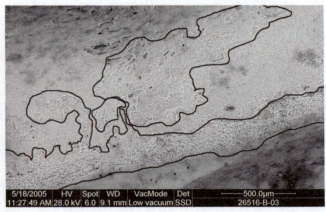

Fig. A.294 SEM microphotograph of a bird bone from the MNCN taxidermic collection prepared in water by natural maceration (see A. 905 and A.907). The bone surface displays linear-like trajectories along the anatomical bone edge of the bone, and corrosion produced by bacterial attack

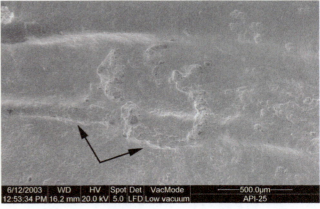

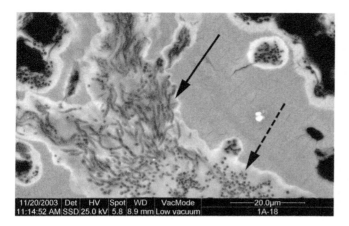

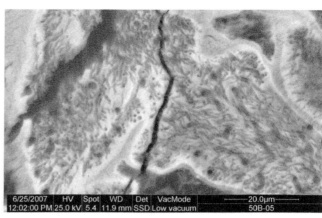

Fig. A.296 SEM microphotograph of a transverse section of the tibia of an adult monitored sheep from Neuadd 1, buried for 6 years in soil (pH 3.5). Bacterial attack may perforate the bone (broken arrow), when the section is perpendicular to the path of the bacteria, or form grooves (black arrow) when the section is parallel to the bacteria trajectory. Bacterial attack is also known as MFD (microscopic focal destruction) and Non-Wedl foci (i.e., Hackett's "linear longitudinal, lamellate and budded foci" Text Fig. 3.8)

Fig. A.299 SEM microphotograph of linear striations in the cut section of a vertebra ND1 from a monitored carcass of sheep from Neuadd 10/4 buried in a manure heap for 14 years (A.799). This photograph shows a predominance of linear longitudinal tunneling made by bacteria

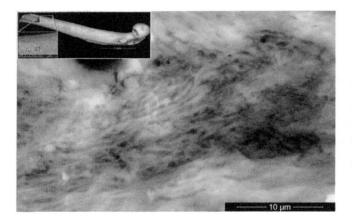

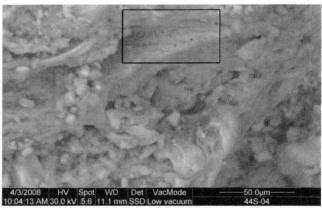

Fig. A.297 SEM microgrophotograph, detail of a small mammal fossil femur from Sima del Elefante, Atapuerca showing tunneling made by bacteria. Small mammal bones are unusually damaged by bacterial attack. This femur comes from a level rich in bat guano (Bennàsar 2010). This specimen is not a cut bone section. Bacterial foci may be distinguished on uneven bone surfaces (as seen in A.300 and A.484) as well as broken edges as shown here

Fig. A.300 SEM microphotograph of a fossil from Azokh Cave, heavily attacked by bacteria on the bone surface. Bacterial tunneling may not only be distinguished in cut sections, but it may also be seen on uneven bone surfaces or fracture edges (A.297), although it is a little more difficult to see. The square is pictured in the following image A.301

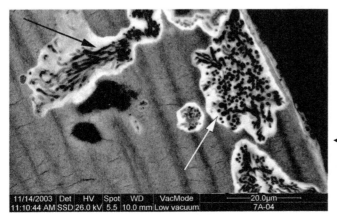

◄ **Fig. A.298** SEM microphotograph of a sectioned metapodial ND5 of an adult monitored horse from Neuadd 41/1 exposed on the ground for 15 years (A.292). The picture shows the linear arrangement of bacterial tubules when the section cuts transversely across their trajectory (black arrow) and perforations when perpendicular (white arrow) as seen in A.296

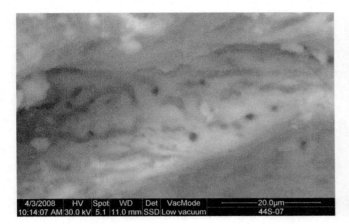

Fig. A.301 SEM microphotograph of a fossil from Azokh Cave, showing linear marks and perforations made by bacteria on bone. Compare it with A.297 uneven bone breakage, and with A.299 on a sectioned bone

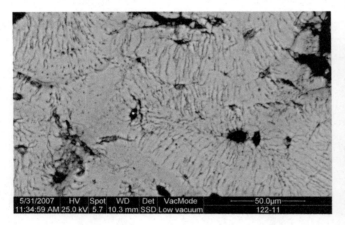

Fig. A.302 SEM microphotograph of a sectioned fossil from Azokh Cave cut transversely and showing another type of tunneling (grey vertical tunnels in this picture) that has not so far been experimentally reproduced. It has strong similarities to enlarged canaliculi (A.275), although width and length are geater and here they have radial distribution (A.303) interrupted by the limit of the osteon (A.304). The agent that caused this remains unknown

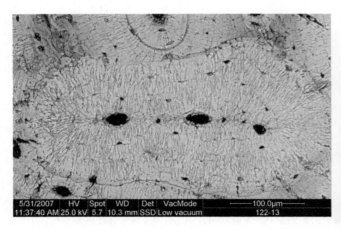

Fig. A.303 SEM microphotograph of a transverse section of a fossil from Azokh Cave. Acid fluids may penetrate bone tissues and enlarge the original canaliculli. This has strong similarities to damage by moss (A.274). This tunneling is not restricted to periosteal cortical layers, as formed by moss, but affects the entire cortex, here affecting primary (on top) and secondary Havers canals (at the bottom). The agent that caused this remains unknown

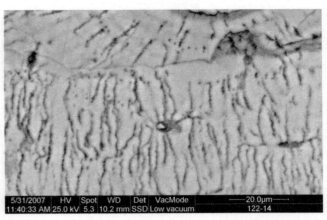

Fig. A.304 SEM microphotograph of a cut section, detail of a fossil from Azokh Cave (A.303) showing a close up view of enlarged canaliculli that are interrupted by the limit between primary and secondary Havers canals. Looking at this tunneling in greater detail, it looks like it was formed by aligned microspheres. The agent that caused this tunneling remains unknown

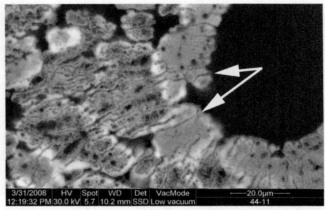

Fig. A.305 SEM microphotograph of a fossil bone from Azokh Cave cut transversely. Bacterial attack can be distinguished at the bottom framed in patches (dark grey spots, see A.296 and A.301), interrupted by aligned microspheres (beaded, white arrows) similar in size to, or even smaller than, bacteria perforations, but with a different dispersal pattern. The agent that caused this tunneling remains unknown

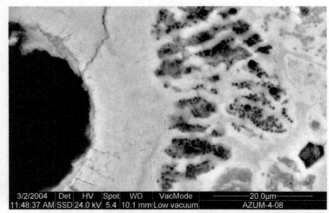

Fig. A.306 SEM microphotograph of a fossil bone from Azokh Cave cut transversely. This shows a closer up view of bacterial paths (distinctive bacterial perforations can be distinguished, see A.298) interrupted by *white* sinuous radial tunneling. Cultured bacteria or fungi in the laboratory, however, have not reproduced this shape and dispersal patterns. Further studies are needed

Chapter 4
Pits and Perforations

Pits are defined as superficial marks on the surfaces of bones, and perforations as marks that penetrate into the underlying tissue of the bones. Pits and perforations have lengths less than 4 times their breadth to distinguish them from linear marks (see Chap. 3). They vary greatly in size and depth of penetration depending on the agent that produces them, but they also vary with type of bone, whether diaphysis, epiphysis or articular surface, from young or old individuals, or mammals versus lower vertebrates. Interpretation of the process making the pit or perforation must always take into consideration these parameters. In addition, pits and perforations produced by different taphonomic agents are more difficult to distinguish than are linear marks, although in some cases they may be associated with linear marks, in which case they may be identified by association, and in other cases they may be identified by location.

Agents and Processes

Inorganic processes:

Marks made by movement of stone against bone

Abrasion (wind, water) A.344 and A.351
Percussion marks made by humans A.308
Trampling A.338

Organic processes:

Marks made by animals: tooth, claw or beak impacting bone

Chewing marks by carnivores A.355
Perforations by diurnal raptors A.403
Penetration by termites and other insects A.414

Solution marks made by plants

Root marks made by plants A.423
Penetration marks by lichen, algae A.437, A.452 and A.432

Marks made by microorganisms: bacteria, fungi A.461 and A.941

Characteristics

Pits have no consistent morphology, but perforations can be distinguished between cone-shaped perforations and square bottomed ones, and between regular and irregular cross-sectional profiles. Different taphonomic agents can produce similar marks, particularly where there is little penetration into the bone, and the same agent can produce different marks, for example depending on the location of the marks on bones.

Morphology of Pits and Perforations

We distinguish again between organic and inorganic processes, and these are further subdivided by shape of the marks. Many of the organic processes produce distinctive cone-shaped perforations that are broad on the surface and narrow interiorly, with surface bone depressed into the pit (A.362). Inorganic processes generally produce broad-based perforations with irregular edges, although some agents such as sharply pointed rocks can also produce perforations (A.316, A.317, A.318, A.319, A.320, A.321 and A.322).

Location and depth of pits and perforations follows the classification first proposed by Andrews and Fernández-Jalvo (1997), and the letters identifying the 13 categories below refer to their definitions. The categories are as follows:

© Springer Science+Business Media Dordrecht 2016
Yolanda Fernández-Jalvo and Peter Andrews, *Atlas of Taphonomic Identifications: 1001+ Images of Fossil and Recent Mammal Bone Modification*, Vertebrate Paleobiology and Paleoanthropology, DOI 10.1007/978-94-017-7432-1_4

a = Shallow pits on diaphyses of limb bones (a);

a1 = Deep perforations on diaphyses of limb bones[1]; A.363

a2 = Perforations on the diaphyses of flat bones; A.365

b = *linear marks on surfaces of bones* ⎫

b1 = *linear marks on ends of bones* ⎬ these linear marks (see Chap. 3) are included here to complete our classification of tooth marks

b2 = *linear marks on articular bone* ⎭

c = Deep perforations on articular ends of bones; A.388

d = Deep perforations on the edges of spiral breaks; A.395

e = Deep perforations on the edges of transverse breaks; A.375

f = Deep perforations along edges of split bones; A.373

g = Multiple perforations on the bone surface made by multi-cusped teeth; A.389

h = Deep perforations on bone ends or edges; A.355

i = Double arched puncture marks on crenulated edges; A.383

These categories agree with other types of measurements of tooth marks (for example minimum dimensions) taken by different authors investigating modern carnivore tooth marks (Selvaggio and Wilder 2001; Domínguez-Rodrigo and Piqueras 2003; Pobiner 2008; Delaney-Rivera et al. 2009) as follows:

• pc: puncture marks on compact bone (category a, pits on diaphyses, and punctures on broken edges: categories d, e, f, i)
• pac: puncture marks on cancellous or articular surfaces (category c, pits on epiphyses)
• gc: grooves on compact bone (category b, scores on diaphyses)
• gac: grooves on cancellous or articular surfaces (category b1/b2, scores on epiphyses).

Descriptions will take into account the following criteria, where data are available:

Diameter of marks, taken as the minimum dimension A.387
Multiple or single marks
Depth of marks
Frequency of marks
Location of marks

---

[1]This modification was not recognized in our initial classification because it was not present in our original sample.

Organic Processes Producing Pits or Cone-Shaped Perforations

By far the most common source of pits and perforations in bone is chewing or gripping by carnivores, and this typically produces conical or cone-shaped perforations into or through the surfaces of bones. There are several sources of variation of carnivore marks that make the relationship between these and the predator species difficult to evaluate (Delany-Rivera et al. 2009). Of critical importance are the type of carnivore tooth and the location of the mark on the prey. For example, marks located on the diaphysis of bones are generally shallower than marks made on the softer epiphyseal bone, where tooth perforations may penetrate deeply below the surface. Another factor to take into account is the potential size relationship between predator and prey. If a hyena eats a small mammal such as a mouse, there will be no perforations of any kind, for the whole body is consumed, but if it eats a large bovid such as wildebeest, there may be many pits and perforations on different parts of the skeleton, for example shallow pits on the diaphyses and deeper perforations on epiphyseal bone and on the edges of breaks. On the other hand, small predators such as foxes or genets eating small mammals or scavenging large mammals may leave small pits and perforations on bone surfaces or on the edges of broken bone, and if they consume the whole body, bone surviving in the scats of the predator may preserve chewing marks made by the predator. As a result, small mammal bone is only likely to have small tooth marks since large marks would destroy the bone completely. Large mammal bone may have large tooth marks preserved with or without breakage, and small tooth marks without breakage, but the relationship between tooth mark size and predator is not straightforward.

The relationships between bone size and predator type as seen from the tooth marks on the bone is shown in Figs. 4.1 and 4.2. Small tooth marks on large mammal bone may be produced both by large predators, if the marks are on the diaphysis of the bone or are made by the predator's incisors, or they may be produced by small predators which have smaller teeth. In contrast, large tooth marks on large mammal bone can only be made by large predators (again taking into account the location of the marks). Therefore, when considering tooth mark sizes, mean values are of little use, and the actual size range, and in particular the upper end of that range, provides the data that are most useful for identifying size or type of predator. Similarly, tooth mark sizes on small mammal bone are of greater value when the full range is provided, rather than mean values, and again it is the largest marks that provide an indication of the type of predator. In both cases the possibility that more than one predator has gained access to a carcass has to be considered, in which case there may be large and small marks from two or more different sized predators.

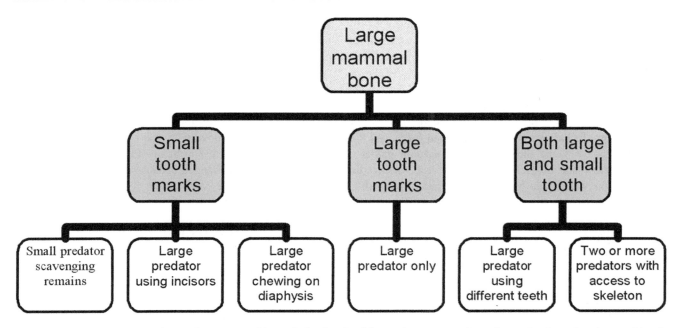

Fig. 4.1 Ranges of tooth mark sizes on large mammal bones indicating the different size ranges as they relate to the sizes of predator making the marks

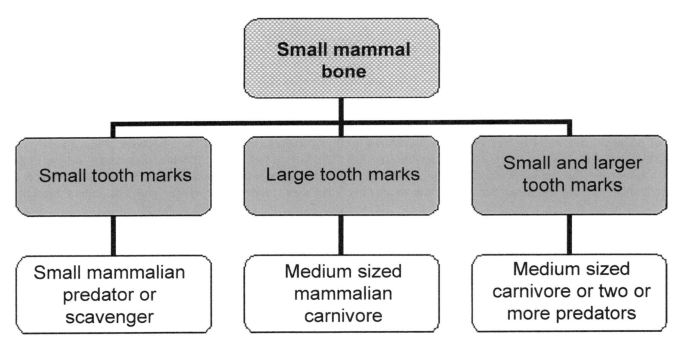

Fig. 4.2 Ranges of tooth mark sizes on small mammal bones indicating the different size ranges as related to the sizes of predator making the marks

Size Distributions of Pits and Perforations

Some data are now available on the sizes of carnivore tooth marks, but no comparisons are currently available between these and other sources of perforations such as from trampling or insect marks. The following discussion is therefore restricted to carnivore marks only, and comparisons of morphology follow in the next section.

Two sets of data are shown in Figs. 4.3 and 4.4. The first shows the size range of pit marks on diaphyses (category a in our classification) of limb bones for two sites, a camel skeleton from Jebel Barakah which was scavenged by jackals and foxes (Table 2.1) and a sample of fox-modified sheep bones from Neuadd (Andrews and Armour-Chelu 1998; Andrews and Whybrow 2005). The modes of the pit sizes for the two samples differ slightly (Fig. 4.3), but the ranges show that most of the marks made by foxes are in the range 0.5 to 1.6 mm, while the Barakah marks range from 1.1 to 3.4 mm. Jackals are slightly larger than foxes, and it is probable that the while the larger sizes of tooth marks at Barakah were made by jackals, it is not clear whether the smaller marks were made by foxes or jackals or by both.

Data in Fig. 4.4 compare the same fox-modified assemblage with the sample of human bones from Sima de los Huesos, Atapuerca (Andrews and Fernández-Jalvo 1997). The size distribution of carnivore pits on diaphyses (category a) show a strong similarity between the two samples at the low end of the range, but the fossil sample has some larger pits as well, up to 7.5 mm in diameter. These marks are beyond the range of fox teeth, and two conclusions are possible. One is that all the marks were made by a medium to large carnivore, which made mainly small marks while chewing the bones, but some larger marks as well; and the other is that there were at least two types of carnivore. If there were two carnivores, one was a small fox sized scavenger, inferred because of the close similarity in tooth mark size at the low end of the range. The second would have been a larger species of carnivore that made fewer marks. The first is inferred to have been a scavenger, because foxes are not generally able to kill humans, and the second could well have preyed upon or scavenged the humans, or at least had primary access to the bodies. Figure 4.4 also shows the distribution of perforations on the edges of spiral breaks, that is those breaks that are made on fresh bone (see Chap. 9). Perforation sizes at the lower end of the size range have

similar distributions for the fox and fossil samples, and the fossil sample has many larger perforations as well. The most parsimonious conclusion is that two sizes of carnivores had access to this sample of human bones, foxes, which are the most common carnivore species in the cave, and an unknown but large carnivore such as a cave lion.

Figure 4.5 compares the tooth mark sizes for the fox assemblage shown above with that of the cave bear (*Ursus spelaeus*) from Arrikrutz in Spain (Pinto Llona et al. 2005). The scale has been changed because the tooth marks on the bear assemblage are very large. The Arrikrutz deposits contain the remains of hyenas and wolves in addition to cave bears, and it is likely that the tooth marks are a combination of all these predators' activities. The range is skewed to the right, with the sizes of both diaphyseal pits and spiral break perforations being larger than in the fox assemblage. The numbers of spiral breaks and incidence of carnivore perforations on the breaks are also much higher in the Arrikrutz sample (Pinto Llona et al. 2005). This is attributed to the variety of large predators at Arrikrutz, and it contrasts with the assemblage at Troskaeta, where the only species present is the cave bear (Fig. 4.6). The distribution of tooth marks on the diaphyses (type a) is wider than on the mixed carnivore assemblage at Arrikrutz, where most pits are at the smaller end of the range, and for most locations of marks the sizes are greater in the assemblage from Troskaeta. Since cave bears are the only carnivores known from Troskaeta, it is likely that they produced the perforation marks at this site, and this in turn suggests that the major sources of the smaller perforation marks on the Arrikrutz bones were the hyenas and wolves rather than the cave bears (Pinto Llona et al. 2005). Similarly, the perforations on articular surfaces of the Arrikrutz bones are generally smaller, with a large number at the small end of the range, and this differs from the

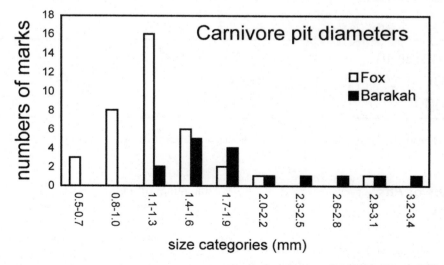

Fig. 4.3 Size distribution of surface pits (type a) for a modern sample of fox-ravaged (*Vulpes vulpes*) bones from Neuadd, Wales (white bars), compared with a sample of bones from Jebel Barakah, UAE (black bars), where the main scavengers are foxes and jackals

Puncture diameters

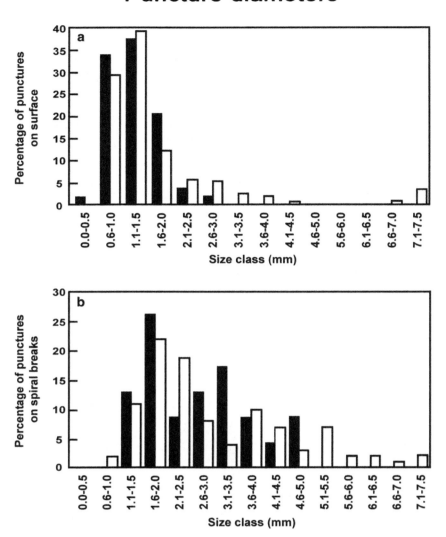

Fig. 4.4 Size distributions of carnivore chewing marks for a modern sample of fox-ravaged (*Vulpes vulpes*) bones from Neuadd, Wales (black bars), compared with a sample of fossil human bones (white bars) from the Sima de los Huesos, Atapuerca. Top: surface pit diameters (type a). Below: perforation diameters on the edges of spiral breaks (type d)

Troskaeta bones, which have higher frequencies of large sized type c perforations. Interestingly, the Arrikrutz bones have greater numbers and larger type d perforations on spiral breaks, as well as a greater number of spiral breaks overall, as opposed to large type c perforations. This could indicate a higher degree of carnivore modification of the assemblage while the bones were still fresh (Pinto Llona and Andrews 2004a, b).

Another factor to take into account when measuring tooth mark size is the skeletal element where the marks occur. Puncture marks on limb bones such as femur, tibia humerus and radius have smaller diameters than those on the scapula or pelvis. Data from the Atapuerca Sima de los Huesos human fossils show that for the major limb elements, the modes of the tooth mark diameters are 0.6–1.0 mm, and this compares with values for the ulna and sacrum of 2.1–2.5 mm, and for the scapula and pelvis of 2.6–4.0 mm (Andrews and Fernández-Jalvo 1997). These differences reflect the varying compactness or density of the bone. Looking just at maximum values, the largest size observed for fox tooth marks were 4.6 to 4.9 mm, while the largest marks on the human fossils from the Sima were 6.8 to 7.5 mm.

These examples serve to emphasize several points when measuring tooth mark size distributions. One is that the size distributions of carnivore marks are different on bones from different parts of the skeleton, and comparisons between modification by different carnivore species is only possible

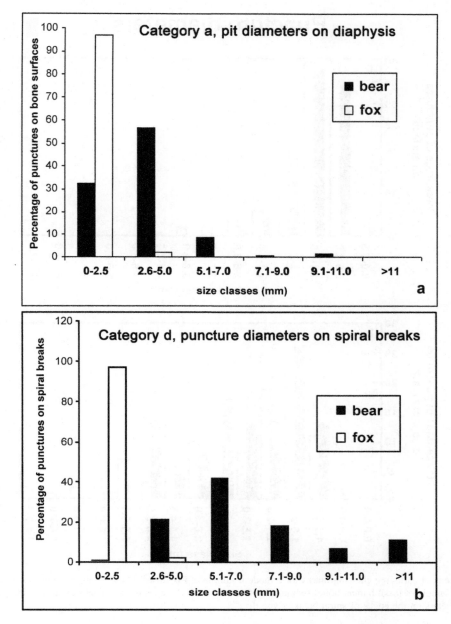

Fig. 4.5 Size distributions of carnivore chewing marks for two samples of carnivore chewing: a sample of fox-ravaged (*Vulpes vulpes*) bones from Neuadd, Wales (white bars), compared with a sample of carnivore-ravaged bones (*Ursus spelaeus*) from Arrikrutz, Spain (black bars). Top: surface pit diameters (type a); Bottom: perforation diameters on edges of spiral breaks (type d)

when comparing like with like, namely tooth mark sizes on equivalent parts of the skeleton. It is also evident that there is considerable overlap in sizes of marks, and all species and all bone parts can include marks at the small end of the size range. In our view, therefore, it is the large end of the size range that is important, for small carnivores cannot make large tooth marks, and in this respect the presence of large

perforations provides evidence for large carnivore activity on the fossils from the Sima de los Huesos (Andrews and Fernández-Jalvo 1997). In general, however, absence of large tooth marks can probably be taken to indicate absence of large carnivore species, although such a conclusion based on absence of evidence could be disproved if larger tooth marks are found later.

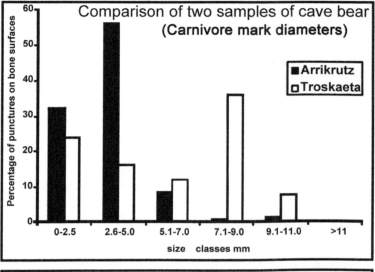

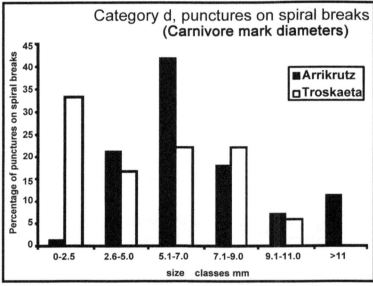

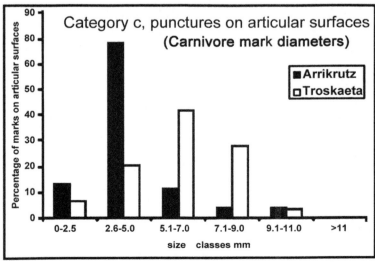

Fig. 4.6 Size distributions of carnivore chewing marks for samples of fossil cave bear (*Ursus spelaeus*) from Arrikrutz and Troskaeta, Spain. Top: pit diameters on the diaphysis (type a). Middle: perforation diameters on spiral breaks (type d). Bottom: perforation diameters on articular surfaces (type c)

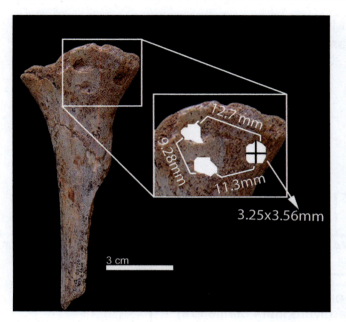

Fig. 4.7 Multiple perforations on the bone surface made by multi-cusped teeth (category g). The three cusp impressions can be seen on the left hand figure, while the measurement of the multi-cusped impression is shown on the right

and displacing surface bone, but deeper perforations penetrate into underlying tissues, and they generally displace surface bone down into the interior of the bone. This displaced bone is not always evident on fossils, as the fragments of bone are easily lost during the process of preservation and fossilization, or even during laboratory fossil preparation, but on freshly perforated bone they are usually evident. Perforations made by crocodiles may penetrate deeply into bone, leaving conical holes (A.371), but they also make smaller marks (Njau and Blumenschine 2006; Baquedano et al. 2012). Multiple perforations on the bone surface may be made by multi-cusped teeth (category g). It is common for the perforations to differ in depth penetrated, for multi-cusped carnivore teeth have one cusp higher than the others (Fig. 4.7).

Insect Damage

Carnivore tooth marks are formed when the teeth of the carnivore penetrate the surface of the bone of its prey, whether as a result of predation or scavenging (Blumenschine 1986). Shallow pits penetrate a little way, crushing

Deep conical perforations may also be the result of insect attack. This can cause extensive pitting of bone, but the perforations differ morphologically from marks made by larger animals. There is usually grooving leading into the perforations, which differs from anything produced by larger animals, and the perforations penetrate deep inside the bone (A.414). The perforations have no floor as the insects penetrate into and consume the inner bone.

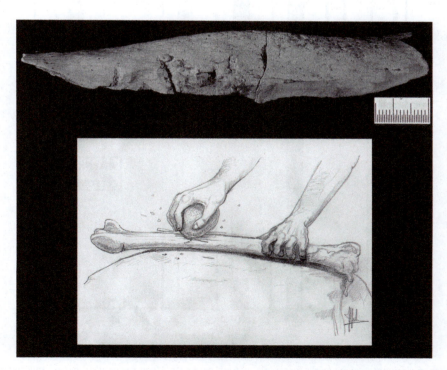

Fig. 4.8 Impact notches or percussion marks consist of pits of variable sizes and depths. These are produced when the bone is held on a hard solid surface and smashed with a stone as shown in the drawing below. (Drawing by Javier Fernández-Jalvo, copied with permission)

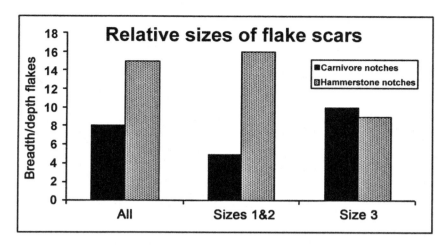

Fig. 4.9 Relative sizes of flake scars (breadth/depth) comparing carnivore notches and hammerstone notches. Data from Cappaldo and Blumenschine (1994). Sizes 1 and 2 come from large mammals weighing less than 115 kg (250 pounds) and Size 3 come from animals weighing ca. 115–340 kg (250–750 lb)

Plant Roots

Plant roots also produce conical perforations with smoothly rounded edges, in addition to the linear marks described in Chap. 3. Roots penetrate into bone and sometimes cover both the exterior and the interior of the bone in a thick mat. Similarly, but at a much smaller scale, rounded conical regular perforations have been observed on teeth from sediment accumulated in calm water. The exact origin of this modification has not been identified yet. We can confirm that it is not bacterial attack, and it may be related to micro-organisms forming a biofilm (community of microorganisms encapsulated within a self-developed matrix and adherent to a living or inert surface) that is homogeneously structured and that superficially modifies osseous surfaces (more specifically enamel (A.495), see Chap. 8). Although early stages of bacterial attack may also produce rounded conical regular perforations, more intense bacterial damage penetrates deeply into the bone producing individual perforations that are irregular in shape.

Inorganic Processes Producing Broad-Based Perforations

Broad-based perforations are made by some abrasion processes caused by agents such as trampling, and human butchery or dismemberment of carcasses. Pits and perforations produced by percussion notches tend to be broader than carnivore marks (Blumenschine and Selvaggio 1988, 1991; Capaldo and Blumenschine 1994), and they are more variable in size compared to marks made by carnivore teeth, which tend to be narrower, more semi-circular and have a perpendicular release angle (Fig. 4.8). Their relative sizes are shown in Fig. 4.9 (data from Cappaldo and Blumenschine 1994). Impacts or percussion marks on the surfaces of bones may be superficial or penetrate deeply into the underlying tissues (Fig. 4.8), depending in part on their location. On the diaphyses they are equivalent to category 'a' or 'b' in our tooth mark classification, and 'c' if they are on articular ends. They also occur on bones broken to gain access to bone marrow (see Chap. 9), but in this case only part of the impact remains on the broken edges of the bone. These are equivalent to category 'd' and 'e' on our tooth mark classification scheme. Impact marks may be associated with spiral breaks since the bone has to be fresh to retain marrow (Capaldo and Blumenschine 1994). The use of hammerstone and anvil is illustrated in Fig. 4.8, and the size comparison with carnivore tooth marks is shown in Fig. 4.9. Human-made percussion marks are irregular but broad-based perforations in the surface of bones, and similar marks may be produced by trampling, simulating the effects of percussion during butchery. The main difference between them is that trampling marks are generally more superficial (A.77 and A.124) and more irregular and do not penetrate far into the bone. They also tend to be more numerous and scattered over the bone surfaces and do not bear any relationship with the anatomical parts of the bone. In other words the marks are scattered randomly over the surface of the bone. Also, trampling marks are associated with rocky or otherwise hard, rough substrates, when substrate is dry, as is the case with linear marks that mimic cut marks (Chap. 3), and in both cases the marks result from the action of bone being rubbed against stone by animals trampling over them. In damp conditions, trampling has been seen to produce punctures on bone surfaces at Concud (A.335) and Senèze (A.336), and this has also been reproduced in the laboratory (A.334). Some impact marks may retain the shape of the stone making them, and they could be the result of natural damage in a cave, if a bone is pressed into a sharply pointed rock, or if a rock falls on to the bone on the cave floor. Context of the bone may indicate if this is the case. Alternatively, it could be the result of human action,

Asymmetric Perforations

Some perforations have distinctive morphology, being short, penetrating deeply into the bone, and having a strongly asymmetric profile, with one edge being almost vertical and the other with a pronounced slope. All of the agents described above may also produce asymmetric perforations. This morphology is apparent on the styloid process of a human skull from Çatalhöyük in Turkey (A.325) (Andrews et al. 2005). The chop mark (A.326) has penetrated deep into the bone with a smooth sloping side showing the direction of the perforation. On this bone there is also a superficial mark just above the chop mark that had failed to penetrate the bone. The styloid process was associated with a skeleton from which the skull had been removed, and the chop mark on the styloid process was probably inflicted during removal, indicating that some force was used. Chapter 3 gives another example of skull removal at Çatalhöyük, but in this case there were cut marks on the atlas vertebra to show how the operation was carried out.

Organic Processes Producing Broad-Based Perforations

Perforations made by large birds of prey tend to be large and irregular. Vultures and eagles can penetrate thin bone with their beaks to produce perforations, or in the process of stripping muscle and tendon from the bone they may tear off irregular strips of bone. The perforations are restricted to very thin bone. Such is the case with the example of vulture beak penetration through the thin bone of the skull and eagle predation on monkeys, in which strips were torn off the scapular blades by the eagles' beaks (Andrews 1990). The resulting marks are sometimes hard to distinguish from the action of mammalian carnivores, such as fox scavenging, but their restriction to thin bone is indicative if not diagnostic.

Lichen can produce a variety of perforations, from round to irregular and diffuse. Plant roots can also produce broad-based perforations as well as or instead of conical ones. These have the potential to be confused with carnivore tooth marks, which may also be broad-based (Domínguez-Rodrigo and Barba 2006). Other types of plant may also produce broad-based perforations in particular circumstances, such as algal growth on bone preserved under water.

In some cases it is not certain if the perforations are caused by algae or due to water corrosion (see Chap. 8).

A final example of broad-based perforations is the product of wind erosion (A.340 and A.341). Some of the perforations have a distinctive shape, oval with one end more pointed than the other, and with flat bottoms. The marks are shallow and uniform in depth, often filled with sediment. It is probable that the elongation of the perforations indicates wind direction

Perforations from Chemical Attack

Several types of perforation that do not fit easily into the classification used here can be mentioned briefly. One is cave corrosion, which consists of erosional perforations that thin out and penetrate bone to such an extent that the surface begins to collapse around the perforation. Cave corrosion is a modification commonly seen in caves, but the process by which it forms is currently unknown. It may have to do with the high levels of humidity common in closed cave environments, and with a mildly acid substrate this could lead to dissolution of bones. In some cases, perforations of bones in caves can be attributed to bacterial attack.

A second type of perforation resulting from chemical attack is the result of digestion, either by enzymes or by a strongly acid environment. Digestion of compact bone may have the effect of removing the surface bone and producing multiple pitting. The pitting is due, however, to penetration of the surface bone and exposure of the weaker bone beneath the surface. These modifications by digestion are described in more detail in Chap. 8. The third type of perforation, which is not strictly speaking a taphonomic modification, is the formation of pitting on the occlusal surfaces of teeth formed during the ingestion and chewing of food (A.352). The frequency and relative sizes of microwear pits is related to type of food ingested (King et al. 1998). Fourthly, it has been proposed (Fernández-Jalvo 1992) that diatoms may produce perforations on bones immersed in water. In fact the modification by diatoms consists mainly of perforations following a lineal trajectory where water energy is relatively high. Diatoms link to each other as a chain to resist the water current (Raven et al. 1986), and where they anchor themselves to the bone, perforations occur. In our experience, this is quite common in bones lying in water and this was also observed by Bodén (1988) on calcareous sand grains. In some cases, bones in water showed an extreme diatom attack with linear areas completely "eaten" away and diatoms incrusted into the bone (A.444). This is surprising because diatoms are siliceous and bone is made up of phosphate carbonate, so they may not penetrate bone for metabolic reasons. The physiological role of this corrosion remains unknown.

Finally, the action of bacteria can be briefly mentioned. Colonies of bacteria have a rounded, sometimes conical shape (A.478) and have an outer fringe of higher density (brighter areas) composed of calcium phosphate, like bone bioapatite. Waves of bacteria can be distinguished as colonies overlapping each other (see Chap. 4 (A.460), and this may provide criteria to distinguish seasonal stages (Fernández-Jalvo et al. 2010).

Individual bacteria perforate the bone forming microscopic tunnels visible in transverse sections. The diameter of the tunnels which has been measured by mercury porosimetry between 0.1 and 1.2 micron (Jans et al. 2004), although these perforations may reach 2 microns in diameter (Fernández-Jalvo et al. 2010) measured on the SEM. Wedl tunneling attributed to fungi has been described by several authors (Hackett, 1981; Bell et al. 1996; Millard 2001) to measure >10 μm in diameter. However, microscopical focal destruction (MFD) produced by both bacteria and fungi appears through porosimetry analyses as increases in pore volume in pores of diameter ~0.1–10 μm while non-microbially mediated collagen loss appears as an increase in pore volume in pores of <0.1 μm diameter (Nielsen-Marsh and Hedges 1999; Smith et al. 2002; Jans et al. 2004).

Atlas Figures

A.307–A.502

Fig. A.307 Impact notches or percussion marks on a fossil human femur shaft from Gran Dolina, TD6 Atapuerca. They consist of pits of variable sizes and depths and are produced when the bone is held on a hard solid surface and smashed with a stone to gain access to the marrow (see Text Figs. 4.8, 9.1, A.331). These pits may be mistaken for carnivore tooth marks. Differing from tooth marks, these pits have an irregular or too flat shape to match with tooth cusps. Photo M. Bautista

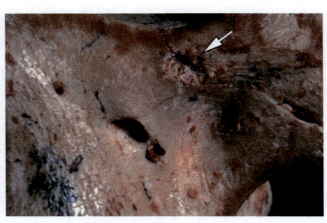

Fig. A.309 Human fossil malar bone from Atapuerca, Gran Dolina TD6 (A.52, A.98, A.99 and A.307) showing cut marks and punctures on the bone surface (arrow points a notch mark). Pits can be found on more fragile bones, such as craniofacial bones, as well as on long bones. This is a detail of the holotype of *Homo antecessor* ATD6-69

Fig. A.308 Impact pits on a fossil tibia of bovid from Gran Dolina, TD6 Atapuerca. This fossil was found together with cannibalized human remains. Impact scars are similar to those seen in A.307 of a human femur and A.331 of a bone fragment from Tianyuandong. Marrow extraction is the purpose of this heavy damage on these bones

Fig. A.310 SEM microphotograph of a fossil bone from a middle sized mammal from Vanguard Cave (A.108). Percussion pits are usually accompanied by abrasions and scratches underneath and around the perforation. The accompanying scratches have been described by Turner (1983) as hammerstone-anvil, caused by friction of the bone against a block or the anvil surface where the bone was resting when it was struck

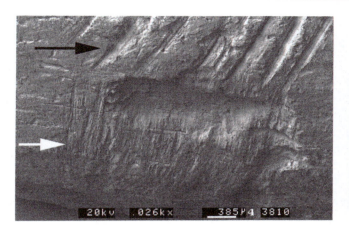

Fig. A.311 SEM micrograph of fossil bone from Vanguard Cave. Detail of an impact notch or percussion mark accompanied by fine scratches (white arrow as in A.310) mainly underneath and also around the notch perforation as result of hammerstone-anvil effect. These scratches show that the bone was already clean of meat when it was struck. The perforation interrupts a set of cut marks (black arrow) indicating the butchering sequence of meat extraction followed by bone breakage to extract the marrow

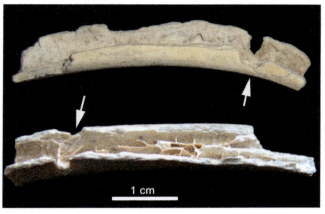

Fig. A.313 Fossil bone fragments from Tianyuandong, Zhoukoudien site 27 (compare with A.312). Impact marks on the edge of the broken bone fragments showing penetration across the shafts (white arrows). The marks have the appearance of having been made by burin-shaped tools

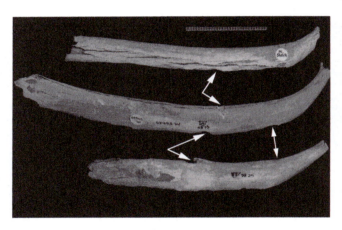

Fig. A.312 Sequence of three associated ribs from a fossil human skeleton from Gough's Cave (A.100, A.102 and A.103). These ribs were found in approximate anatomical position, and they have percussion/chop marks on the edges of the ribs: rib 4 on the posterior edge, rib 5 on the superior edge opposite the mark on rib 4; similarly, rib 5 has another mark posteriorly, and rib 6 a superior mark opposite the rib 5 mark. Ribs 5 and 6 have a second set of marks, all shown by white arrows

Fig. A.314 SEM microphotograph of a fossil from Abric Romani. Sometimes hammerstone-anvil scratches (A.310 and A.311) may be hard to distinguish (black arrows). The presence of a fine groove pattern (white arrows) is associated with the impact mark where the bone rested against the stone or anvil surface when it was struck. These traits differentiate percussion marks from other taphonomic modifications

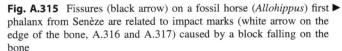

Fig. A.315 Fissures (black arrow) on a fossil horse (*Allohippus*) first ▶ phalanx from Senèze are related to impact marks (white arrow on the edge of the bone, A.316 and A.317) caused by a block falling on the bone

Fig. A.316 Punctures on the edge of a spiral break (white arrow) of a fossil horse rib from Senèze resulted from falling blocks. These may mimic puncture marks by carnivore teeth. Spiral breaks are usually associated with breakage of fresh bones. Impact marks may differ from carnivore tooth marks (A.374 and A.396) by the presence of thin cracks (black arrows) associated with the mark, indicating that the bone was free of meat when the impact took place

Fig. A.319 Fossil sesamoid of *Eucladoceros* from Senèze bearing a mimic tooth mark and grooves on the articular surface. The shape of this puncture is not an inverted cone as produced by tooth cusps (see A.355), but instead the puncture has an irregular shape. This puncture mark is associated with a deep crack, probably due to compression against sedimentary clasts. Arched scratches on the left side are linked to shallower puncture marks, indicating these are trampling marks

Fig. A.317 Right femur of a fossil cervid (*Eucladoceros*) from Senèze. It has cracking that is related to the lateral breakage (bottom of this picture) giving an arched shape to the impacts (black arrows). Blocks of falling stone may hit a skeleton that is free of meat, skin or any other soft tissue, and this may cause impact marks (A.374 and A.396) similar to carnivore puncture marks on the bone surface (compare with A.355) on the broken edge (white arrows)

Fig. A.320 Fossil bone. Coccyx from Senèze bearing a multi-cusped tooth print mimic that was, however, produced by percussion (see A.321 and A.322). This time the punctures are conical, like marks made by carnivore teeth. However, a circular flattened puncture mark (black arrow) does not match with a carnivore tooth print mark. Looking at this puncture in more detail, the top puncture has a rather flat bottom shape similar to the puncture pointed to by an arrow

◄ **Fig. A.318** Distal metapodial of an immature individual of fossil horse (*Allohippus*) of Senèze. It has shallow puncture marks on the articular surface produced by pressure of gravel. Tooth marks on articular surfaces are larger and damage is heavier than on shafts because articular ends are soft, and still softer when the individual is juvenile, as in this case (A.357 and A.358). The lack of heavy damage suggests that this is the effect of another taphonomic agent, apparently independent of the nutritional value of the bone

Fig. A.321 Fossil bone. Coccyx bones were found in anatomical connection at the site of Senèze. A rounded percussion mark (white arrow) shows that the cancellous tissue is not compressed (A.322) as would be expected in a tooth mark (A.355). This was probably a result of cortical flaking by a light impact, because otherwise these soft bones would have entirely collapsed

Fig. A.322 Fossil bone. Detail of A.321, percussion mark on a coccyx from Senèze. The rounded edge of this mark could mimic the shape of a tooth. The cancellous bone, however, is not compressed, but just exposed underneath a piece of missing cortical bone. The shape of this mark and the recovery of this bone in anatomical connection with the rest of the coccyx, also bearing percussion marks, excludes its identification as a tooth mark on A.321

Fig. A.324 SEM microphotograph of a fossil human vertebra from TD6, Atapuerca A.323. This inverted pyramid-shaped mark is 6 mm long, and it was most likely made by a pointed stone. The sediment of unit TD6 of Atapuerca-Gran Dolina site is clayish with gravel and limestone clasts. Limestone clasts may have sharp edges, which might penetrate thin cortical bone like a vertebra by compression under trampling action. There are no accompanying scratches

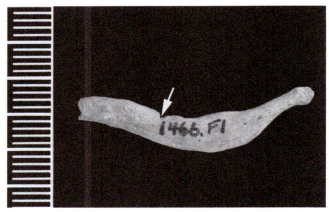

Fig. A.325 Chop mark on fossil human styloid process from Çatalhöyük leaving a scar made by a sharp stone tool edge striking the bone surface (A.326 and A.327). The skull had been removed after burial and partial decay of the body, and the rest of the skeleton survived intact

◄ **Fig. A.323** SEM microphotograph of a fossil human vertebra from TD6, Atapuerca. The triangular perforation penetrates the surface of the vertebra. The perforation has sharp-cornered triangular edges, and the fragments of bone are compressed into the perforation. The inverted pyramid-shape of this puncture mark is unusual for a tooth mark (A.360 and A.362). Note the top left grey stain is the glue used to label the fossil

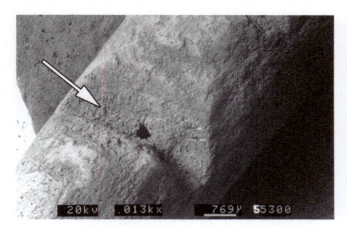

Fig. A.326 SEM microphotograph of a fossil bone, detail of a chop mark on the human styloid process from Çatalhöyük (A.325). The mark is a deep and wide V-shaped scar. The action is related to cutting strong muscle attachments or dismembering. The absence of scratches suggests that the bone was still bearing flesh, in contrast to percussion or impact marks

Fig. A.329 SEM microphotograph of a fossil bone, detail of a chop mark (compare A.327 with A.330) on a long bone shaft of an unidentified small sized mammal species from Vanguard Cave. A piece of stone from the tool edge (black arrow) has been encrusted in the bone surface

Fig. A.327 SEM microphotograph of a fossil bone, detail of a chop mark made when striking a bone surface with the sharp edge of a stone tool. The resulting mark is a deep and wide V-shaped scar (A.326). The action is related to cutting strong muscle attachments or dismembering the bones of the skeleton

Fig. A.330 Impact deformation by a falling block on a fossil radio-ulna of a horse (*Allohippus*) from Senèze. The rapid action of the impact compresses the cortical layer where the block struck the bone (compare with A.327)

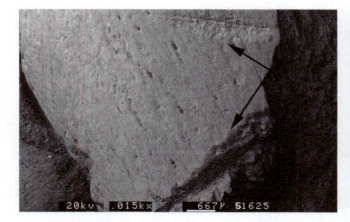

Fig. A.328 SEM microphotograph of a fossil bone, detail of chop marks on the diaphysis of a long bone of a small sized large mammal from Vanguard Cave (A.325, A.326 and A.327). Notice the similar shape of the edge of both marks on this fossil (black arrows)

Fig. A.331 Bone fragment from Tianyuandong hammered by percussion marks on the surface, similar to A.307 and A.308 from TD6. Punctures are irregular and imprint the shape of the stone with which the bone was struck to break it and extract the marrow. Courtesy of H. Tong

Fig. A.332 Fossil bone from Tianyuandong showing a large conchoidal scar with an impact mark associated (white arrow). Conchoidal
breakage on the edges of the breaks may also be produced by
carnivores. In this case, the conchoidal scar is too large to be caused by
carnivores (A.396) and is more probably the result of human breakage
(Blumenschine 1988)

Fig. A.335 Fossil from Concud with punctures that are surrounded by
plastically deformed bone (white arrows). In a damp sedimentary
environment, grains of sediment may penetrate into the bone and leave
deep pits when compressed possibly by trampling as experimentally
reproduced in A.334. Courtesy of D. Pesquero

Fig. A.333 Fossil bone. Right mandible of cervid (*Eucladoceros*)
from Senèze. It has linear marks produced by friction against coarse
sediment generated by trampling. These are accompanied by pits or
puncture marks (small square enlarged in the box on top right). These
pits were experimentally reproduced (A.334) in a wet environment and
are surrounded by plastically deformed bone as seen in other fossils
(A.335 and A.336)

Fig. A.336 Pits on a fossil fragment of an unknown species from
Senèze. The bone is plastically deformed around the pits. It is
interesting to compare these marks from trampling with crocodile tooth
marks (A.371, Njau and Blumenschine 2006; Baquedano et al. 2012),
which are similar although smaller in size and non-bisected pits or
traces of carinae shown in A.372. This is because chewing by
crocodiles also occurs after immersion of their prey in water. On this
specimen, the pitting is associated with scratches (black arrow, small
box bottom right), suggesting trampling

◄ **Fig. A.334** Experimental bone compressed after several weeks
immersed in water using static materials testing equipment. The bone
was compressed against coarse sediment (gravel size and stones). The
results of the experiment showed grains of sediment penetrating the
bone shaft and articular surfaces. The main trait is the presence of
plastically deformed bone around the depressions instead of the more
conical shaped perforations produced by carnivore teeth (see A.335 and
A.336)

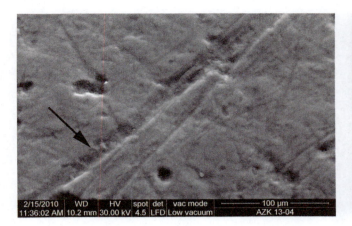

Fig. A.337 SEM microphotograph of a fossil bone showing linear marks and pits from Azokh Cave. These are caused by sediment particles penetrating into the bone when rubbed against the sediment during trampling. Note the irregular shape of the lateral side and bottom of the linear mark (see A.116)

Fig. A.340 Modern large bone fragment collected from the Sahara desert. The bone surface appears irregularly pitted, polished and rounded (A.565 and A.566). E. Aguirre's Collection

Fig. A.338 SEM microphotograph showing trampling of fossil bone in Westbury cave where the sediment is rich in gravel-sized limestone clasts (Andrews and Ghaleb 1999). Sediment particles produced abundant pits on the bone surface by penetration of gravel grains into the bone (A.124). The site is interpreted as an all-male cave bear den, and trampling was probably made by bears. Field width equals 7 mm

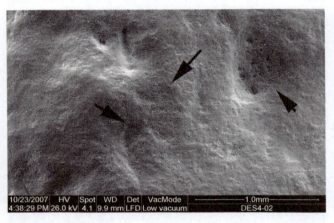

Fig. A.341 SEM microphotograph of a modern bone from the Sahara sand desert. It has been rounded and polished by the action of fine sand and wind. that hits the bone surface leaving a strongly pitted surface (A.566). Many of the pits and irregularities of the bone surface contain grains of sand (arrows)

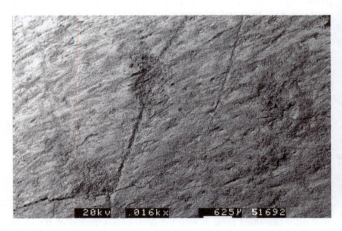

Fig. A.339 SEM microphotograph of a fossil bone from Vanguard Cave. The sediment in this cave is coarse sand and dispersed gravels. Friction of the bone against the sediment caused linear marks and pits on the surface due to trampling

Fig. A.342 SEM microphotograph of a recent bone from the Sahara sand desert that has been polished by intensive aeolian abrasion (A.566). The photograph shows homogeneously dispersed pitting all over the bone surface. This pitting is produced by the impact of sand grains against the bone by the wind

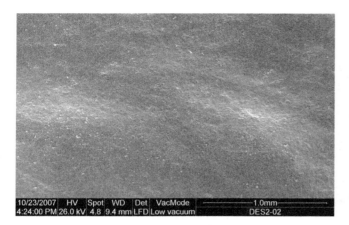

Fig. A.343 SEM microphotograph of the surface of a modern bone from the Sahara sand desert. The surface of the bone has been pitted and polished by sand and wind (A.566). In some concave areas of the bone, pitting is slightly less than in convex areas, but the whole of the exposed surface was affected

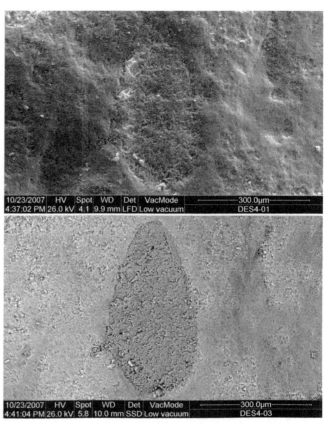

Fig. A.345 Top: SEM microphotograph of the surface of a modern bone pitted by wind and sand abrasion from the Sahara sand desert. Detail of A.581 with pits filled with cemented sand grains in the interior of the pit. Picture taken at SE emission to observe the topography. Bottom: SEM microphotograph of the same area of the same specimen A.341 collected from the Sahara sand desert, observed at the BSE mode to see differences in density. The infill of cemented sediment in the pit was chemically analyzed by Energy Dispersive X-ray Spectroscopy (EDS), and this showed that the chemical composition had a high content of Calcium and Silica together with other elements in minor proportion, such as Aluminum, Magnesium, Sodium, Iron and Sulfur

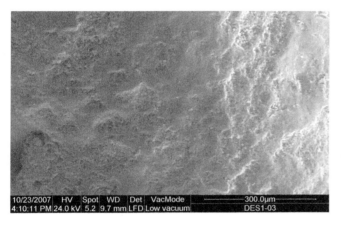

Fig. A.344 SEM microphotograph of a recent bone surface pitted by sand and wind, shown here at a higher magnification than A.341, A.342 and A.343. It shows the surface intensively pitted by sand grains that were blown against the bone surface by wind action

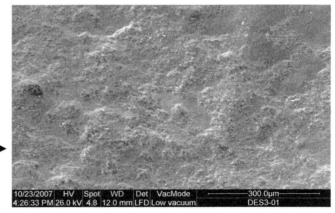

Fig. A.346 SEM microphotograph of the intensively pitted surface by ▶ wind abrasion of a modern bone collected from the Sahara sand desert. Wind polishes the bone which in detail consists of areas of intense pitted surface and some patches of smooth polished bone (top right corner of the picture). Compare with Text Fig. 6.5

Fig. A.347 SEM microphotograph of the pitted surface of a modern bone collected from the Sahara sand desert. There are areas of polished surface that are characteristic of abrasion by wind (compare with A.112 coarse sand in water)

Fig. A.348 Rounding and polishing is also produced by sediment and water action. The aspect of this type of abrasion is similar to the action of wind and sediment (A.566). Microscopic analysis distinguishes between wind and water abrasion

Fig. A.349 SEM microphotograph of recent weathered bone at stage 4 weathering (Behrensmeyer 1978). It has been abraded by water and coarse sands in a stone polishing tumbler for 360 h. The surface shows dispersed superficial pitting where the pits are bigger but more dispersed than evident for wind abrasion (A.346), and the shiny (smooth) polished patches also are larger (see Text Fig. 6.5). Particle size has greater variation in water-carried sediment than wind transported sediment, and water acts as cushion between the bone and grain particles

Fig. A.350 SEM microphotograph of recent weathered bone at stage 4 weathering (Behrensmeyer 1978). It has been experimentally abraded with very coarse sands and gravels (>1 mm grain size) in water, in a stone polishing tumbler for 360 h. Pitting by sand is greater on bone surfaces abraded by wind and shiny (smooth) polished patches are larger in bones abraded by water (see Text Fig. 6.5)

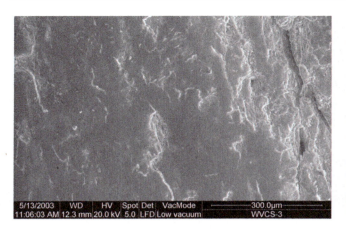

Fig. A.351 SEM microphotograph of fossil bones abraded by water and gravel. The general aspect is of pitting with some random scratches, and the shiny (smooth) polished patches are smaller than in weathered bones. This may be caused by a more homogeneous consistency and hardness of fossils than in modern bones. Sediment and water abrasion on fossilized bone produces greater rounding and polishing than on weathered or fresh bones (see Text Fig. 6.2)

Fig. A.353 Fossil phalanges of *Eucladoceros* from Senèze. Soil corrosion may also perforate the bone cortex. The volcanic sediment at this site was acidic, and body putrefaction under permanent water, together with the action of vegetation (aquatic algae) could cause this strong corrosion on some of the fossils A.807 and A.808

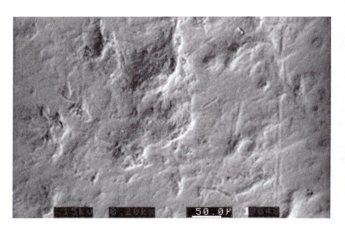

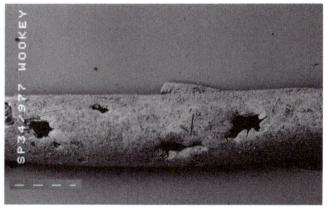

Fig. A.352 SEM microphotograph of a fossil, detail of the occlusal surface of a hominid molar from Paşalar showing irregular pit shapes on the occlusal enamel (compare with A.338). This has been produced by tooth wear during the life of the individual and is related to the type of food eaten by that individual. Microwear pitting is greatest for species specializing in hard object feeding and/or when grit is ingested with the food

Fig. A.354 SEM microphotograph showing deep perforation of recent bone by cave corrosion at Wookey Hole, a limestone cave connecting to Swildon's Hole. The bone is hollowed out from the inside so that the bone surrounding the perforations has started to collapse into the interior. (See other cave soil corrosion types from the same site (A.786) and from other caves (A.788)). Field width equals 5.2 mm

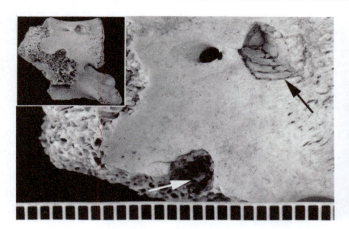

Fig. A.355 Puncture marks on a recent monitored vertebra from Neuadd. Puncture marks are produced by a vertical compression, producing inverted cone impressions – black arrow and arch shaped rim at broken or anatomical edges – white arrow. Carnivore puncture marks are characterized by an inverted cone shape, with depressed cortical bone covering the bottom and walls of the puncture (compare with A.320 and A.322). The small box on top left shows the body of the sheep vertebra with these two tooth mark types

Fig. A.357 Punctures on the epiphysis (A.358) of a fossil hominin proximal humerus from Sima de los Huesos, Atapuerca. The sizes of punctures on articular ends are generally larger than on diaphyses, since the thinner and softer bone of articular surfaces is more easily penetrated (A.157). See text for explanation and classification of carnivore chewing marks by locations on bones

Fig. A.356 Carnivore tooth marks on fossil bone from Galeria, Atapuerca. Carnivore puncture marks differ on different types of bone: broken edges (transversal, spiral or splits), type of bones (long bones, flat bones: A.363, A.364 and A.365) or bone surfaces (articular or intact edges). Each of these cases has different sizes of marks based on types of mark (e.g. puncture, groove), anatomical element (e.g. humerus, scapula) and location (e.g. diaphysis, articular surface). (© Atapuerca EIA)

Fig. A.358 Puncture mark on the epiphysis (A.357) of a fossil cervid from Galeria, Atapuerca, made by a large carnivore (© Atapuerca EIA)

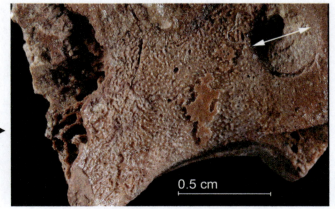

Fig. A.359 Puncture mark on fossil bone from Miocene deposits at ▶ Concud, showing an inverted cone shape (A.355 and A.387). The mean sizes of punctures on surfaces of limb bone shafts are generally smaller than those on the articular ends. Puncture marks on bone surfaces are measured taking the minimum dimension. Courtesy of D. Pesquero. (Pesquero 2006)

Fig. A.360 Puncture marks recorded on the bone surfaces of fossil bone fragments recovered from Galeria, Atapuerca site, probably made by wolves (A.356). Courtesy of I. Cáceres

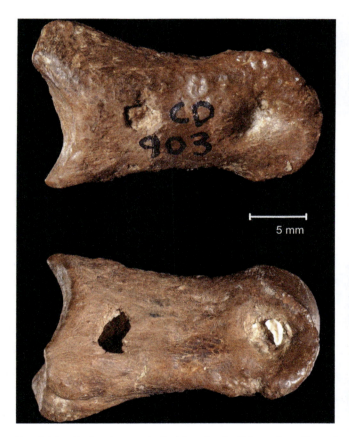

Fig. A.361 Perforations on both sides of a fossil bone from Concud made by carnivore chewing. The outer surface of bone has been depressed into the interior. When the compact bone is thin (such as in phalanges), punctures may perforate the cortical bone into the interior rather than forming a conical shape depression (A.355 and A.387). Courtesy of D. Pesquero

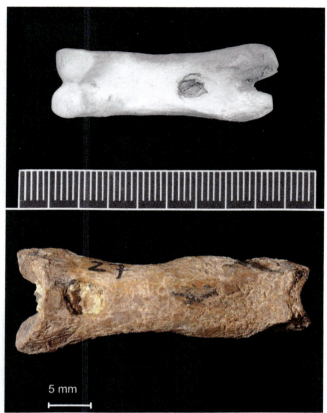

Fig. A.362 Top, single puncture made by a fox from a monitored modern carcass from Neuadd. The minimum diameter of the puncture is 4.2 mm, which appears too large for it to be made by a fox until its location penetrating soft trabecular bone is considered. Isolated punctures are less diagnostic than tooth prints, but they may still be indicative of the size of the carnivore involved. Bottom, a similar mark on a fossil from Concud (see A.361). Courtesy of D. Pesquero

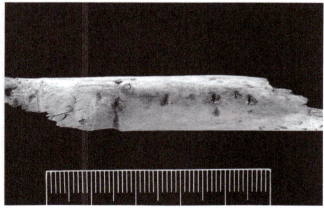

Fig. A.363 Recent monitored bone fragment from Neuadd with fox puncture marks (compare with A.362 and A.364). The sizes of punctures on surfaces of limb bone shafts are generally smaller than those on the articular ends and range between 1 and 2 mm measuring the minimum dimensions. See text for explanation and classification of carnivore chewing marks by locations on bones

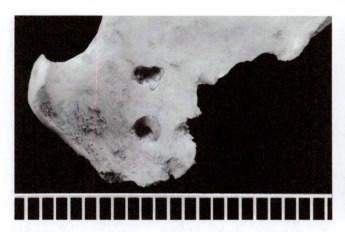

Fig. A.364 Recent monitored bone from Neuadd 11/1. Puncture marks on the surface and on the edge of a rabbit pelvis were made by a feral cat. See text for explanation and classification of carnivore chewing marks by locations on bones

Fig. A.367 Detail of experimental modern human chewing marks making a puncture mark near the broken edge of a recent bone. This mark was made with the premolars and has a triangular shape (A.368)

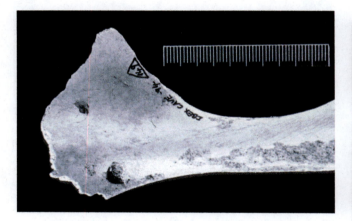

Fig. A.365 Puncture marks on a fossil pelvis of ibex from Gibraltar were probably made by a fox. Perforations in thin bones (A.364) made by mammalian carnivores are larger than those on limb bone diaphyses, in this case 3–4 mm in minimum diameter. See text for explanation and classification of carnivore chewing marks by locations on bones

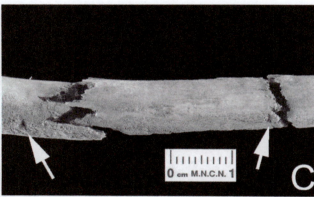

Fig. A.368 Human tooth marks near broken edges on a modern rib that was chewed experimentally. Thin bones are chewed to suck the marrow content in bones. Human puncture marks (white arrows) may be associated with bending of the bones by holding them between the upper and lower cheek teeth and pulling down the free end with the hands. This action provides a consistent pattern of bent ends (see Text Fig. 9.4). This bent shape has also been seen on bones chewed by chimpanzees

◀ **Fig. A.366** Modern young pig rib experimentally chewed by humans. Human puncture marks (black arrows) are rare, except in areas near breakage or anatomical ends of light and thin bones. Humans use tools to break bones, but they may chew anatomical ends or fragment edges of bones like ribs or vertebrae, or other bones after they have been broken to extract the marrow. The shape of these puncture marks is triangular (see A.368)

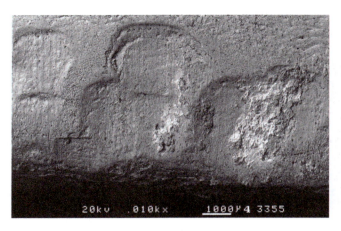

Fig. A.369 SEM microphotograph showing crescent pits on a rib from TD6, Atapuerca. These have been identified as made by hominins when compared with experimental marks made by human incisors. These marks are deeper than gnawing marks observed from both modern humans (experimental A.134) and Neanderthals (Gorham's Cave, A.145, A.148 and A.151). Note the central mark seems almost to be an image of the flattened occlusal surface of the incisor that made the mark

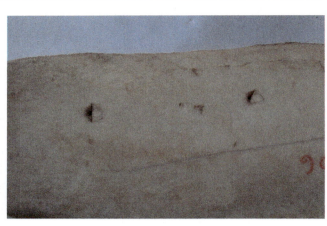

Fig. A.372 Modern bones. Detail of crocodile tooth marks showing sharp puncture marks. Similarities with trampling marks in wet environments are evident, and it may be difficult to distinguish them if the sediment gravels and clasts have sharp edges, such as in Senèze (see A.336). On the other hand, perforations from trampling may contain some of the sedimentary clasts forming the perforations

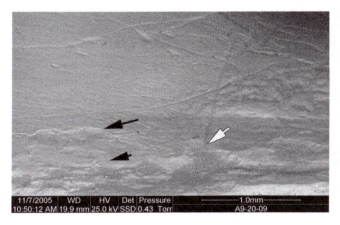

Fig. A.370 SEM microphotograph of a fossil rabbit pelvis surface. Pits similar to those experimentally made by humans (see A.134) have been recognized on small animal bones from Gorham's Cave. According to these experiments, shallow crescent pits (black arrow) are associated with shallow linear marks (short black arrow). These chewing marks interrupt previous incision marks made by a stone tool (white arrow). See A.145 and A.151

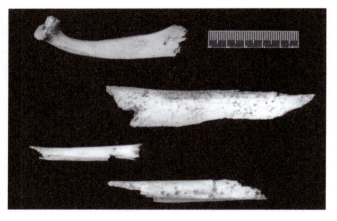

Fig. A.373 Chewing marks on modern monitored sheep bones scavenged by foxes at Neuadd. From top to bottom perforation of the end of a rib shaft, perforation of the broken (left) end of a bone fragment, perforation on lateral surface of bone diaphysis, and multiple perforations on the shaft of a metapodial. The sizes of punctures on spiral breaks are larger than punctures on transverse breaks (A.363 and A.364). See text for explanation and classification of carnivore chewing marks by locations on bones

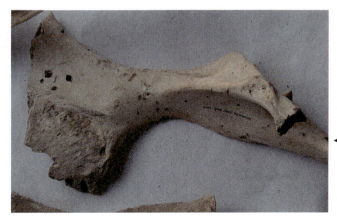

◄ **Fig. A.371** Modern bones. Puncture marks made by crocodiles (compare with A.336). Crocodiles have rows of unicuspid conic teeth, each of them having the shape of canine teeth of mammalian carnivores. Crocodiles do not chew their food but grip their prey strongly when catching it and when dismembering carcasses. Courtesy of F. Njau. (Njau and Blumenschine 2006)

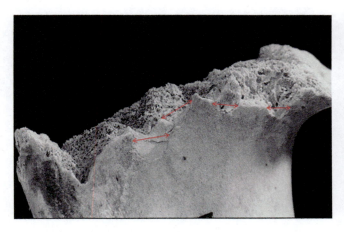

Fig. A.374 Edge punctures on a transverse break on a fossil limb bone from Atapuerca. The double ended red arrows show how arch-shaped puncture marks are measured. Punctures exposed in the broken ends of bone shafts are generally larger than the surface pits, particularly when they occur near the articular ends. See text for explanation and classification of carnivore chewing marks by locations on bones. (Compare A.316). Courtesy of I. Cáceres

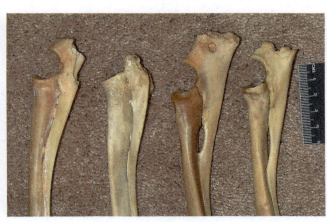

Fig. A.376 Modern bones (compare with A.375). Heads of sheeps' ulnae chewed by domestic dogs. Sheep were given to Koi people in South Africa for their consumption, and after the carcasses were eaten, the Koi gave the remains to their dogs. In all cases the olecranon process has been chewed with puncture marks along the edges. C.K. Brain's collection. (Brain 1981)

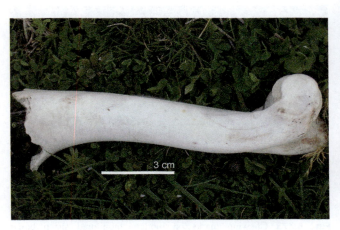

Fig. A.375 Pit marks on a modern monitored deer humerus from Riofrío. Tooth marks near the articular ends (A.376) and at transverse breaks may be related to marrow extraction. The sizes of pit and puncture marks on transversal broken edges are generally smaller than on spiral breaks. See text for explanation and classification of carnivore chewing marks by locations on bones

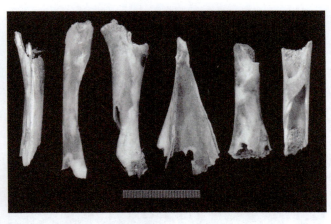

Fig. A.377 Modern bones. There may be great variety of sizes and shapes of edge punctures in a single assemblage, as seen here on monitored sheep bones from Neuadd scavenged by foxes (*Vulpes vulpes*). The shape and size of the perforations on bones like the scapula are similar to those produced by eagles or vultures (A.408), although similar marks are present also on more robust bones in the same assemblage. These tooth marks are larger on flat bones than pits on long bone diaphyses made by the same predator (A.363)

Fig. A.378 Modern bone. Perforations in thin bone on monitored ▶ sheep bones at Neuadd scavenged by foxes. Punctures on thin and soft bones, such as sacrum, pelvis and scapula, are consistently larger and deeper (penetrating the cortical bone) than those on more compact bone (see A.375). See text for explanation and classification of carnivore chewing marks by locations on bones

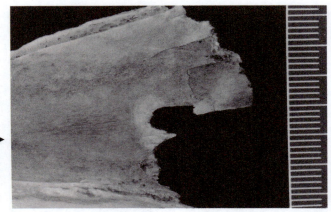

Fig. A.379 Tooth mark on a hominin pelvis at Sima de los Huesos, Atapuerca, after disarticulation. Carnivores frequently gain access to the carcass through the pelvic area, resulting in heavy damage to the pelvis. Punctures measured on thin soft bones, are larger than those on other more compact bones (e.g. limb bones, A.392). See text for explanation and classification of carnivore chewing marks by locations on bones

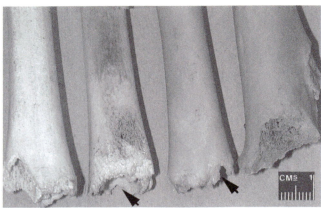

Fig. A.382 Modern bones. Sheep chewed by Koi people. Long bone ends of sheep have been chewed by Koi people in South Africa. Note the arch shaped rim of the broken edges (A.383). Koi boiled the sheep before eating the carcass as described in Brain (1981)

Fig. A.380 Modern bone. Lateral and dorsal processes of monitored vertebrae of a fallow deer carcass from Riofrío chewed when still in anatomical connection (A.381)

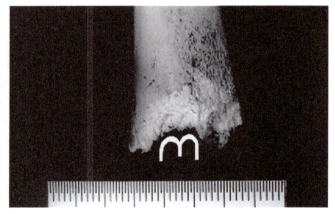

Fig. A.383 Modern bone. Experimentally chewed long bone by modern humans. Looking at the crenulated edges in detail, a double arched puncture can be distinguished in our experimental collection and can be compared with bone chewing by Koi people (Brain 1969). This double puncture results from human chewing using the molar teeth and appears to be a characteristic of human chewing (A.382 and A.384) and differs from any other carnivore tooth mark

◄ **Fig. A.381** Modern bone. Dorsal processes of sheep from Barranco de las Ovejas (Pyrenees Huesca Spain) have been chewed while still in anatomical connection (A.380). This may be common practice for small sized carnivores scavenging large prey. The scat left by the carnivore for marking its territory has been provisionally identified as coming from a small sized carnivore that could scavenge these carcasses

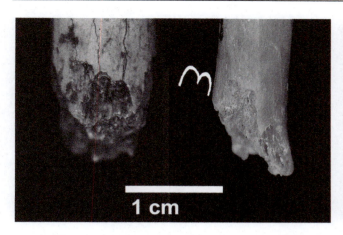

Fig. A.384 Fossil bones. Double arched punctures on fossils from TD6 Atapuerca. These characteristic marks made with human molar teeth have been identified on a human bone (right side) and on a small sized mammal (left side)

Fig. A.385 Double tooth marks on a recent monitored sheep mandible from Neuadd. Associated puncture marks from a multicuspid tooth such as this are described as a tooth print. Punctures are measured individually as well as measuring the distance between them (A.387). The first measurement measures the cusp sizes and the second measures the size of the tooth size. See text for explanation and classification of carnivore chewing marks by locations on bones

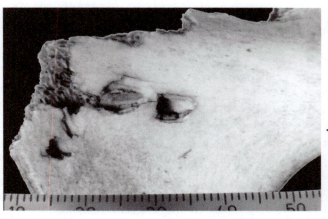

Fig. A.387 Two tooth impressions are seen here on a fossil bone from Concud. The length and breadth of the separate puncture marks are first measured, and then the distance between the center of each puncture provides an estimate of the separation of the cusps on the tooth. The outer rim distance of the punctures provides a closer estimate of the overall size of the carnivore tooth that made the marks, although it will always underestimate the size (see A.389). Courtesy of D. Pesquero

Fig. A.388 Fossil bone. Astragalus of *Sivatherium*, a late Pliocene/early Pleistocene short necked giraffe from Langebaanweg. Two carnivore puncture marks can be seen on the surface, probably made by an hyenid (A.389). There are five species of hyena known at Langebaanweg site

◀ **Fig. A.386** Double tooth marks on a recent monitored sheep bone from Neuadd. Tooth prints are not frequent, but if there are several bones in an assemblage with them present they provide good evidence of the carnivore involved, in this case a fox. Tooth prints provide an estimate of the distance between tooth cusps (A.387), and hence tooth size, of potential carnivores so as to identify the species involved (A.389)

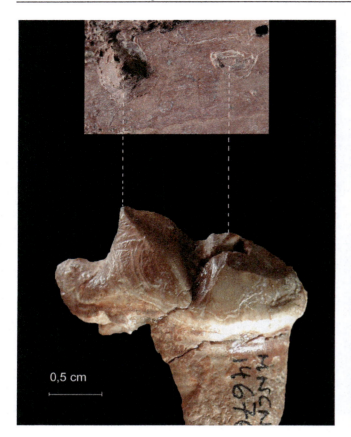

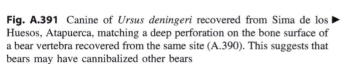

Fig. A.389 Fossil bone. Top, part of the surface of a vertebra of *Hipparion* from Concud. The two punctures are formed from a tooth print, with one conical perforation penetrating the surface of the bone, and a second shallow pit that indicates this is a multiple mark made by a single tooth. The tooth print is compared (at the same scale) to a tooth of *Lycyaena chaeretis*, an extinct hyena found at the site (Pesquero 2006), which shows how the shape and size of the two tooth cusps match the shape and size of the tooth print. The distance between the outer rims of the punctures provides a estimate of the overall size of the carnivore tooth that made the marks (A.387), although it will always underestimate the size because of tooth ridges and cingula extending the alveolar dimensions of the tooth. (Courtesy of D. Pesquero)

Fig. A.390 Fossil bone. A vertebra of *Ursus deningeri* from Sima de los Huesos, Atapuerca, bearing a multicuspid tooth print. From its size, the tooth print was probably made by a cave bear (A.391) molar. This is indicated by the large size of the tooth mark and the three lateral cuspids present on the articular surface of the vertebra, with the separate perforations nearly 1 cm in diameter and the total with of the mark nearly 3 cm

Fig. A.391 Canine of *Ursus deningeri* recovered from Sima de los ▶ Huesos, Atapuerca, matching a deep perforation on the bone surface of a bear vertebra recovered from the same site (A.390). This suggests that bears may have cannibalized other bears

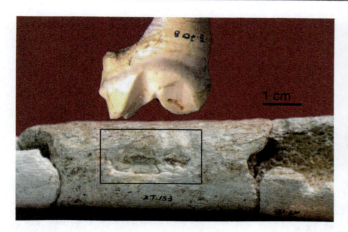

Fig. A.392 Tooth print recorded on the bone surface of a fossil human long bone recovered from of Sima de los Huesos, Atapuerca. The shape and size is different from other tooth prints seen here (A.387 and A.389), matching better with a carnassial of *Panthera leo*, which is also recorded at the site (shown above)

Fig. A.393 SEM microphotograph of a fossil bone, detail of a tooth print (see Text Fig. 4.7) made by a fox-sized carnivore premolar on a small sized mammal bone fragment from Atapuerca

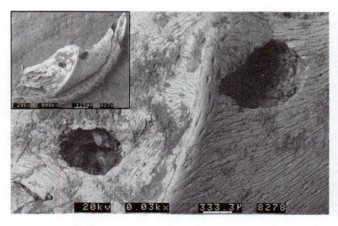

Fig. A.394 SEM microphotograph of tooth print on a recent rodent mandible from a scat of a monitored white tailed mongoose, *Ichneumia albicauda* (see also A.173) from Meswa Bridge. The rodent mandibular body is shown in the top left box, with the details of the puncture marks shown below. The distance between the midpoints of the two perforations is 2.3 mm, which indicates that the marks were probably made by the two mesial cusps of the M1 (Andrews and Evans 1983)

Fig. A.395 Two puncture marks on spiral breaks on a fossil from Concud. Puncture marks on spiral breaks are larger than the ones on transverse breaks. See text for explanation and classification of carnivore chewing marks by locations on bones. Compare with A.316 and A.396. Courtesy of D. Pesquero. (Pesquero 2006)

Fig. A.396 Measurement of an arched shaped puncture mark on a broken edge of fossil bone from Concud. Note that the tooth mark impact has struck off a flake of bone (compare with A.316). Courtesy of D. Pesquero

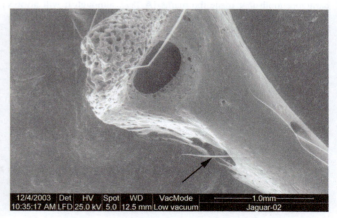

Fig. A.397 SEM microphotograph of an experimental modern rodent distal humerus digested by a jaguar. The thin bone of the articular surface has been penetrated by the digestive action (A.617). Digestion provides direct evidence of carnivore involvement, producing penetration of bone surfaces, sometimes reaching the marrow cavity. Evidence left by digestion may be more diagnostic of predator damage than chewing marks. Courtesy G. Gómez

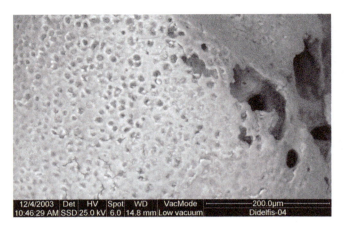

Fig. A.398 SEM microphotograph of the articular surface of an experimental modern rodent bone that has been digested by *Didelphis*. Chemical digestion by acid secretions is here producing perforations into cancelous tissues (far top right) as well as into articular bone (A.397). Modern collection of G. Gómez

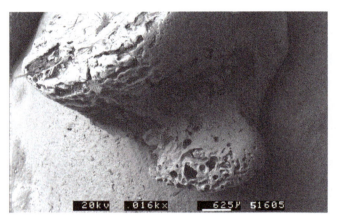

Fig. A.399 SEM microphotograph of the distal end of a monitored recent bird humerus digested by a jackal. Note the surface is perforated by the action of the acid juices of the mammalian carnivore (A.398)

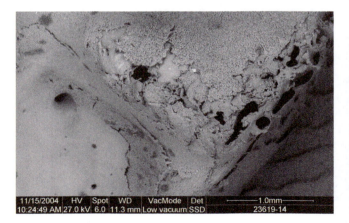

Fig. A.400 SEM microphotograph of a modern bone from the taxidermic collection of the MNCN. Detail of damage done during bone preparation. This has produced perforated surfaces through use of enzymes to prepare the bone for museum display. In this case, the destructive activity of enzymes could be due to either prolonged exposure to the enzymes or to the fact that treatment took place during the summer when temperatures were higher than usual

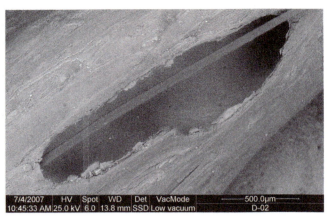

Fig. A.401 SEM microphotograph of a modern bone from Tswalu (South Africa). Detail of a perforation on a bone fragment digested by modern hyena. Perforation is the result of the highly acidic gastric juices of hyenas (compare with A.402)

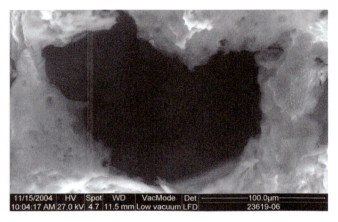

Fig. A.402 SEM microphotograph of a from the taxidermic collection of the MNCN. Detail of bone showing the effects of enzyme action. Perforations by enzyme activity are similar to bones those produced by predator digestion (see A.397 and A.401)

Fig. A.403 Vulture beak marks perforating the thin bone of a bovid skull collected from Olkarian Gorge (A.202 and A.205). One of the beak perforations is indicated by an arrow, but the edge has several other beak puncture marks

Fig. A.404 Perforations of a recent deer bone scavenged by vultures. This specimen was monitored by park guards at Riofrío, and they confirmed that this fallow deer carcass was exclusively eaten by vultures (A.403, A.405 and A.658)

Fig. A.406 Tibia of sheep pecked by vultures (black arrow) from the Barranco de las ovejas (Huesca Pyrenees, Spain). Carcasses are abandoned by the local shepherds and scavenged by Lammergeier or Bearded Vulture, *Gypaetus barbatus* (compare with A.403 and A.405)

Fig. A.405 Top: modern mandible of fallow deer from Riofrío that has been heavily damaged along the inferior border by vultures' beak marks (white arrow). This specimen was monitored by park guards at Riofrío, and they confirmed that this fallow deer carcass was exclusively eaten by vultures (A.403 and A.404). Bottom: modern red deer maxilla from Riofrío heavily damaged by vultures. Numerous puncture marks were made by penetration of the beak into the bone, some of which are indicated by white arrows. Vultures can be highly destructive and break fragile bones of small to large sized mammals

Fig. A.407 Magpie beak puncture marks are present on an experimental bone covered by lichen. The dark matter surrounds the area pecked by the crow (A.406)

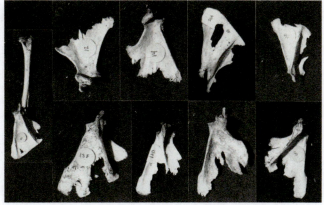

Fig. A.408 Modern bones. Monkey scapulae with perforations in the blade and stripping of the vertebral edge, both carried out by eagles. The monkeys were killed by the eagles, and all or parts of their bodies had been carried into the eagle's nest (A.377). When the bones had been stripped of their flesh, they were discarded and dropped to the foot of the tree below the nest where they accumulated. (Andrews 1990)

Fig. A.409 Fossil hedgehog mandible from Olduvai Bed I (FLKN5). This damage consists of a triangular regular pitting, the pits all similar in size, covering the mandible surface. These pits are isolated or formed at the ends of linear marks. The size and shape of the pits fit with the tips of spines may have been made by humans (A.128, A.129 and A.130)

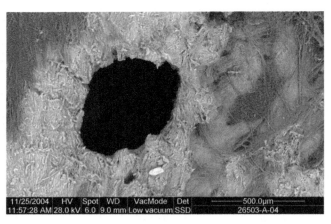

Fig. A.412 SEM microphotograph of a modern bone, from the taxidermic collection of the MNCN. Detail of a perforation made by dermestids. Note the irregular edge of the perforation due to the powerful dermestid's jaw (see A.207 and A.410)

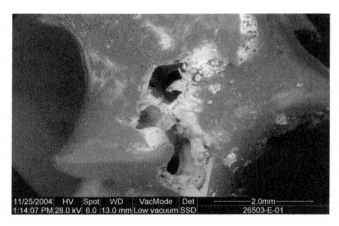

Fig. A.410 SEM microphotograph of a vertebra of *Milvus* from the MNCN taxidermic collection prepared using dermestids. Note the perforations made by the dermestid larvae and deep scratches at the centre of the specimen (see A.207 and A.208)

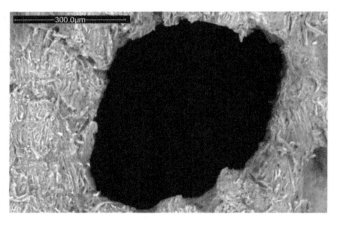

Fig. A.413 SEM microphotograph of a modern bone, from the taxidermic collection of the MNCN. Detail of a perforation made by dermestid beetle larvae, showing grooving around the sides of the perforation and deep penetration into the bone. The irregularity of the edge could be a response to the powerful jaws of the coleopteran larvae as they penetrate the bone (see A.207 and A.410)

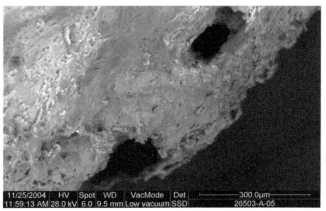

◀ **Fig. A.411** SEM microphotograph of a modern bone from the taxidermic collection of the MNCN. Detail, of perforations made by dermestids due to a prolonged exposure to dermestid larvae (see A.207 and A.410)

Fig. A.414 SEM microphotograph of a modern vertebra monitored and collected from a termites' nest in Tanzania. Termites have strong jaws that may cut into bone. Several perforations were distinguished on the bone surface, and all have the characteristic sinuous radial grooves around the hole as seen in this picture (compare with A.207 and A.410). A.K. Beherensmeyer's collection

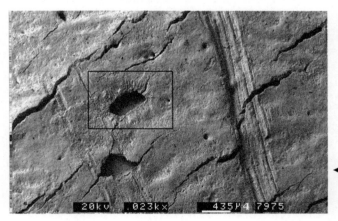

Fig. A.416 SEM microphotograph. Detail of the boxed hole from A.415 from Atapuerca, Trinchera Elefante. Traits of this perforation and surrounding scratches are highly similar to modern termite perforations seen in A.414. Field width equals 1.2 mm

Fig. A.417 SEM microphotograph of a modern bone, detail of perforations of bone by insect larvae on human remains from the Spitalfields burial crypt. These perforations have striations along the edges, slightly different from those made by termites (A.414) or dermestides (A.413)

◄ Fig. A.415 SEM microphotograph of perforations on a fossil from Trinchera Elefante (ATA'98, TEi 31, n.1,), Atapuerca. Features of termite chewing are characteristic and can be also identified in fossils One of the perforations is boxed and shown in close up in A.416

Fig. A.418 SEM microphotograph of a modern bone, detail of conical perforations made by a species of insect in human bones in the Spitalfields crypt. These are volcano-like craters with radial striations around the hole. Differences in shape and size between these perforations (A.413, A.414 and A.218) suggest that different insects were responsible for these modifications

Fig. A.421 SEM microphotograph of a modern bone, detail of the circular depressions on the human skull from Spitalfields crypt (A.420). One explanation for these marks is that they were made by Diptera. Two species of Diptera have been identified in Spitalfields' coffins, *Conicera tibialis* and *Ophira leucostoma* (Molleson, pers.com). Diptera as well as arachnids lack powerful jaws, so they secrete digestive fluids to soften their food and absorb the pre-digested food

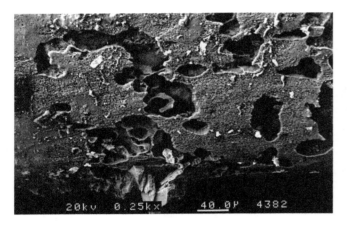

Fig. A.419 SEM microphotograph of a modern bone, detail of perforations made by a species of insect in human bones in the Spitalfields crypt. Scratches are present along the edges of the holes (A.218), different from other insect damage on the human bones

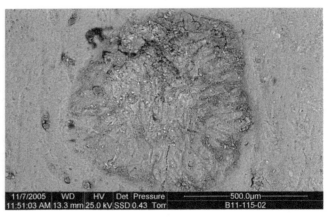

Fig. A.422 SEM microphotograph of a fossil bone from Gorham's cave, showing two types of modifications. Chemical digestive fluids secreted on the bone surface have softened the bone, producing a circular corrosive scar (compare with A.421). Within this area, physical penetration has produced linear marks converging into the area. This type of damage could be due to arachnids (see A.223 and A.224), as they cannot eat solid substances, but this needs to be verified

◄ **Fig. A.420** SEM microphotograph of a modern human skull from Spitalfields crypt. The circular depressions (bottom right of picture) could have been made by insects. The smooth surface inside the circular depressions (A.421) discards coleoptera species or termites. Damage by insects needs further experimental studies to distinguish different insect activity which would provide invaluable environmental information

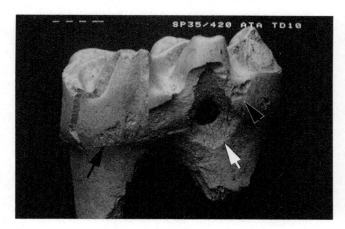

Fig. A.423 SEM microphotograph showing a fossil rodent molar from Atapuerca-TD10 damaged by vascular plant roots (see A.778). Perforation by roots has etched both the enamel and the dentine (white arrow). Characteristic cracking texture can be distinguished in the interior of the root mark around the perforation. The enamel is split off and linear marks etched into the dentine (black arrow on the left) and into the enamel (black arrow on the right). Field width equals 2.8 mm

Fig. A.426 SEM microphotograph of a root mark that has perforated (A.423 and A.884) a fossil small mammal long bone from Atapuerca. Corrosion of the bone is present around the perforation. Differences in the type of corrosion observed in root marks (perforations, grooves, internal cracking or surface corrosion) may indicate the type of plant that made the mark. Further experimental work is needed to obtain this information

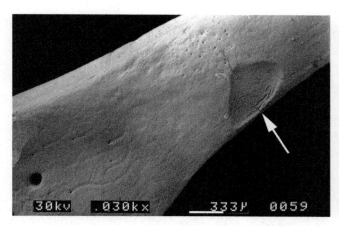

Fig. A.424 SEM microphotograph of a fossil small mammal long bone from Atapuerca showing an incipient root pit (white arrow). Note the cracking texture in the interior of the root mark (A.425 and A.430). Linear grooves at the bottom left side of the bone surface are vessel imprints

Fig. A.427 SEM microphotograph of a small sized fossil animal bone from Gorham's Cave showing perforation by roots (A.423 and A.430). Note the cracking in the interior of the root mark

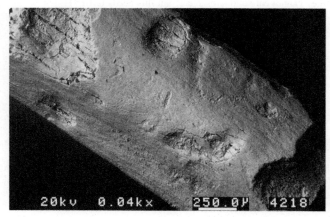

◀ **Fig. A.425** SEM microphotograph of a fossil small mammal long bone from Atapuerca showing deep root pits (A.430 and A.882). Cracking is present in the interior of the root mark that is not affecting the bone surface outside these pits. Cracks in the interior of root marks are not always present, but where present they indicate root mark origin of the mark

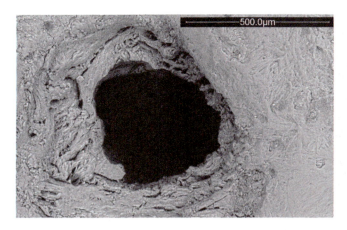

Fig. A.428 SEM micrograph of a root perforation, on a fossil small mammal long bone from Atapuerca (A.423 and A.430). Note the cracking texture in the interior of the root mark around the perforation

Fig. A.431 SEM microphotograph showing pits on a modern monitored horse scapula (ND21, Neuadd 25, see A.264) preserved in calm water for 10 years (measured pH 5.1). The surface was covered by a dense film of algae (seen at the bottom left side), and the surface perforations were probably produced by the algae (see A.432 and A.435). Algae are no longer classified within the Kingdom Plantae, but they have been included here as traditional classification

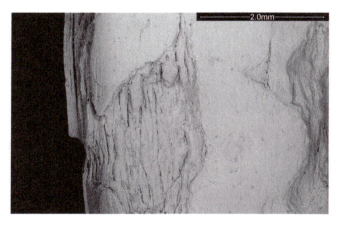

Fig. A.429 SEM microphotograph of a fossil from Gorham's Cave at the cave entrance. The bone surface has been degraded by root marks, identified by the internal cracking (A.240 and A.430) at the bottom of the root mark and absent on the bone surface outside the damaged area

Fig. A.432 SEM microphotograph of the pits on a modern monitored horse scapula (ND21, Neuadd 25, A.264) preserved in calm water for 10 years (measured pH 5.1). Close up view of boxed area at A.431 to show in detail the pit marks

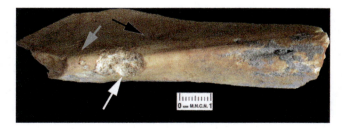

Fig. A.430 Circular root mark mimicking an impact mark on a fossil from Cueva Ambrosio (white arrow). The bottom of the mark has an irregular surface indicating it was made by a chemical agent. It connects to a branched linear mark (black arrow), with a brown-reddish halo around the mark. This reinforces its biochemical origin and indicates that the mark is a recent root mark. The grey arrow indicates a fossil root mark (compare with A.227 and A.249). Specimen provided by S. Ripoll

Fig. A.433 SEM microphotograph of another pitted area on a modern monitored horse scapula (ND21, Neuadd 25, A.431) preserved in calm water for 10 years (measured pH 5.1)

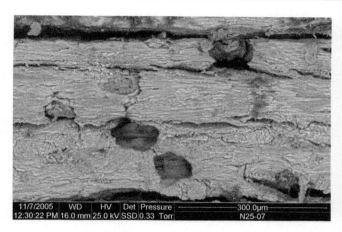

Fig. A.434 SEM microphotograph of the pits in a modern monitored horse scapula (ND21, Neuadd 25, A.431) preserved in calm water for 10 years (measured pH 5.1). This shows details of perforations which we believe have been produced by algae

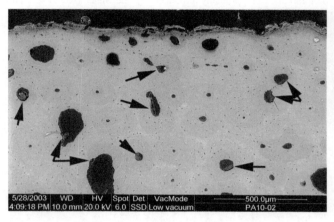

Fig. A.435 SEM microphotograph of a cut section of the modern monitored horse scapula (ND21 Neuadd 25 A.264) preserved in calm water for 10 years (measured pH 5.1) and covered by algae. Canals of Havers contain remains of algae (black arrows) indicating that algae have penetrated the bone throughout the Haversian system (see A.265)

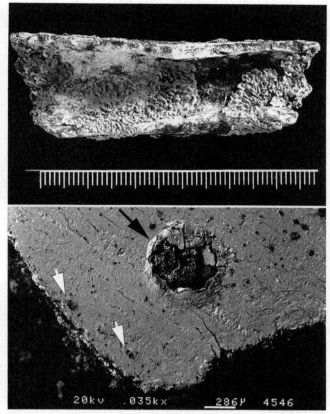

Fig. A.437 Top: recent long bone shaft fragment heavily covered by lichen from A.J. Sutcliffe's subarctic collection. Bottom: SEM microphotograph of the cut and polished section of the bone above, showing a segment of lichen penetrating (A.262) the medial cortical layer (black arrow). This penetration is independent of any histological trait. The smaller black spots indicate more incipient lichen penetration, some of which is indicated by white arrows. A.J. Sutcliffe's collection

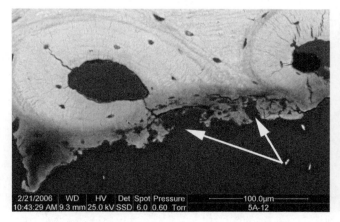

Fig. A.436 SEM microphotograph of a cut section of the modern monitored horse scapula (ND21 Neuadd 25), detail of the cortical surface. Algae have heavily etched the bone surface (PCL) making perforations as seen in A.265. In cross section these holes are seen as tubular shapes (white arrows). The most important penetration, however, is made through the histological traits (Havers canals) of the cortical bone (see A.435)

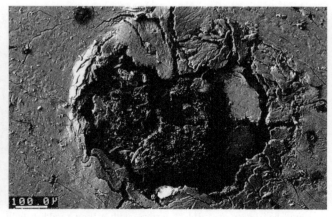

Fig. A.438 SEM microphotograph of a recent long bone shaft fragment heavily covered by lichen from A.J. Sutcliffe's subarctic collection. This is a detail of A.437 showing lichen perforating the bone cortex and including remains of lichen in the interior of the perforation

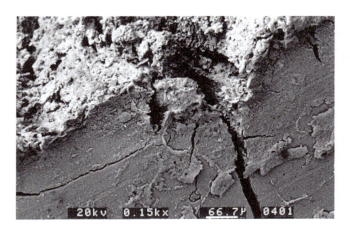

Fig. A.439 SEM microphotograph of a recent long bone collected from subarctic environments. The surface of the bone is entirely covered by lichen. This is a cut section of a bone fragment showing lichen penetration cracking the bone surface of thick cortical bones. The bone surface is also etched and damaged (A.440 and A.896). A.J. Sutcliffe's collection

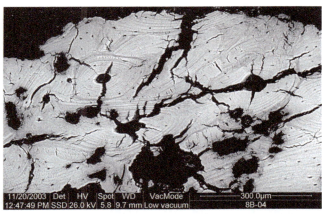

Fig. A.442 SEM microphotograph of a recent monitored axis vertebra from Neuadd 2 (ND12) which was exposed on the ground (pH 4.0) for 18 years. The lichen penetration shows as cracks in the cortical layer and perforations surrounded by damaged (grayish) bone tissue. The lichen penetration is independent of histological structures (A.262), and it is apparent that the Havers canals are not used by lichen as done by algae (A.435)

Fig. A.440 SEM microphotograph of localized damage on a recent bone produced by lichen on the bone surface. The lichen was removed but some pieces of lichen remain on the surface (black arrow) see A.896. A.J. Sutcliffe's collection

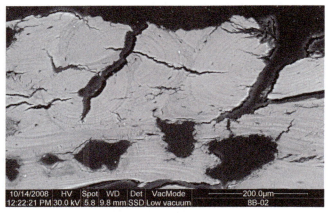

Fig. A.443 SEM microphotograph of a recent monitored axis vertebra from Neuadd (ND12) which was exposed on the ground (pH 4.0) for 18 years. The lichen penetration (A.262) shows as cracks in the cortical layer (top of the photograph) and perforations surrounded by damaged (grayish) bone tissue at the bottom of the photograph. (Note that the holes and cracks are filled by the resin used to prepare this cut section.)

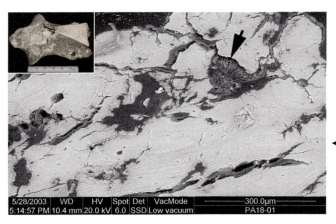

◄ **Fig. A.441** SEM microphotograph of a recent monitored horse vertebra from Neuadd 2 (ND12 top left box). The vertebra was exposed on the ground (pH 4.0) for 18 years and entirely covered by lichen. This has penetrated into the bone breaking the bone surface (see A.262, A.437 and A.895). Remains of the lichen are present in some of these holes (black arrow)

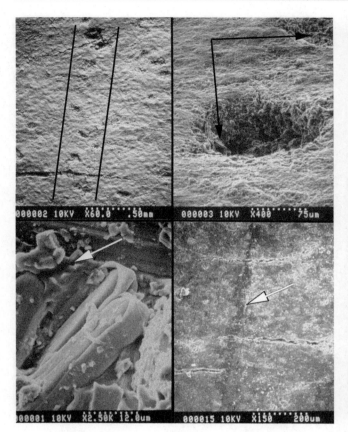

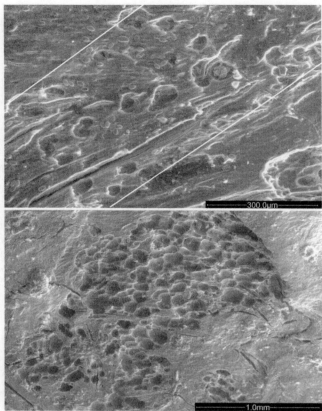

Fig. A.444 SEM microphotographs of a modern bone fragment from East Fork river, Wyoming, USA This is a high energy stream during the spring snow melt season. The bone surface shows pits lineally arranged (top left see A.447). Inside these pits, there are abundant diatoms (top right some of them indicated by black arrows). These diatoms are covered by a mucilaginous substance (white arrow at the bottom left photograph). The siliceous composition of diatoms questions whether diatoms may corrode bone. Boden (1988), however, described perforations made by diatoms on calcareous sedimentary clasts. When conditions turn unfavorable, diatoms may synthesize mucilage (A.448 and A.449) that sticks diatoms cells together so that they form long chains attached to the substrate in high energy water systems (Raven et al. 1986). This could be the origin of the smooth shiny track observed at the bottom right photograph pointed by a white arrow. A.K. Behrensmeyer's collection

Fig. A.445 SEM microphotographs taken with SE emissions of a recent fallow deer mandible collected from a seasonal water stream at Riofrío. The bone surface shows a slightly linear arrangement of pits (top). Diatoms are dispersed over the surface, possibly because the bone was collected in summer when the water flow was slow and calm. More rapid seasonal water flow in spring could force diatoms to line up and attach themselves to the bone (A.446). Boden (1988) proposes that the mucilage produced by diatoms must have an acid and/or corrosive substance that attacks carbonates, although he did not analyze it chemically to verify his hypothesis. This type of pit is distinguished from natural histological traits of the bone, e.g. bone remodeling. Bone remodeling is shown in the bottom picture and is three times bigger and the arrangement does not follow any linear distribution

Fig. A.446 SEM microphotograph at SE emission mode of a fallow ▶ deer mandible collected from a seasonal water stream at Riofrío (see A.445). A slightly darker and more organic substance covers the bone surface, forming as linear tracks (marked between white lines) each with small pits along the interior (some pointed by white arrows) which are also lineally arranged

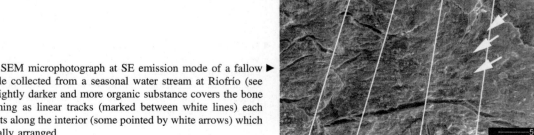

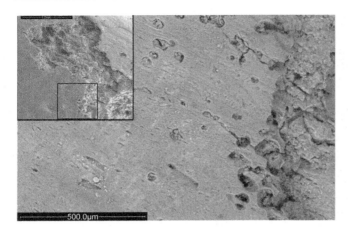

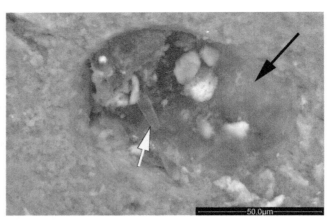

Fig. A.447 SEM microphotograph at BSE mode of a modern sheep radius monitored in a spring river at Neuadd for several years. It is heavily damaged on the surface (A.892). The top left photograph shows a linear arrangement of pits coalescing to form the damaged area. The area of the small box is shown enlarged in the main photograph. Diatoms cover the bone and this is the only apparent cause of this damage

Fig. A.449 SEM microphotograph of a perforation on a recent sheep's radius monitored from a spring river at Neuadd photographed at BSE mode. Backscattered electrons penetrate into the surface. Note the organic layer covering the perforation (black arrow). Diatoms are covered by mucilage, but backscattered electrons allow us to distinguish partially covered diatoms (white arrow) which are more superficial (A.444)

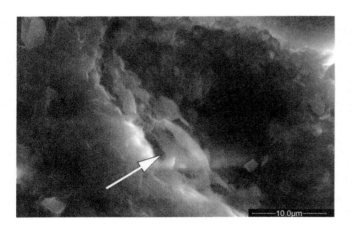

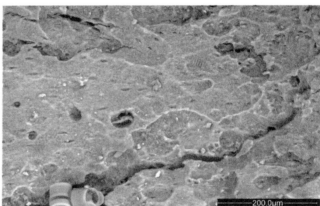

Fig. A.448 SEM microphotograph of a perforation on a recent sheep radius monitored in a spring river at Neuadd for several years. Observations of this bone at SE emission mode shows a dense organic layer covering the damaged areas. A diatom can be distinguished underneath the organic layer or mucilage (white arrow see A.444)

Fig. A.450 SEM microphotograph of perforations from the sheep's radius monitored from a spring river at Neuadd photographed at BSE mode. Deep and incipient perforations are filled by diatoms. The shape and distribution of these perforations differ from modification in bone structure as a result of bone remodelling (see A.445 bottom)

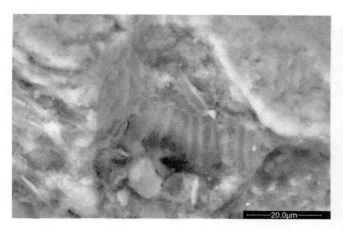

Fig. A.451 SEM microphotograph of a perforation on a recent monitored sheep's radius from a spring river at Neuadd at BSE mode. Incipient perforations show this alignment of diatoms (see A.450) of this distinctive diatom species. Backscattered electrons have a limited penetration capacity and only superficial layers can be observed. Depth and shape of the pit bottom cannot be seen

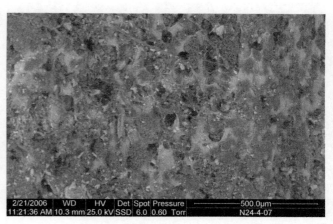

Fig. A.453 SEM microphotograph of surface pits on a modern monitored sheep's femur (ND15 from Neuadd 24.) The surface of the bone is covered by diatoms, algae and plant roots. The carcass was resting in calm water and bog vegetation for 12 years. In this case, perforations may also be made by colonial algae (see A.431 and A.434) as well as unicellular diatoms

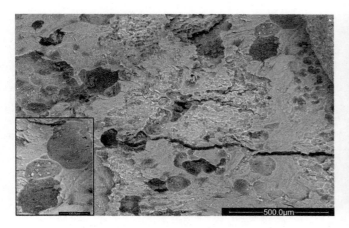

Fig. A.452 SEM microphotograph of perforations and detail of perforations (bottom left) from a modern monitored sheep's radius from Neuadd at BSE mode. Perforations are more randomly dispersed than the pitting resulting from bone remodeling (A.445). Some of these perforations have a regular shape (bottom left box), but others have an irregular edge, like the bottom pit in the small box. All are covered by mucilage (grayish substance) from diatoms

Fig. A.454 SEM microphotograph of surface pits on a modern monitored sheep's femur (ND15 from Neuadd 24). The surface of the bone is covered by diatoms, algae and plant roots. The diatoms (black arrow) are attached to the bone and partially covered by mucilage (A.444), with plant roots superimposed on the diatoms (double head white arrow)

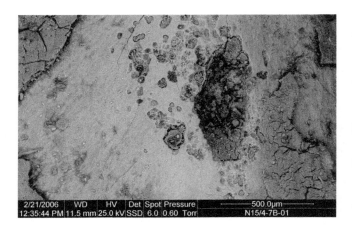

Fig. A.455 SEM microphotograph of a modern monitored sheep's tooth (ND26 from Neuadd 15/4). The carcass was 3 years in a seasonal water stream and covered by algae and diatoms (A.458). Notice the surface pits on the enamel

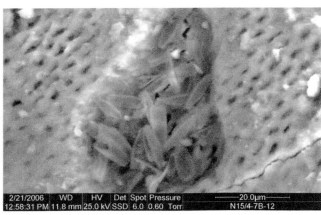

Fig. A.457 SEM microphotograph of a modern monitored sheep's tooth (ND26 from Neuadd 15/4). The carcass was 3 years in a seasonal water stream. A close up of one pit shows that it is filled with diatoms, but mucilage seems to be absent (see A.449). The perforation may have been formed first as a result of water corrosion, algae (A.431 and A.434) or a bacteria-fungal biofilm (A.500), and diatoms were just be lodged inside

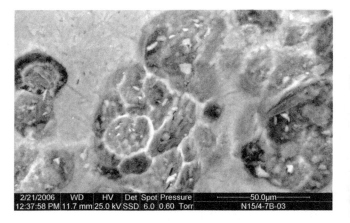

Fig. A.456 SEM microphotograph of a modern monitored sheep's tooth (ND26 from Neuadd 15/4). The carcass was 3 years in a seasonal water stream. Pitting is here arranged as a honeycomb-like shape, slightly more similar to remodeling by osteoclasts (A.445 bottom), but not possible because enamel lacks osteoclasts. The shape and pattern suggest an organic origin, possibly a biofilm, but further experimental work is needed (see A.493)

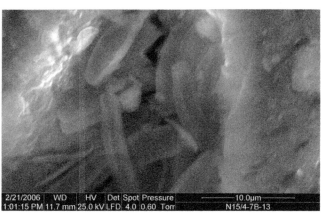

Fig. A.458 SEM microphotograph of a modern monitored sheep's tooth (ND26 from Neuadd 15/4). The carcass was 3 years in a seasonal water stream. Detail of A.457 taken at SE emission mode. Note diatoms are not covered by mucilage, they are individually and randomly piled up inside the hole

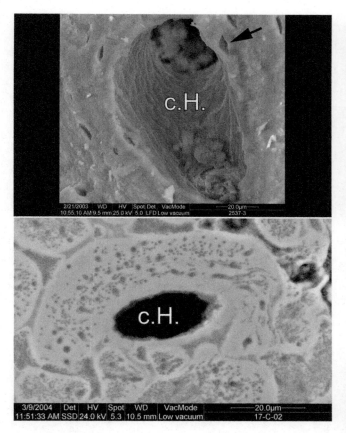

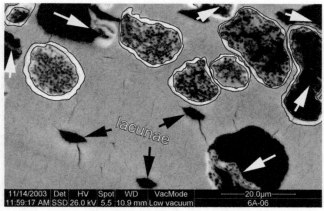

Fig. A.461 SEM microphotograph of a cut section of a modern monitored bone from Neuadd. White arrows point to some shallow bacterial colonies distorted during cutting (A.478). Typical terrestrial bacterial attack has rims around individual colonies. The rim is highly electron dense (brighter at BSE mode) indicating hypermineralized zones. EDS analysis returned a calcium phosphate chemical composition for both the bone and the outer ring

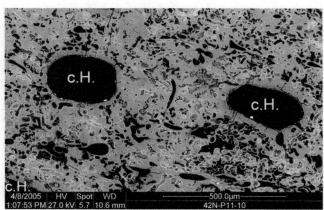

Fig. A.459 Top: SEM microphotograph of a fossil bone from Concud. The photograph shows a detail of an osteon, where the central canal is the canal of Havers which contains the bone's nerve and blood supplies. The Canal of Havers is surrounded by lacunae (black arrow), the space that an osteocyte occupies when osteoblasts become trapped in the bone matrix they secrete. This fossil has excellent histology, usually rare. Bottom: SEM microphotograph of a transverse section of a bear fossil fragment from Azokh Cave. Colonies of bacteria (brighter areas–MFD), are located around a Havers canal (A.475-top, A.919 and A.927). Pits inside these bright areas are perforations of individual bacteria. Osteon features (see above) have been destroyed by bacterial attack, except for the canal of Havers (c.H.). Bacterial attack is also known as Non-Wedl foci

Fig. A.462 SEM microphotograph of a transverse section of a fossil bone from Azokh North. Havers canals (c.H.) are the only histological traits that can be recognized. Bacterial colonies are dispersed all over with a chaotic arrangement. The low intensity of bacterial attack, and the randomly dispersed bacterial foci distributed independent of histology, may indicate a late bacterial attack (compare with A.475)

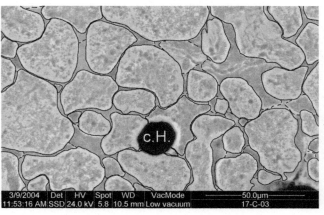

◀ **Fig. A.460** SEM microphotograph of a fossil bone fragment from Azokh Cave. The bone has been heavily attacked by bacteria (A.476). Solid lines delineate individual colonies of bacteria. Underneath these colonies, previous ones (see A.472) can be distinguished (dashed lines), although some boundaries are uncertain. MFD distribution around histological features (c.H.) suggests invasion from the interior of the bone (A.927)

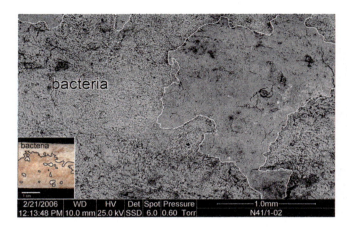

Fig. A.463 SEM microphotograph of the diaphysis of a monitored metapodial (ND5) of a modern horse from Neuadd 41/1 (A.292). The bone is heavily attacked by bacteria, and the white rounded areas are bacterial foci. This is apparent on the SEM as intense pitting in contrast to non-affected cortical bone, here outlined in solid white line

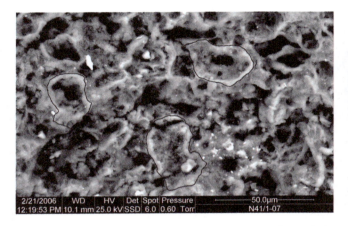

Fig. A.464 SEM microphotograph of the bone surface of the diaphysis of a metapodial (ND5) of a monitored modern horse from Neuadd 41/1. Close up view of A.463 that shows perforations produced by bacterial colonies, some of them outlined in solid black line. This is not a cut section, but the bone surface and bacterial colonies can be distinguished on the surface

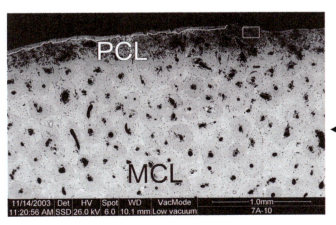

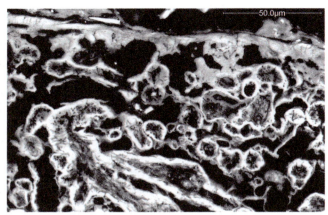

Fig. A.466 SEM microphotograph of the cut section of a monitored horse metapodial ND5 (A.463 and A.292) from Neuadd 41/1, detail of A.465. The link between post-mortem modificationmodification and peripheral distribution of MFD has been already described (Bell and Elkerton 2008; Bell et al. 1996, 2009). According to them, endogenous bacteria disperse throughout marrow cavities and Haversian system during putrefaction

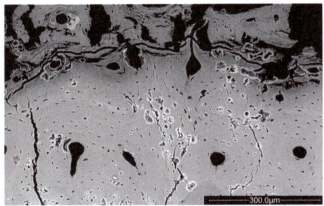

Fig. A.467 SEM microphotograph of modern bone cut, polished and mounted in resin, of a tibia of a monitored adult sheep from (ND16 from Neuadd 1 (A.925). Scattered distribution of destructive foci by bacteria (scattered, dispersed, non-histologically arranged and/or affecting periosteal and endosteal layers, but not medial layers of cortical bone) suggest to us a late bacterial attack probably from the soil. Bacteria could follow cracks or cracks could be formed through MFD weaker areas

◄ **Fig. A.465** SEM microphotograph of the cut section of a monitored horse metapodial ND5 (A.463 and A.292) from Neuadd 41/1. Bacterial attack is concentrated at the PCL (periosteal cortical layer, see Text Fig. 2.2). The boxed area is displayed in A.466. The white solid line outlines relatively undamaged cortical bone islands drawn in A.463. The cut section shows the MCL (medial cortical layer, see Text Fig. 2.2) is mildly attacked by bacteria

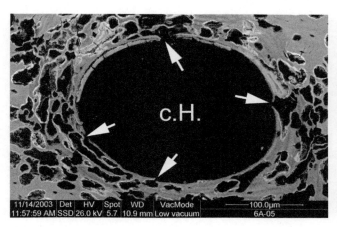

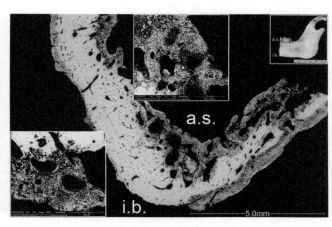

Fig. A.468 SEM microphotograph of a sectioned bone of a monitored adult horse radio-ulna, ND6 from Neuadd 41/1. The bone was exposed on the surface of the ground for 25 years in open upland conditions. Bacteria colonies are arranged concentrically around the canal of Havers (c.H., compare with A.459) indicating that they were dispersed through histological features of the bone. Invasive colonies broke the thick wall of the Havers canal (white arrows) and dispersed throughout the bone

Fig. A.471 SEM microphotograph of a sectioned bone of a modern monitored sheep's mandible left in a stream for 3 years (specimen ND20 from Neuadd 15/4). The intact bone surface is shown in the small box on top right. Bacterial attack has affected the inferior border (i.b.) and the alveolar sockets (a.s.), but the medial cortical layer (see MCL Text Fig. 2.2) was not reached by bacterial action. Details of bacterial damage to the inferior border are shown at bottom left and to the alveolar sockets, center top

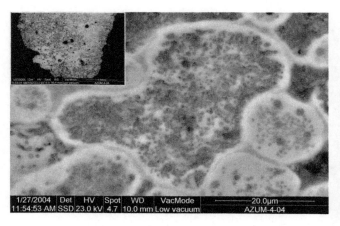

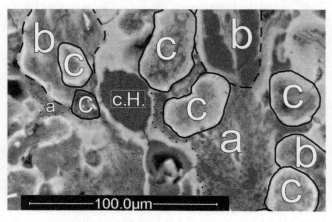

Fig. A.469 SEM microphotograph of a sectioned fossil bear bone cut transversely from Azokh Cave. The fossil is intensively affected by bacterial attack, shown in the small square at the top left side. A close up view of this specimen (right) shows the bacterial colonies and perforations made by bacteria (MFD) are arranged around histological features (c.H.). MFD invaded from the interior of the bone during decay (A.460)

Fig. A.472 SEM microphotograph of a sectioned bone of a modern monitored sheep's mandible left in a stream for 3 years (specimen ND20 from Neuadd 15/4) showing bacterial attack. Secondary bacterial foci appear superimposed on earlier foci, suggesting successive generations of bacteria. Intermittent water flow during summers could have affected the preservation of the bone, producing the successive generations of bacteria from a, the earliest to c, the latest (see A.460)

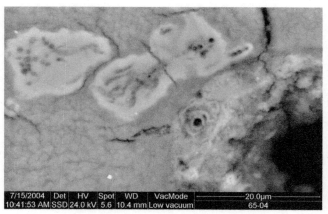

◄ **Fig. A.470** SEM microphotograph of the transverse section of a fossil from Concud. The incidence of bacterial attack is low (A.467), and this may be due to carnivore activity, reducing putrefaction by the complete consumption of muscle and other soft parts, which serves to prevent endogenous bacteria from invading bone tissues (Bell et al. 1996)

Fig. A.473 Heavily damaged bone surface on an immature fossil bear metapodial from Sima de los Huesos, Atapuerca (Coll. MNCN). The bone is heavily pitted as a result of bacterial attack (see A.475 and A.916)

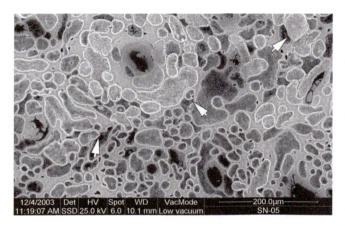

Fig. A.474 SEM microphotograph of a sectioned fossil adult bear ulna from Sima de los Huesos, Atapuerca (Coll. MNCN). Bacterial colonies appear as bright areas around the Havers canal (as in A.469). Although bacteria colonies may appear interrupted, a detailed study shows that boundaries adapt to each other except for rare instances (white arrows). This suggests that darker colonies are the older, primary colonies, and they are interrupted by brighter and later colonies

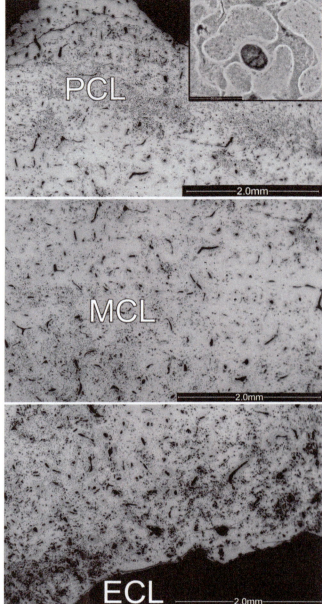

Fig. A.475 SEM microphotograph of a sectioned fossil adult bear ulna from Sima de los Huesos, Atapuerca (Coll. MNCN). Bacterial attack is equally intense at the PCL, MCL) and ECL (see Text Fig. 2.2). MFD are dispersed around Haversian systems as seen in A.474. All these traits suggest endogenous bacteria were active, since decay processes continued to affect the bone. Constant environmental conditions (humidity and temperature) and a soil highly rich in bacteria, might favor intense bacterial activity. An extensive study of bone surface, bacterial dispersion through or independently of histological features, and information of the site environment, are all needed to obtain conclusive results on exogenous/endogenous bacteria. So far, we start to distinguish a particular bacterial arrangement around Havers canals on specimens that decayed naturally. Bacteria arrange around the canal of Havers, forming a "flower like" shape. This has been seen for instance in this photograph (small box top right), A.459 bottom, A.474, A.919 and A.927 (white arrow). However, more extensive studies and observations are needed to see if it is possible to find the appropriate pattern to distinguish exogenous from endogenous bacteria

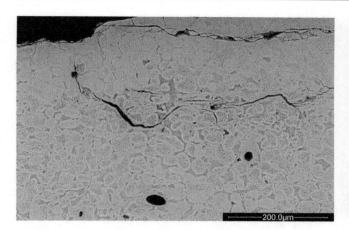

Fig. A.476 SEM microphotograph of a fossil bone from Azokh Cave that has been cut transversely. Intensive pitting is present as a result of bacterial attack that has destroyed osteons and bone histology (A.460). This intense bone bacterial attack is described by Hedges et al. (1995) as stage 0 of Oxford Histological Index (OHI). The only recognizable histological features are canals of Havers (empty black holes)

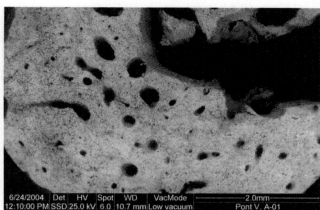

Fig. A.479 SEM microphotograph of a transverse section of a fossil bone from Pontvallain (Pays de la Loire, France) (A.480). ^{14}C dating of one bone sample gave an age of $3,204 \pm 56$y. Fossils analyzed are from a single individual of aurochs (*Bos primigenius*) that fell, died and decayed in a deep karstic crevice. Fossils are heavily attacked by bacteria, but this did not affect preservation of DNA (see Pruvost et al. 2007)

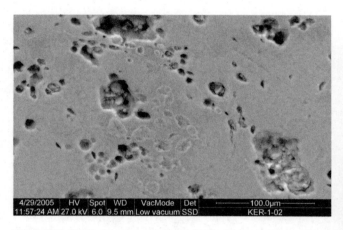

Fig. A.477 SEM microphotograph of a detail of the transverse section of a Neolithic bone from Tell el Kehr (Syria). Bacterial attack is mild and superficial. The black stains are empty holes previously filled by bacterial colonies, which were superficial enough to be detached when cutting the bone (see A.461 and A.478). Specimen provided by E.M. Geigl

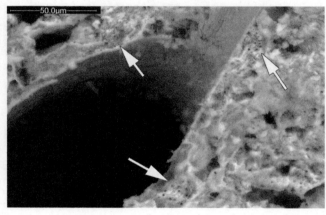

Fig. A.480 SEM microphotograph of a transverse section of a fossil bone from Pontvallain (Pays de la Loire, France) (A.479). The section was polished but not embedded in resin. Evidence of bacterial attack is indicated by white arrows. DNA analysis showed that two fossils excavated in 1947, washed and stored in the local museum since that time, contained no DNA, and three fossils dug in 2004 and kept in aseptic conditions, did contain DNA (see Pruvost et al. 2007). Specimen provided by E.M. Geigl

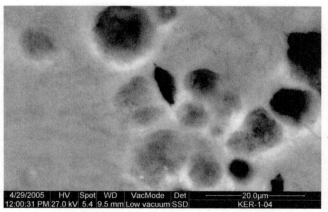

◄ **Fig. A.478** SEM microphotograph of a transverse cut section of a Neolithic bone from Tell el Kehr (Syria) detail of A.477. Bacterial colonies were removed during preparation of the cut section (as seen in A.461). Removal of these shallow and incipient bacterial colonies shows that the perforations penetrating into the bone have a rounded conical shape. Specimen provided by E.M. Geigl

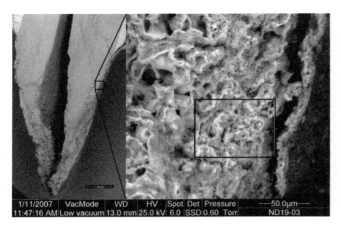

Fig. A.481 Left SEM microphotograph, photo-composition of the distal roots of a modern monitored sheep tooth ND19 from Neuadd 15/4. The tooth root was inside its alveolar cavity in a mandible (sample ND20) that remained for three years in a seasonal water stream. The mandible shows bacterial attack along the inferior border and in the alveoli sockets (see A.471 and A.472). Right The tip of the tooth root, in contact with the alveolar socket, has also been heavily affected by bacterial attack. Square enlarged in A.482

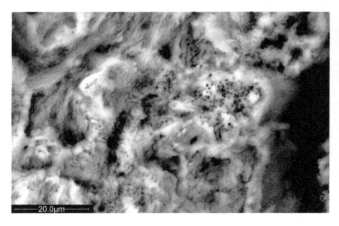

Fig. A.482 SEM microphotograph of the tooth root of a monitored sheep ND19 from Neuadd 15/4. This detail of bacterial attack on the tooth root (boxed area in A.481) shows simultaneous perforations and tunneling by bacteria. Bacteria were restricted to the apical area of the roots in contact to the mandible sockets, but the top of the root and the tooth crown (cementum, dentine and enamel) were not damaged. This indicates early diagenetic bacterial attack independent of decay

Fig. A.485 SEM microphotograph of a transverse section of a fossil ▶ from Concud showing diagenetic breakage. The bone surface appears heavily corroded on the surface and corrosion can be seen to affect the bone peripherally (white arrows). This corrosion is a new type of taphonomic bioerosion which is strongly linked to aquatic environments and more specifically to lacustrine environments (Pesquero et al. 2010). See also A.910, A.911 and A.912

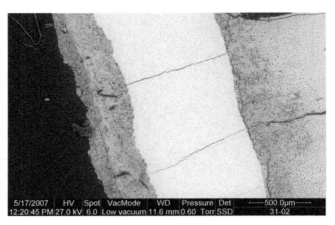

Fig. A.483 SEM microphotograph of a transverse cut section of a tooth. The enamel (the white strip) is unaffected by bacteria. In contrast, bacteria have intensively attacked the dentine (on the right of the enamel) and cementum (on the left). Apparently the enamel acts as a barrier to bacterial attack, for enamel is rarely invaded by *post-mortem* bacteria (see also A.909). Specimen provided by E.M. Geigl

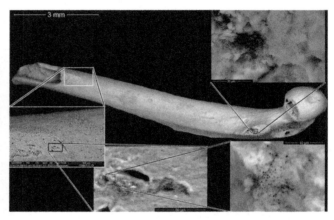

Fig. A.484 SEM microphotograph of a fossil small mammal femur from Sima del Elefante, Atapuerca showing bacterial attack at both distal and proximal ends (see A.297). Accumulations of small mammals are usually produced by predators, and it is likely that digestion would hinder bacterial action. This femur, however, was in a soil rich in bacteria, probably from the bat guano covering this level of the fossil site (Bennàsar 2010)

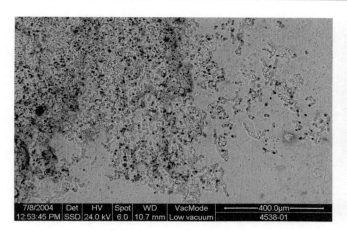

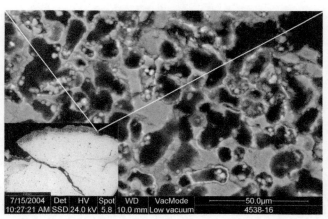

Fig. A.486 SEM microphotograph on a fossil bone surface (A.910 from Concud). Some areas appear affected by intense bioerosion pitting (A.487), increasing the bone porosity, in contrast to some other areas that are mildly damaged (A.488). Surface distribution is, therefore, discontinuous, which may suggest colonial microorganisms. Bioerosion penetrates the bone peripherally from the surface inwards. These traits indicate an exogenous origin for microorganisms

Fig. A.489 SEM microphotograph of a transverse section of fossil from Concud. The small box at the bottom left shows that the pitting is restricted to the external edge of the PCL (see Text Fig. 2.2). This bioerosion penetrates about 200–300 microns into the bone, with no histologically-orientated organization (A.912)

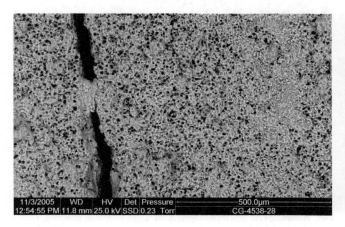

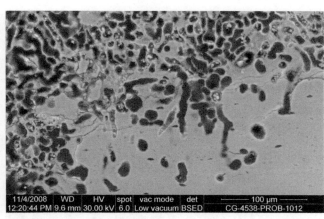

Fig. A.487 SEM microphotograph on the fossil bone surface from Concud showing very intense surface pitting. Aquatic bioerosion may intensively affect the bone surface, where the texture and structure of compact bone cannot be recognized. The distribution of these patches of intense pitting suggests a colonial organic origin. Experimental work, especially in lacustrine environments, is needed to identify the origin of this type of modification and differentiate it from other environments (see A.486 and A.910)

Fig. A.490 SEM microphotograph of bioerosion on the transverse cut section of a fossil from Concud. The distribution of this microboring in the bone sections is peripheral (A.912). The diameters of these tightly clustered microtunnels ranged from 7 to 18 microns (N = 73; mean = 9. 84 ± 2.6), their dimensions and morphology remaining constant

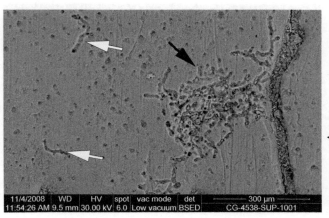

◄ **Fig. A.488** SEM microphotograph on the fossil bone surface of bioerosion described for some Concud fossils. In areas where this invasive microboring is less intense, individual grooves (white arrows) can be distinguished, compared with other areas where there is progressively increasing density and where these channels form a ramified arrangement (black arrow) also in cross section (A.490)

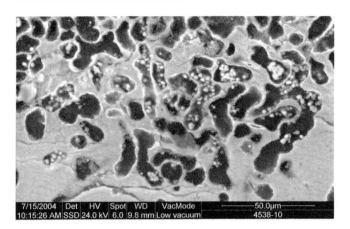

Fig. A.491 SEM microphotograph of bioerosion on the transverse cut section of a fossil from Concud. The microtunnel walls show a rim of highly electron dense material (A.489 and A.492) with a width of 0.1–1.5 microns (N = 45; mean = 0.81 ± 0.25). EDS analysis showed the chemical composition of the unaltered bone and these rims to be calcium phosphate in identical chemical element proportions

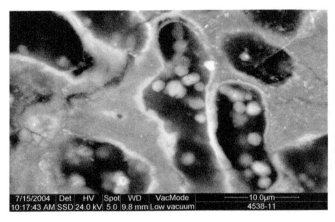

Fig. A.492 SEM microphotograph of bioerosion on the transverse cut section of a fossil from Concud. BSE-SEM images show the rim (A.491) to have a bright white color, indicating it to be denser and more compact than unaltered bone. Microboring contained highly electronic dense (bright color) spherical aggregates or 'microspheres'. EDS analysis showed the chemical composition of these microspheres to be calcium phosphate (Pesquero et al. 2010)

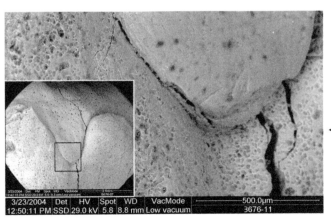

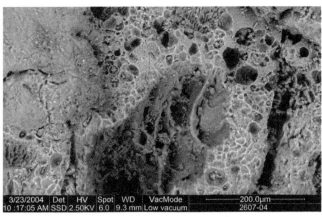

Fig. A.494 SEM microphotograph of the heavily pitted enamel surface of fossil tooth n.2607). This intense pitting, in contrast to common bacterial attack, only affects the enamel (see A.483). This has been observed on several teeth from Concud and other sites with calm water environments (see A.943, A.944 and A.945). Further observations and experiments are needed to identify the specific microorganism that produces this modification

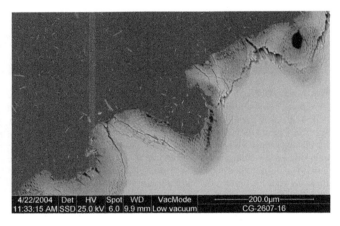

Fig. A.495 SEM microphotograph of a sectioned tooth from Concud. Rounded conical and regular perforations penetrate a few microns peripherally into the enamel. No microboring or distinct trait is observed, except for a shallow chemical corrosion of the enamel surface (grey color). The pattern of this corrosion is distinctive and is reminiscent of algal penetration (see A.265 and A.436) suggesting biochemical corrosion

◄ **Fig. A.493** SEM microphotograph of the distribution of another type of surface damage of unknown origin. This fossil is from Concud, and the modification only affects the tooth enamel. This modification has been found at other sites (lakes and calm water streams, see A.498 and A.943). Traits suggest a colonial biotic agent, probably linked to aquatic environments, like the new type of taphonomic bioerosion described in Concud (Pesquero et al. 2010) A.487

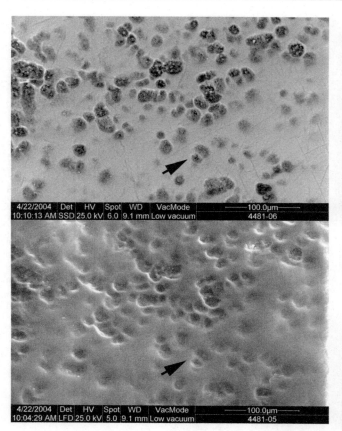

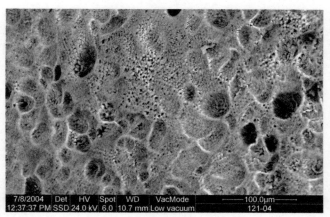

Fig. A.498 SEM microphotograph from Charcognier (Neolithic, France) of superficial pitting on enamel. A more intense stage of this modification has a honeycomb shape (compare with A.456). Specimen provided by E.M. Geigl

Fig. A.496 SEM microphotograph of superficial pitting on the surface of a horse tooth from Concud. This is similar to the modification on A.455 affecting the enamel. The incipient stage of modification on this specimen shows a distinctive pattern, where three or two joint-lobes can be distinguished (black arrows). This pattern is illustrated at SE emission mode (bottom figure), which shows better the topographic traits. No outer hypermineralized rim is distinguished around perforations, as seen for bioerosion on bones and described by Pesquero et al. (2010)

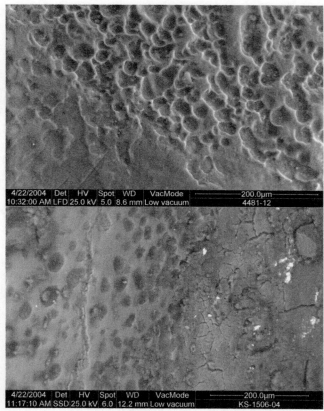

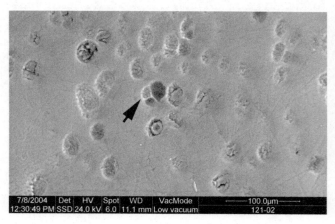

Fig. A.499 SEM microphotograph of superficial pitting on enamel. A more intense stage of this modification has a honeycomb shape. Top horse molar from Concud. Bottom a horse molar from Los Casiones (Teruel, Spain), another Miocene paleo-lakeshore site having similar modification on tooth enamel. Certain similarities are also found with modern cases like Neuadd (ND26) A.455, A.456, A.457 and A.458, although the intensity of modification is milder in Neuadd. The environmental conditions at Neuadd 15/4 correspond to seasonal flooding, where calm periods alternate with high energy episodes

Fig. A.497 SEM microphotograph of a horse tooth from Charcognier, a riverside bank site (Neolithic, of France). The pitting on the enamel is similar to the bioerosion found in Concud (A.496). The incipient stage of the modification of this specimen shows similarities with the distinctive pattern observed on Concud specimens, where two/three joint-lobes can be distinguished (black arrow). Specimen provided by E.M. Geigl

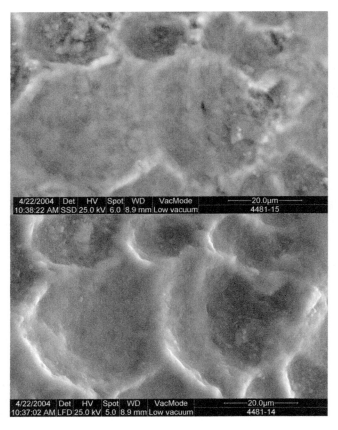

Fig. A.500 SEM microphotograph of superficial pitting on the enamel of a fossil tooth (n. 4481) from Concud. Top BSE mode; Bottom SE emission mode. Detail of individual pits. The shape and distribution of the pitting is similar to erosion found at Neuadd (ND26, A.456) in continental aquatic environments (see A.498 and A.499)

Fig. A.502 SEM microphotograph of the surface of a fossil horse incisor from Concud. This superficial and homogeneous modification attacks the enamel (in contrast to *post-mortem* bacteria, A.483), but the dentine is almost intact. The general appearance of this surface erosion suggests that microorganisms could produce this modification. At the present stage of knowledge, we may speculate it to be related to a biofilm (an aggregate of free-floating microorganisms) that attached to a hard substrate (enamel)

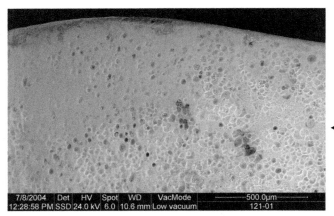

◄ **Fig. A.501** SEM microphotograph of a tooth from Charcognier (Neolithic, France) of superficial pitting on enamel (see A.496, A.497 and A.502). Distribution, dispersal, shape and size of this modification are similar to those on specimens from Concud, and slightly similar to specimens from Neuadd 15/4. They share aquatic conditions of still water and rich in microorganisms. Further experimental work and more extensive observations of other situations are needed

Chapter 5
Discoloration and Staining

Changes in color may indicate speed of burial, or exposure to environmental factors such as humic acids, oxygenated environment, burning or water. It can also be misleading, for bones can change color very quickly within and between sedimentary horizons. Color also varies according to the type of bone and the type of animal, young or old. Probably the most useful source of information is color differences on a single bone, for this can say much about the history of that bone. For example, half buried bones may have the exposed end lighter in color than the buried end, or bones resting in mud in flowing oxygenated water may be differentially stained black by manganese dioxide. For both of these eventualities there is a very clear taphonomic signal with information about the local environment.

Agents and Processes Affecting Bone Color

Inorganic processes:

Leaching effects of water, A.503 and A.507
Deposition of minerals, A.504, A.505, A.510 and A.529
Wildfire, A.511, A.523 and A.524

Organic processes:

Organic acids in soil, A.531, A.532 and A.534
Human activity, A.539 and A.540
Cooking by humans, A.522
Carbon deposition, A.519 and A.520
Root growth, A.536 and A.538
Microbial attack, A.521

Characteristics

Organic acids in soil cause brown staining, and in oxygenated soils rich in iron there may be a reddish color to the bone. Fungal attack may blacken the surface of bone. Exposure to water may lighten the color, but if the water is newly oxygenated it may promote crystal growth of manganese dioxide, staining the bone black. The presence of manganese deposition is related to environmental condition characterized by wet, mildly alkaline and oxidizing, as well as the involvement of bacteria (López-González et al. 2006). Manganese dioxide is insoluble and tends to form crusts and coatings in caves. Burial in soils may lighten the color of bone if the soils have impeded drainage (gleying), or darken it in soils rich in organic matter. Cooking or burning may alter the color through a sequence of stages. Carbon deposition may produce black deposits on the surface. Root growth on bone surfaces also alters the color of bone, sometimes lightening it in the linear marks made by the roots, or sometimes darkening the color. There may be differential coloration on bones, for example in bones partly buried in soil, with exposed parts being bleached by the sun and the buried parts stained by organic acids.

Discoloration may be accompanied by other modifications of bone, but in this chapter we consider the color of the bone surfaces and the mineralogical composition of either surface depositions or deep penetration of the bone. The most straightforward way to consider these issues is to categorize by color rather than agent, and so we will distinguish black staining, brown staining, preservation of original light colors or lightening of bone. We will further distinguish between ubiquitous staining and localized staining.

© Springer Science+Business Media Dordrecht 2016
Yolanda Fernández-Jalvo and Peter Andrews, *Atlas of Taphonomic Identifications: 1001+ Images of Fossil and Recent Mammal Bone Modification*, Vertebrate Paleobiology and Paleoanthropology, DOI 10.1007/978-94-017-7432-1_5

Inorganic and Organic Modifications

Black Staining

This is one of the most common features of fossil bone assemblages. It can be produced by several different agents and by several different processes. Staining by manganese dioxide is common (A.504). It may appear as overall black surface staining, which could indicate total immersion in water or wet sediment, but more commonly the staining is patchy (A.503). The distribution of staining may be confined to one side of a bone, which is likely to be due to the bone resting on a wet or damp surface and periodically immersed in water (A.504). It may be present on diaphyses and meta-physes and absent on articular surfaces (A.506), suggesting that manganese staining occurred when skeletal elements were in anatomical connection (López-González et al. 2006). Manganese may also be precipitated dry by bacterial growth on bone (A.508) or form massive accretions (A.509). Finally,

one of the most characteristic phases of manganese deposition is the formation of dendritic patterns, branching patterns that resemble plant growth on the surfaces of bones and stones. The crystalline structure of the manganese deposition is seen clearly in SEM images (Fig. 5.1).

Another form of black staining may come about from deposition of carbon (A.519) on bone (Molleson et al. 2005). This is usually a special case: for example, the skeletons of old individuals from Çatalhöyük, Turkey had accretionary carbon deposited on the inner surfaces of their ribs due to smoke inhalation during life. The houses of this Neolithic town had no ventilation, so when heating or cooking took place inside the the single roomed dwelling, the room would have filled with smoke. Constant breathing this smoke over many years accumulated inert carbon residues in the lungs of the inhabitants, so that in old age the lungs would have had heavy carbon residues. On death and after the lungs decayed, the inert carbon remained and accumulated on the inner surfaces of the ribs, mainly on the lower side of the body but also to some extent on the upper side. This feature was observed on 7 out of

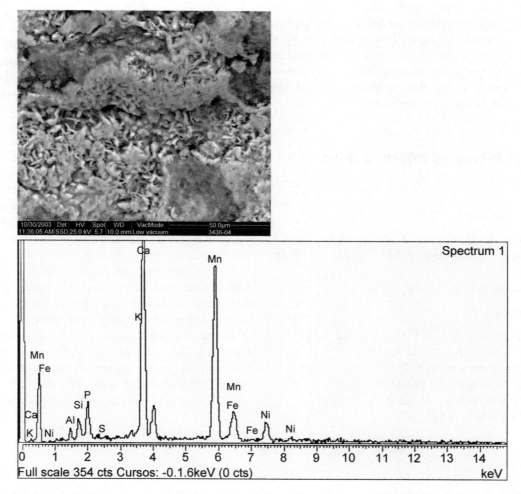

Fig. 5.1 SEM microphotograph of manganese seen by BSE detectors showing the shape and crystallographic growth. Below is the EDS analysis of the metallic deposit, which shows peaks for manganese (Mn) as well as bone minerals (P and Ca) together with other minor elements, such as Iron, Nickel, Aluminium, Silica and Sulphur, contained in the sediment

10 old individuals in the 1995–99 excavations at Çatalhöyük, but it was absent from young/middle-aged adults and juveniles (Andrews et al. 2005).

Black staining has also been observed on human burials at Çatalhöyük as a result of fungal attack (A.521) (Andrews et al. 2005). Primary burials at this site are distinguished by their pristine white color, but graves with multiple burials, where the grave has been reopened to make room for later burials, contained bones from the earlier interments that had been disturbed by the later burials. Exposure to the air, perhaps while still rich in organic matter, has resulted in fungal attack by fungus, imparting a grayish black color evenly distributed over the whole surfaces of the bones. Graves reopened only once have scattered and weak fungal staining on bones from the earlier interment, but graves opened several times (3 to 12 times in some cases) had bones with heavier degrees of staining. The relationship between degree of disturbance and extent of fungal staining is so great at Çatalhöyük that it forms an independent check on the number of disturbance (see Chap. 10).

Another source of black staining comes from fire (Stiner 1995) (A.522). The effects of fire usually produce more than one color change as well as variable effects on crystallinity, shrinkage, weight and cracking (Shipman et al. 1984, see Chapter 7) on different parts of the bone, depending on its position relative to the fire (Fernández-Jalvo and Perales 1990). Experiments on fire-induced modifications of bone have shown color changes between white, brown and black (see below: Fig. 5.2). Bones associated with hearths remain hot for a longer time period than wood, and they provide a good fuel to keep the hearth hot for longer and with almost no smoke. Such usage better explains the presence of burnt bone fragments inside hearths rather than a simple aim of cleaning the site as previously proposed by some authors (e.g., *toss zones* Binford 1981).

Brown and Black Variable Staining

Agents responsible for variable staining are similar to those described above, with variations in process or duration accounting for the variations in color. Skeletons from single individuals may have elements differing widely in color if the bones were preserved in different conditions. For example, a skeleton that is only partly buried in soil could have some skeletal elements stained brown by organic matter such as humic acids in the soil and others that remain on the surface either unchanged or even lightened in color if exposed to weathering (A.531 and A.532). This difference in color has little taphonomic significance, since even bones from the same skeleton may differ in color, but in the fossil record such a difference might be accorded greater

Fig. 5.2 Unburnt bones start off with the original bone color; stage 1 burning is marked by limited areas of brown staining; stage 2 is seen after early exposure to heat or burning, when the bone is brownish, is still intact and is not cracked. Stage 3 burning shows charring and a dull black color; stage 4 the bone is dark grey in color, with extensive cracking of the surface, shrinkage of the bone and some remineralization; stage 5 is the final stage, with bright white color of the bone which had become calcined; calcined bones are powdery and easily destroyed (Stages defined by Cáceres 2002). The maximum temperature acquired during the experiment was 855 °C (Data provided by I. Cáceres)

significance than is warranted. In contrast, rapid burial in sediment increases the probability that the bones would survive into the fossil record. Sometimes a color change is seen on a single bone, for example if the bone was half buried. The buried end may be stained a darker color than the exposed end, which may also show signs of weathering.

Bone fragments subjected to burning at different temperatures show big differences in color change in relation to fire temperature (A.522) (Schmidt and Uhlig 2012). Color change is only partially related to temperature, however, because there are several other parameters that may affect the bone, such as wind, environmental temperature and relative humidity (Stiner 1995), or even water dripping (in caves). Also, the type of wood used and the type of bone may be important, particularly in terms of anatomical elements, and the presence of soft tissues can insulate the bone and prevent its direct exposure to fire. An experimental project using fire at recorded temperatures (Fig. 5.2) was performed by Cáceres (2002) and Cáceres et al. (2002). This experiment showed that there is a form of heat pre-treatment that produces early transformation due to heat (identified as stage 1) either resulting from the heating of bones still bearing soft tissues (cooking) or when clean bones free of meat are directly exposed to fire embers to facilitate breakage. This heat pre-treatment facilitates bone breakage to extract the marrow, which takes on the consistency of gel facilitating its extraction (Cáceres 2002). It has also been found that breakage of previously heated bones shows a mixture of fracture angles (see Chapter 9). Stage 1 heating of bones causes them to lose moisture rapidly, so that when they are broken they behave as if they were dry rather than green (predominance of mixed fracture angles as described by Villa and Mahieu 1991).

The five stages of color change due to fire proposed by Cáceres (2002) (Fig. 5.2) are as follows: stage 1, unburnt bones with dispersed spots of brown color (stage 1); stage 2, bone is brownish and is still intact, without cracking; stage 3, bones with charring and a dull black color; stage 4, bones dark grey in color, with extensive cracking of the surface, shrinkage of the bone and some remineralization, and the surface of the bone may become shiny; stage 5 is the highest grade of fire exposure and it produces a bright white color of the bone. These are calcined bones,which have become powdered and are easily destroyed. Stage 5 is also characterized by heavy cracking and exfoliation of the bone, with further shrinkage and remineralization and with a shiny bone surface (Cáceres 2002, Cáceres et al. 2002). Using energy dispersive spectrometry (EDS) and SEM-BSE (backscattered electron mode), bones blackened by manganese can be easily detected and distinguished from bones exposed to fire (Fig. 5.3) (Schmidt and Uhlig 2012; Schmidt et al. 2012; Fernández-Jalvo and Avery 2015).

Brown staining, or the reverse, lightening of color, may be the result of a number of organisms in an active soil. Bacterial attack on the surfaces of bones may produce a patchwork of corrosion (see Chap. 8) that is lighter in color than the rest of the bone. Modern roots may also stain fossils black, and at some sites where evidence of fire is present, this can cause confusion. Such was the case at the top series of Gran Dolina (Atapuerca). The presence of a dark blackish stained long bone diaphysis fragment (and a modern root underneath) was thought at first to be evidence of fire. However, the inverse sequence of coloring (black outside-grey inside) showed this was not the result of fire. In all these cases, the color change is associated with other taphonomic modifications that make their interpretation more straightforward, and to a large extent this renders color interpretation redundant since the other modifications such as corrosion, bacterial attack or root marks provide the necessary information.

Red Staining

Staining of bones a reddish brown color may occur in iron rich soils. It is characteristic of oxygenated and biologically active soils and is a good indication of such conditions at time of burial (A.528). Red staining is also seen on human burials where the body, or the bones of the skeleton, have been covered with red ochre (A.539). This is a common funerary ritual and is readily apparent both from the color and the context of the burial (Molleson et al. 2005). Cinnabar (mercuric sulphide) applied to the face of skulls has also been recorded on human skulls from Neolithic burials, giving a red color that may be indistinguishable at naked eye from red ochre (Molleson 1990; Molleson and Andrews 1996).

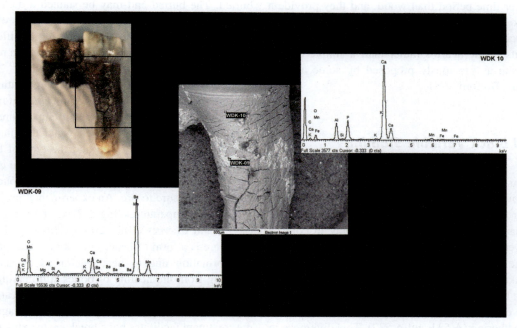

Fig. 5.3 Rodent molar from Wonderwerk Cave (South Africa). The fossil is completely blackened. Differences between staining by burning and manganese deposition can be observed at the SEM-BSE. Using EDS, manganese could be detected (spectrum WDK-09, left bottom) on small whitish patches dispersed on the fossil surface. This is in contrast to the homogeneously grey surface of the tooth root. The EDS analysis (spectrum WDK-10, top right) shows a calcium phosphate composition of the tooth root (bioapatite)

Atlas Figures

A.503–A.540

Fig. A.503 Fossil mandible from middle Pleistocene deposits in Atapuerca with black staining. This has formed through one of the most common forms of staining of fossil bone, from manganese dioxide precipitating out on bone surfaces (A.510). Courtesy of I. Cáceres

Fig. A.505 Fossil mandible of *Metacervoceros* from Senèze. The side is heterogeneously stained by manganese, having a gap in the center, caused may be by the presence of a stone or another bone that hindered the manganese deposition (see A.504)

Fig. A.504 Fossil right mandible of *Eucladoceros* from Senèze. Manganese may affect bones heterogeneously (A.505). The buccal side (top) is completely covered by manganese with extensive scratches produced by friction against the sediment. In contrast, the lingual side (bottom) is unstained

Fig. A.506 Fossil second phalanx of *Metacervoceros* from Senèze. Shafts of limb bones are sometimes completely covered in manganese, but articular ends are not. This pattern has been considered to indicate that bones were in anatomical connection when the manganese was deposited (López-González et al. 2006). Compare with A.504 and A.505

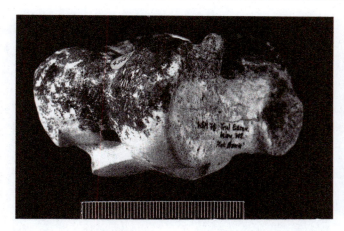

Fig. A.507 Differential manganese staining on one side (as in A.504) of a fossil ruminant astragalus from Westbury Cave, unit 11/2. The bone was resting in the sediment in the orientation shown here, and while the bottom was unstained, the upper surface was heavily stained with manganese. The presence of oxygenated water on the surface of the sediment may have brought about this difference in staining

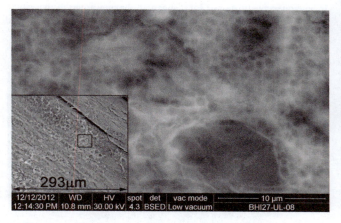

Fig. A.508 SEM microphotograph of a modern monitored bone from a rodent carcass buried for a year in sandy soils. Manganese was detected by BSE-SEM as whiter deposit (small box bottom left) and identified by energy dispersal spectrometry (EDS, as in Text Fig. 5.1 and A.511). A closer up view of areas covered by manganese deposits shows rounded casts that could be caused by bacteria (see A.905)

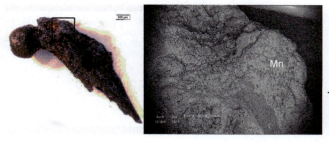

Fig. A.510 Right: SEM microphotograph (squared box on the left figure) of a fossil bone from Wonderwerk Cave (South Africa) showing a detail of manganese dendritic crystal growth (pirolusite, as in A.503). This type of deposit of manganese is frequently observed on fine sediments and in cave deposits

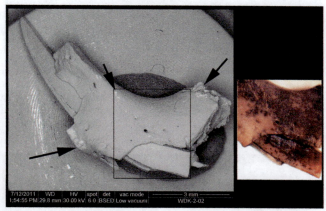

Fig. A.511 Fossil rodent mandible from Wonderwerk Cave. The color image on the right shows apparent manganese staining, but the SEM microphotograph on the left has no higher electron emissions (whiter color, black arrows) as would be expected for a metallic deposit such as manganese (A.513 and Text Fig. 5.3). The black staining is attributed to fire in the context the Wonderwerk site

◀**Fig. A.509** Left: blackened small mammal fossil (proximal end of femur). Square at the top is enlarged in the right figure. Right: SEM microphotograph of massive manganese accretion on a fossil bone from Wonderwerk Cave (South Africa). Compare with Text Fig. 5.1

Fig. A.512 Two fossil fragments from Sima de los Huesos, Atapuerca have differential manganese distribution on the two parts. This may show differences in depositional microenvironments within one level of a site, and when the fragments can be refitted it suggests that the two fragments were not adjacent during diagenesis. Compare with A.503, see A.564

Fig. A.515 Mandible of a fossil bovid from middle Pleistocene deposits in Atapuerca. The body of the mandible is heavily stained by manganese, while the teeth are unstained, as in A.503, A.504 and A.514. Courtesy of I. Cáceres

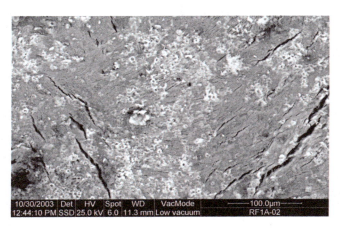

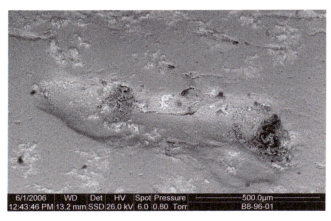

Fig. A.513 SEM microphotograph of a modern mandible of a fallow deer from Riofrío collected from a seasonal river. Observations at the SEM provide different color perception compared with light microscopes. Manganese has a high density because it is a metallic element, and BSE images show manganese deposits as bright (white spots) as in A.510, A.513 and A.516. The grayish areas show the surface of the bone, which is less dense because is more porous and organic

Fig. A.516 SEM microphotograph of pirolusite, oxide of manganese, an arborescent form (see A.510) deposited on a fossil bone fragment from Gorham's Cave. The depression has been identified microscopically as a smooth root mark as in A.247

◄ **Fig. A.514** Bones and teeth may fossilize with completely different colors, as in this case of a cave bear *Ursus deningeri*, from Westbury Cave. Both the maxilla and the dentine of the teeth (exposed as wear pits on the first molars) are stained black with manganese while the tooth enamel is unstained as in A.503, A.504 and A.515. Teeth are usually not covered by manganese, except for cracks or broken edges

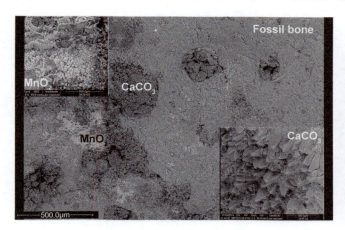

Fig. A.517 SEM microphotograph of a fossil bone from Concud that has dispersed manganese dioxide. The fossil is heavily mineralized, and histological features (canals of Havers) are filled in calcite crystals. With natural light, calcite crystals appear as whitish shiny crystals and the manganese shows up as black stains. The reverse is seen under SEM-BSE mode with manganese in white and calcite in grey, even darker than fossil bone (calcium phosphate). Compare with A.513

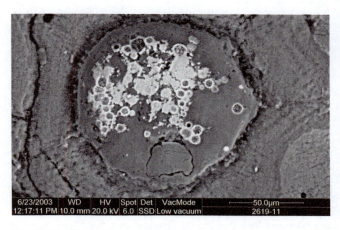

Fig. A.518 SEM microphotograph of a fossil bone from Concud. The substance that covers the canal of Havers and the bright circled deposits in the interior has been analyzed using EDS. This element analysis spectrometry showed that the substance in the interior of the Canal of Havers is massive calcite covered by nodules of iron cut and polished when sectioning the fossil. Compare with A.517

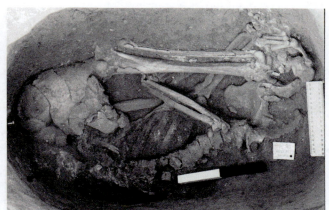

Fig. A.519 Top: human burial from Çatalhöyük. Black staining is present on the inner surfaces of the ribs of this old individual, and it was also present on seven of the nine individuals of similar age from the same site. The staining is due to carbon deposition, and it is considered to be the result of inhalation of smoke during life due to inadequate ventilation in ancient houses. The deposition of carbon is more concentrated on the lower side of the body. Carbon deposition is only seen on very old individuals at Çatalhöyük and is absent on juvenile and mature adults. Bottom: carbon accretion on the inner surfaces of the ribs of a very old individual from Çatalhöyük. (Andrews et al. 2005) as in A.520. Compare with A.503

Fig. A.520 Rib fragment from a human skeleton from Çatalhöyük. The ▶ carbon deposition on bones forms a thick accretionary mass in extreme cases where layers of carbon have been deposited. Note the bright shiny texture of black carbon stains, as in A.519. Compare with A.503

Fig. A.521 Bone fragment of a human skeleton from Çatalhöyük with dark staining as a result of fungal attack when the burial was re-opened and the bones exposed. This specimen came from a grave with multiple burials. Compare with A.503

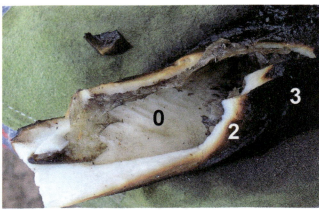

Fig. A.524 Experimental bone broken after burning by direct exposure to fire flames (A.522). Three categories of fire/heating are present on this bone, visible because it has been broken post-fire. It shows a color sequence from the outside, which was in contact with flame and higher temperatures and is at stage 3, to the inside, which is protected to some degree from fire and remains at stage 0. If this sequence is inverse, the cause is due to another taphonomic agent. Photo I. Cáceres

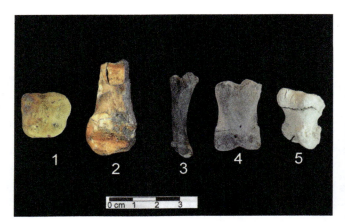

Fig. A.522 Bones experimentally exposed to fire. Stage 1 burning is marked by small spots of brown-reddish staining; stage 2 is reached after early exposure to heat or burning where the bone is brownish and is still intact, without cracking; stage 3 shows charring and a dull black color; stage 4 is dark grey in color, with extensive cracking of the surface, shrinkage of the bone and some remineralization; stage 5 has bright white color of the bone and the bone is calcined so that it is powdery and easily destroyed (see Text Fig. 5.2)

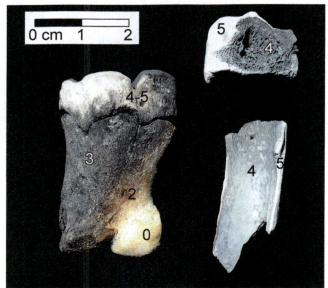

Fig. A.525 Experimentally burnt metapodial and phalanx of young pig showing simultaneous presence of stages 2 to 4 discoloration in the phalanx (on the left). The metapodial (on the right) was broken when collecting the bone after the experiment. The bone fragment shows a sequence from stage 4 (grey) in the interior to 5 (white) at the exterior (see A.522 and Text Fig. 5.2)

◀ **Fig. A.523** Burnt animal bone from Çatalhöyük. The color of burnt bone corresponds, in general terms, to temperature reached, and it varies gradually from brown-black-grey to white. According to experimental sequence (Cáceres et al. 2002), this bone was directly exposed to fire and reached stage 3 (see Chap. 5 and Text Fig. 5.2)

Fig. A.526 Fossil bone fragment from Cueva Ambrosio showing stage 1 burning (see Text Fig. 5.2 and A.522). Stage 1 is only evident as small brown spots of burnt bone dispersed on the unburned bone surface. These dispersed spots occur when the bone is still covered by flesh or because the greasy bone was heated over fire embers. These spots are where the hot stones, sediment grains or pieces of wood are in contact with the bone. Specimen provided by S. Ripoll

Fig. A.529 Human phalanges from a Roman town where skeletons were buried with ornamental objects of copper that stained the bones (bottom), while those close to but not in direct contact with jewelry were not stained (top). EDS chemical element analysis (Text Fig. 2.6) detected a high copper peak on the bone surface Specimens provided by A. Rosas

Fig. A.527 Fossil bone fragments from Cueva Ambrosio showing simultaneously two grades of burning (see Text Fig. 5.2 and A.522). Burning is a gradual process, as also happens with other taphonomic agents (e.g. weathering or digestion)

Fig. A.530 SEM microphotograph of iron nodules deposited on a fossil from Concud. This suggests that iron not only can be dissolved in water and impregnate the fossil bones (A.518), but that nodules of iron are also forming in the sediment. Chemical analysis of the nodules made with EDS detectors at the SEM confirmed the ferric nature and composition of these nodules

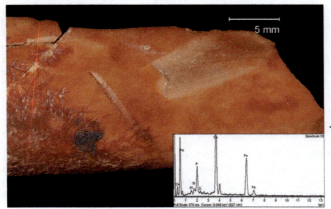

◄ **Fig. A.528** Bone fragment from Concud. EDS chemical element analysis (Text Fig. 2.6) detected a high abundance of iron which provides the reddish color to the fossil. Minerals such as iron, copper, nickel may be derived from the sediment or through soil minerals in contact with bones after burial. These minerals may stain the bones and provide a variety of characteristic colors (e.g. red, green, blue, black). Courtesy of D. Pesquero

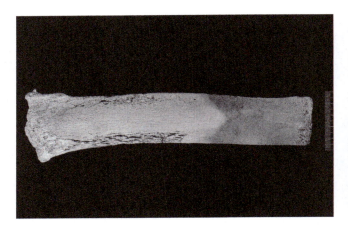

Fig. A.531 Modern bone. Color change on a single bone from Jebel Barakah. The right hand end of the bone was found buried in the dry sandy soil, and the left hand end was exposed to weathering. There is both a color change and difference in weathering stage on this single bone, as in A.532, A.533 and A.535

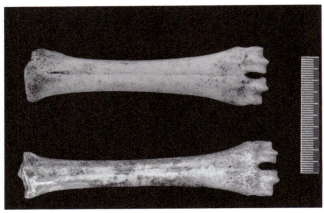

Fig. A.534 Modern bones. Two sheep metapodials from the same individual from Neuadd 1. There is a difference in both color and texture between the two bones. Parts of this skeleton were buried by natural processes within three years, and the rest remained on the surface. After six years, the skeleton was excavated and the buried metapodial (above) was stained brown by humic acids and the surface metapodial at the bottom was only slight stained but showed signs of very early stages of weathering, as in A.535

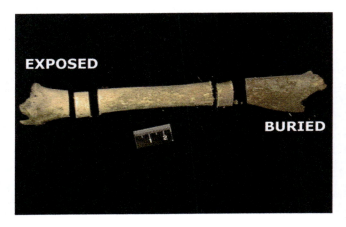

Fig. A.532 Modern bone. Color change on a monitored sheep tibia from Neuadd 25/2. The right hand end of the bone was buried in a wet clay-rich soil, and the left hand end had been left exposed to weathering for six years, as in A.531 and A.533

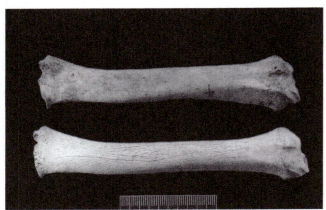

Fig. A.535 Modern bones. Two sheep radii from the same individual from Neuadd 1. There is a difference in both color and texture between the two bones, as in A.534. Parts of this skeleton were buried by natural processes within three years, and the rest remained on the surface. After six years, the skeleton was excavated and the buried metapodial (above) was stained brown by humic acids and the surface metapodial at the bottom was only slight stained but had reached stage 1 weathering

◄ **Fig. A.533** Color change of a fossil bone fragment recovered from fluvial deposits. This fossil shows color contrast between both ends, probably related to differential burial, with the right side buried in soil while the left side was more exposed (as in A.532 and A.533). Courtesy D. Pesquero

Fig. A.536 Miocene fossil from Concud with recent root marks (see A.537) etching the bone surface and staining the bone a dark reddish brown. Courtesy of D. Pesquero

Fig. A.539 Red staining on a human skull from Çatalhöyük. This color comes from the application of red ochre to the skin of this individual after death. The staining became transferred to the bone on decay of the skin (as in A.540)

Fig. A.537 Recent root marks of plants have affected the surface of this fossil specimen from Concud (see A.536), etching both old surfaces and recent breakage (whiter surface on the right side of the picture) when the fossil was still buried in sediment. Roots have been able to penetrate the fossil bone surface and have left a characteristic reddish color (dark red-brownish in this case) which is different from the color of the fossil. Courtesy of D. Pesquero

Fig. A.540 Red ochre staining on a human rib from Çatalhöyük, probably from red ochre applied to the skin of this individual, with the staining transferred to the bone on decay of the skin (as in A.539)

◀ **Fig. A.538** Fossil rabbit bone from Cueva Ambrosio. This rabbit fossil has been heavily damaged by modern roots (reddish color, as in A.537) that even perforated the cortical bone at the mid shaft. The difference in color between the root marks and the rest of the bone is what identifies these marks as recent. Specimen provided by S. Ripoll

Part II
Modifications Affecting Shape

Chapter 6
Abrasion and Rounding

Abrasion is a general term that includes any degree of rounding or polishing of bones, whether broken ends of bones, processes that protrude from the general level of bone, or the complete bone. Abrasion can occur at varying times post-mortem, for example when bone is used as a tool by hominins, but it most commonly occurs on older post-mortem or even fossilized bone as a result of transport or trampling. The rate at which rounding occurs, and its degree, depend to a large extent on the condition of the bone being modified. Experiments have shown that the effects of abrasion are related to the type of bone, whether fresh, dry, weathered, or already fossilized. For example, Behrensmeyer (1991; and see Fernández-Jalvo and Andrews 2003) has shown that weathered bone is more vulnerable to abrasion. Andrews (1995) gives an example of the interplay between abrasion and weathering, and Martill (1990) provides a possible explanation for the resistance to abrasion in fresh bone. Abrasion also varies according to the type of sediment (gravel, coarse sand, fine sand, clay and silt) (Olson and Shipman 1988; Fernández-Jalvo and Andrews 2003).

Agents

Inorganic processes:

Water transport of bones in sedimentary environments, A.541, A.569 and A.570
Transport of sediment impacting bones, A.543, A.547, A.562, A.563, A.576 and A.582
Wind erosion, A.565, A.578 and A.579
Trampling by large animals, A.584 and A.585

Organic processes:

Carnivore and human action, A.584, A.587, A.590, A.600, A.606, A.614 and A.629
Trampling in organic substrate, A.596 and A.597
Bioturbation: plant roots, A.267 and A.636
Aquatic bioerosion, A.637 and water corrosion, A.577

Characteristics

The principal characteristic of abrasion is the degree of rounding, which is categorized as slight, moderate or great, and at its most extreme, fossils that become rounded pebbles. The location of the abrasion may be restricted to ends of bones or bone processes, or bones that are rounded all over. Abrasion on all bone surfaces is generally characteristic of transported bones or when moving sediment particles impact partially stabilized bones. Abrasion on one side of the bone may occur with wind erosion if the bone is stabilized on the land surface, when it may take the form of polishing, and it is more localized on digested bone, where it occurs as rounding, and trampling.

There is little difference in the morphology of abrasion between different agents, and the context of the bone is needed to identify the agent causing abrasion. For this it is important that specimens being excavated are marked with a permanent marker to show how they were lying and which side was facing the underlying substrate. Recording size, shape and orientation of fossils in the site are also important when identifying sources of abrasion (see Chap. 2, methods). The main differences regarding abrasion are seen in comparisons of broken edges versus other bone surfaces, the presence or absence of polishing, the presence of use-related striations on rounded surfaces in bone tools, and the location and distribution of the abrasion. Abrasion will therefore be described in terms of degree of abrasion, polishing and location.

The sequence of abrasion with weathering and breakage can be informative. For example, broken edges that are rounded or weathered show that breakage occurred before the other two processes; while unabraded broken edges on an abraded bone show that breakage formed after the abrasion. Similarly, abrasion before and after weathering can be distinguished from each other.

© Springer Science+Business Media Dordrecht 2016
Yolanda Fernández-Jalvo and Peter Andrews, *Atlas of Taphonomic Identifications: 1001+ Images of Fossil and Recent Mammal Bone Modification*, Vertebrate Paleobiology and Paleoanthropology, DOI 10.1007/978-94-017-7432-1_6

Inorganic Processes

It is possible to relate degree of abrasion to the nature of the agent, the type of bone being affected, and to the time of exposure to the abrading agent (Faith et al. 2009; Gaudzinski-Windheuser et al. 2010; Thompson et al. 2011).

Degree of Abrasion

Degree of abrasion of mammalian bone is affected by the type of bone, whether fresh, dry, weathered or fossilized, the associated sediment type, and by the duration and the strength of the agent causing it (A.541, A.569 and A.570). In order to test these variables, an experiment was set up using a series of tumblers with four different grades of sediment acting on the four bone types for progressive time periods up to 360 h, shown here divided into four time stages (Fig. 6.1). This tests one process only, abrasion of bones by sediment movement in water. The results from the experiment with

different types of bone and fossil (Fernández-Jalvo and Andrews 2003), are as follows:

After 72 h abrasion, degrees of rounding are related to sediment type:

Gravels produced maximum rounding, as follows:

fossil bone	moderate rounding
weathered bones	moderate rounding
dry bones	slight/moderate rounding
fresh bones	slight rounding

Coarse sand showed little rounding on most bones:

fossil bones	slight/moderate rounding
weathered bones	slight/moderate rounding
dry bones	very slight rounding
fresh bones	very slight rounding

Fine sands produced a pattern similar to coarse sands, although abrasion rates were slightly higher than in the coarse sands, and fresh bones are slightly rounded.

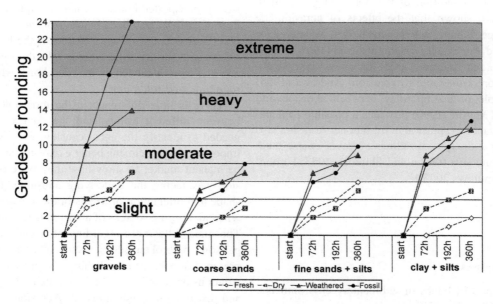

Fig. 6.1 Grade of abrasion. The grades of abrasion are shown on the vertical scale on a scale of 0–24, and on the horizontal axis are shown the four sediment grades used in the abrasion experiment. Each plot is shown divided into the four time stages by which the degrees of abrasion were assessed, from the beginning of the experiment up to 360 hours. Finally, the four types of bone used in the experiment, fresh, dry, weathered and fossilized, and these are shown separately for each sediment grade

Clay and silt produced more abrasion than sands (coarse and fine), as follows:

fossil bones	moderate rounding
weathered bones	moderate rounding
dry bones	slight rounding
fresh bone	no rounding

After 8 days (192 h), abrasion was more evident but still following trends similar to those above. Gravels remained the most abrasive sediment, but by this stage fossils were the most rounded specimens (heavily rounded), more than weathered bones (moderate-heavy rounding). It may happen that use of tumblers distorts this result, for the lighter-weight bones may stay "afloat" in the sediment rather than being entrained within the moving sediment. Fossils, because of their greater density, move within the sediment and thus have a greater degree of rounding. Clay sediments produced less abrasion than gravels, but more so than coarse or fine sands, with fossils still less rounded than weathered bones, and dry and fresh bones much less rounded than fossils. Fine sands produced more abrasion on fresh bones than on dry bones, coarse sands have almost equally rounded fresh and dry bones (very slightly).

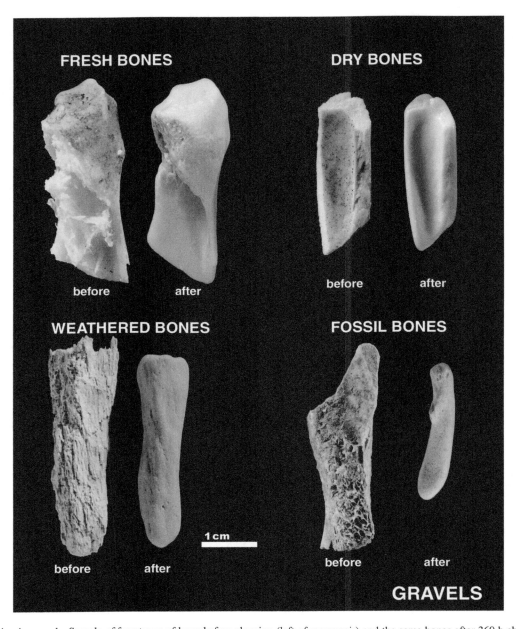

Fig. 6.2 Abrasion in gravels. Sample of four types of bone before abrasion (left of every pair) and the same bones after 360 h abrasion in gravel substrate. Gravel is the most abrasive sediment and produces high rates of polishing and brightness, followed by weathered, then dry and fresh bones (see Fernández-Jalvo and Andrews 2003, for further details). The water flow was constant and equivalent to a horizontal speed of 15 cm/s, simulating fluvial abrasion

Fig. 6.3 Abrasion in coarse sands (above) and fine sands and silt (below). Sample of four types of bone bones after 360 h in the two substrates. Abrasion was progressive with time, so that after 1 h, coarse and fine sands produced little rounding on most bones, with fresh and dry bones equally but only slightly rounded. Fossil bones were more rounded and weathered bones the most. After 360 h both types of sands (coarse and fine) produced more rounding on fresh bones than on dry bones, and slightly higher in fossils than in weathered bones (see Fernández-Jalvo and Andrews 2003)

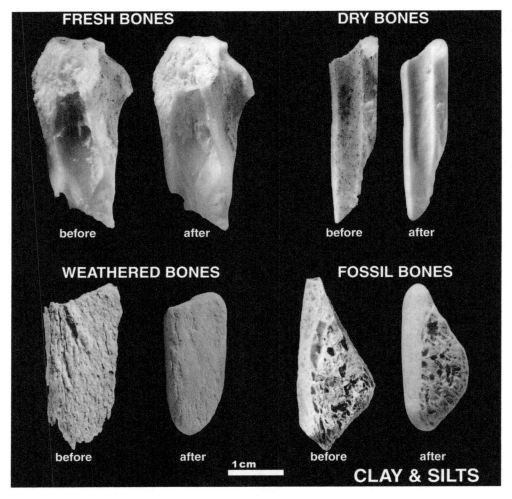

Fig. 6.4 Abrasion in clays and silts. Sample of four types of bone before abrasion (left of every pair) and the same bones after 360 h abrasion in clay and silt substrate. After 72 h, clay and silts produced more rounding than sands (coarse and fine) on weathered bones and fossils. Fossils were slightly more rounded than weathered bones, and dry and fresh bones much less rounded than fossils (see Fernández-Jalvo and Andrews 2003)

After 15 days (360 h), most specimens showed significant changes:

Gravels produced the most extreme rounding (Fig. 6.2)

fossil bones	extreme rounding
weathered bones	heavy rounding
dry bones	moderate rounding
fresh bones	moderate rounding

Coarse sand showed less rounding on most bones (Fig. 6.3).

fossil bones	moderate rounding
weathered bones	moderate rounding
dry bones	slight rounding
fresh bones	slight/moderate rounding

Fine sands produced a similar pattern to coarse sands, although abrasion rates were slightly higher than in the coarse sands, being moderate in almost all cases except dry bones.

Clay and silt produced more abrasion than sands (coarse and fine) (Fig. 6.4)

Fossil bones	heavy rounding
weathered bones	moderate–heavy rounding
dry bones	slight/moderate rounding
fresh bones	slight rounding

The experiment was extended to teeth (A.542 and A.561), but almost no change in the enamel was seen except for some rounding on broken edges and salient ridges. Dentine and roots, however, were affected to a degree similar to that of bones. Corroded bones were also included, but they only survived the first observation, and soon afterwards they disintegrated.

No clear characteristics of scratches, grooves and notches on bone surfaces with different sediment grades or duration of modification could be observed on the different

specimens. There is, however, an increasing extent of smoothed areas and brightness correlated with increasing particle size of the sediment, with fossil and weathered bone showing the greatest brightness (Fig. 6.5).

The interaction of abrasion and weathering was investigated on a fossil assemblage from the Miocene of Turkey (A.583) (Andrews 1995). The fossil assemblage was transported to the site at Pasalar in a single massive mud flow, and many of the fossils showed evidence of abrasion. Some of these had also been weathered (see Chap. 7) and it was evident that they were rounded by abrasion subsequent to the weathering. All stages of weathering are represented in the fossil assemblage, and degrees of abrasion and rounding are proportional to the degree of weathering: the fossils that showed low incidence of abrasion also showed little evidence of weathering; while the fossils that showed the highest

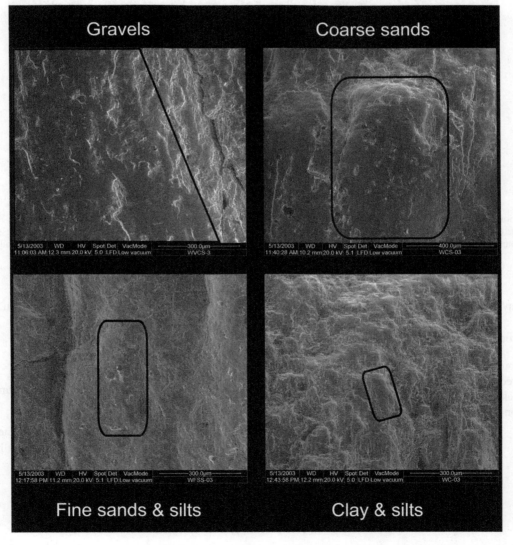

Fig. 6.5 ESEM analyses of the bone surfaces from the experiment shown in Figs. 6.1, 6.2, 6.3 and 6.4 (see Fernández-Jalvo and Andrews 2003) provided little additional information. Scratches, grooves and notches could be observed on the different specimens, but no clear correlation between frequency and intensity could be established. This requires further analysis. The most interesting observation was the presence of isolated polished and flattened patches on the bone surface (marked by black boxes) which increase in size in line with the size of the sediment grain

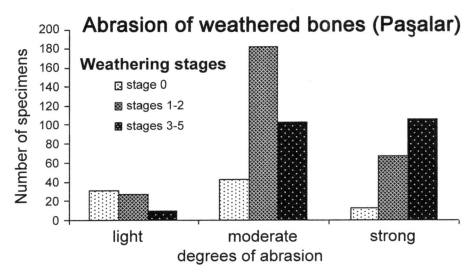

Fig. 6.6 Abrasion and weathering on the sample of fossil bones from Paşalar, Turkey. Three stages of abrasion are shown on the horizontal axis, light, moderate and strong; numbers of bones affected are shown on the vertical axis. Independently of this, the sample is divided between three weathering stages (Behrensmeyer 1978), with the weathering pre-dating abrasion in all cases. Fossils with no or light abrasion are least well represented in the fossil assemblage, and they generally show no or light degrees of weathering. Moderate abrasion is only slightly more frequent on unweathered fossils but is much more frequent on those that show evidence of prior weathering. The strongest degree of abrasion is associated with the highest degrees of weathering: Chi square = 251.54, p = 0.000

incidence of abrasion also had the highest degree of weathering (Fig. 6.6). Survival of bone at this site could therefore be shown to be primarily a function of the weathering, since it was mainly the heavily weathered bones that became abraded and were destroyed.

Abrasion of amphibian limb bones is only indirectly comparable with that of mammalian bone (Pinto Llona and Andrews 1996) because the structure of their bone is so different. An abrasion experiment was undertaken, but the experimental protocol entailed different periods of exposure to abrasion and slightly different sediment grades (Pinto Llona and Andrews 1999). The methodology of the experiment was the same, however, with amphibian bones abraded in a rotary tumbler for varying periods with mixtures of sediment and water in four grades: fine sand, coarse sand, gravel and pebbles. The results are as follows:

After 92 h, degree of rounding relative to sediment type:

fine sand	no modification other than slight rounding,
coarse sand	no modification,
gravel	moderate abrasion, ends rounded, some tissue loss,
pebbles	extreme abrasion, bones reduced to tiny splinters

After 168 h (7 days), degrees of rounding relative to sediment type:

fine sand	low abrasion, especially of limb bones,
coarse sand	less abrasion than fine sand,
gravel	more rounded, no more breakage

After 408 h (17 days), degrees of rounding relative to sediment type:

fine sand	no further modifications,
coarse sand	no further modifications,
gravel	pronounced abrasion and rounding of cartilaginous ends and abrasion of broken ends was now extensive.

After 744 h (31 days), degrees of rounding relative to sediment type:

fine sand	broken ends were lightly rounded,
coarse sand	light abrasion,
gravel	extensive abrasion, bone reduction, and loss of tissue broken ends showed extreme abrasion

Rounding of bones is most commonly seen where there is long-term movement of water and sediment, particularly in rivers or on beaches. The experiments described above show that it takes considerable energy to produce significant rounding, but the circumstances involved in this need not include actual transport of the bone. Bones in a beach environment, for example, are strongly abraded and rounded, but they might not be moved more than a few tens of meters from their place of initial deposition, because tidal change would not move the bones in the absence of on-shore currents. On the other hand, bones preserved in high energy river conditions and transported long distances may be extremely rounded. The fossils associated with high energy sands and gravels from unit 1 in Westbury cave, for example, look like rounded pebbles, for they were transported into the cave system from over 10 km away based on the composition of the associated sedimentary particles (A.569) (Andrews and Ghaleb 1999).

Polishing

It has been shown above that polishing of bone can occur as a result of abrasion by sediment in water, e.g., during transport of bone in rivers, but it is seen in most extreme form as a result of wind erosion, particularly on sandy substrates in desert environments, where there is little vegetation to protect bones from the scouring effect of wind-blown sand (A.578). In this case, the polish covers the whole of the exposed surface of the bone, producing a shiny surface, and if the bone is stabilized on the ground, one surface may be polished by wind action and the other only slightly abraded by the movement of the bone on the surface of the ground and with no evidence of polishing (Shipman and Rose 1983; Denys et al. 2007). In addition, the brightness of bone surfaces (A.562) changed during the abrasion experiment described above, increasing with increasing polish on the bone surfaces. In general terms, brightness increases according to the grain size of the sediment, but with wind polishing, fine wind-blown sand produces an even higher degree of brightness than anything seen on water-sediment abrasion (Fernández-Jalvo and Andrews 2003).

The following sequence is observed:

A small experiment using the sediment tumblers was also carried out with small mammals (A.544 and A.568), including mandibles (with teeth in their alveoli) together with long and flat bones (Fernández-Jalvo and Andrews 2003). Rounding was observed on the tips of the rodent incisors, affecting both the enamel and the dentine to similar degrees. In contrast to digestion, where enamel may be removed by digestive corrosion, the dentine is more strongly affected by physical abrasion and may be extremely rounded. Molars were detached from their alveoli after 48 h, and they showed rounded salient angles, again with enamel and dentine being rounded. Limb bones showed rounding and polishing with sands and gravels, but no effect was observed after abrasion with silt and clay. The only modification caused by silts and clays was perforations on scapular blades but no rounding or polishing (A.568). It is likely that small mammal bones may be transported in suspension in water, reducing their contact with sedimentary particles (Korth 1979; Fernández-Jalvo and Andrews 2003), while in contrast, exposure of small mammal bone to large particles in the absence of water is highly destructive (Andrews 1990; Fernández-Jalvo and Andrews 2003).

Location of Abrasion

Localized rounding may form on one part of the bone only (A.585 and A.598). This may occur, for example, on the end of a limb bone exposed to an abrading agent, and it also occurs on bone processes or other protuberances that project from the general level of the bone. Localized abrasion is distinguished from abrasion by transport or wind, both of

Brightness

	(MIN)			(MAX)
(MIN)	Fresh bones	Clay + silt − Fine sand + silt − Coarse sand − Gravels
	Dry bones (1)	Clay + silt − Fine sand + silt − Coarse sand − Gravels
	Weathered (4)	Clay + silt − Fine sand + silt − Coarse sand − Gravels
(MAX)	Fossils	Clay + silt − Fine sand + silt − Coarse sand − Gravels

which tend to abrade whole surfaces. Localized rounding is produced by trampling, human action or digestion. For the third of these, it appears that avian and reptilian predators produce little rounding of the bones of their prey during ingestion and digestion. Rather their strong digestive processes penetrate bone tissue to produce corrosion of the surface (Fisher 1981a, b, c; Andrews 1990). Rounding has however been recorded on the bones of the prey of *Bubo bubo*, and *Circus cyaneus* (see also Chap. 8). Heavy rounding can be observed more commonly on bones recovered from scats of mammalian carnivores (A.591 and A.610). The rounding capacity of digestion is due more to the effects of digestive juices and enzymes in combination with stomach acids (Denys et al. 1995). It appears that crocodiles with their low enzyme concentrations but low pH do not produce much rounding but almost completely destroy bones and teeth that they have ingested (see Chap. 8). Abrasion and rounding due to digestion processes has been observed for the following mammalian predators:

Vulpes vulpes, slight rounding of broken ends of limb bones A.619

Genetta genetta, slight/moderate rounding of broken ends of limb bones A.633

Mustela putorius, moderate rounding of dentine A.629

Puma concolor, extreme rounding A.610

Panthera onca, abraded bone from zoo animal A.615

Felis wiedii, heavy abrasion of broken ends of limb bones A.848

Martes martes, slight/moderate abrasion of broken ends of limb bones

Canis latrans, moderate to heavy rounding of broken ends of limb bones A.616 and A.630

Crocuta crocuta, heavy abrasion of bone flakes A.589 and A.606

Hyaena hyaena and *Hyaena brunnea*, heavy abrasion of bone flakes A.748

Carnivores may also produce abrasion due to the enzymatic activity of saliva while licking bones as well as chewing. This is especially frequent among young individuals in carnivore dens.

Localized abrasion has also been observed as a result of trampling. If a bone is partly buried and fixed in position in soil or sediment, movement of animals over the most exposed part of the bone, either in direct contact with it or by friction of sediment over the exposed end, can produce extreme abrasion on the end of the bone exposed to this process (A.598). For example bones just beneath the surface of the soil, which is being trampled by frequent passage of animals over the surface, may become polished in whole or in part. On the other hand, some fossil assemblages have high proportions of edge-rounded specimens in the absence of any obvious sedimentary conditions that could have caused it. We propose that in these instances the rounding is due to in situ sediment movement and compaction of the sediment (Andrews et al. 1999; Cáceres 2002).

Localized distribution of rounding on bones may be a feature of human use (A.587) (Brain 1967; Lyman 1984a, b, 1994a). It has been suggested, for instance, that use-related abrasion should be expected to be at or close to a broken end of a bone (Lyman 1994a). This type of rounding, however, also can be produced by carnivore action or even trampling (Andrews 1990; Barham et al. 2000). Use of bones as tools produces localized abrasion on the ends of bones that has superficial similarity to the abrasion from carnivore digestion or trampling. The broken ends of bones may be rounded and polished, but detailed examination may show the presence of a specific pattern of striations related either to the manufacture of the tool or to the use to which it was put. A series of bone points from Mumbwa cave include one undoubted bone tool (Barham et al. 2000), (see Chap. 3), but also includes a number of rounded bone points that bear no indication of human manufacture. The one undoubted bone tool has longitudinal polishing lines and a once-rounded tip now damaged, possibly as a result of use (Barham et al. 2000), but the other abraded bones with rounded tips lack any consistent pattern of abrasion and are probably the product of natural abrasion.

Atlas Figures

A.541–A.638

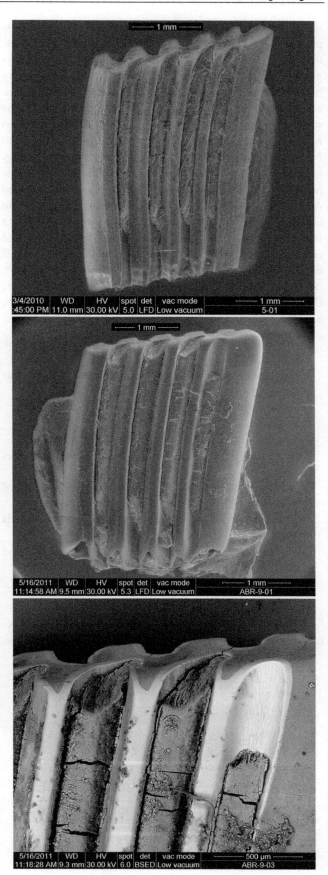

Fig. A.541 SEM microphotographs of modern abraded molars of an ▶ arvicolid. The teeth were experimentally abraded for 192 h in a tumbler with gravels and water (see also A.542, A.543 and A.544). Top: molar before experiment. Middle Same molar after abrasion photographed using secondary electron (SE) detector. Bottom: detail of the occlusal end of the same abraded molar photographed using backscattered electron (BSE) emission. The photograph at the bottom shows a brighter color for the enamel distinguishing it from the dentine (grayish). This photograph shows that both enamel and dentine are equally rounded on the salient angles of the tooth. This trait distinguishes this physical effect of abrasion from digestion, where the enamel is firstly rounded by digestion. The middle image shows that the whole tooth has been affected by rounding, and not just the salient angles. These images show the different properties of BSE and SE emission modes

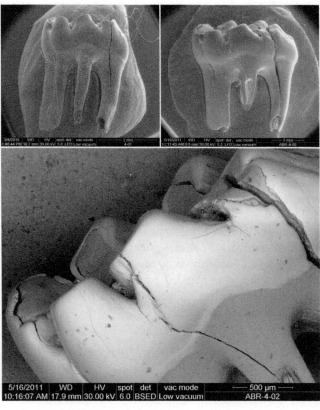

Fig. A.542 SEM microphotographs of modern abraded molars of a murid photographed using secondary electron detectors. The teeth were experimentally abraded for 192 h in a tumbler with gravels and water. Top left: molar before experiment. Top right: same molar after abrasion Bottom: close up view of the abraded tooth using backscattered electron (BSE) emission. Note that both the enamel surface and the roots are smoothed and rounded due to physical abrasion by sediment and water. This is in contrast to the pitting present on the enamel of digested teeth and modification of the enamel/dentine junction of murins and soricids (see A.767, A.768, A.848 and A.849)

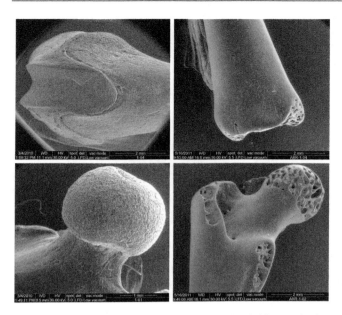

Fig. A.543 SEM microphotographs of modern abraded femur of rodent photographed using secondary electron detectors. The bone was experimentally abraded for 192 h in a tumbler with gravels and water. Top left: distal femur before experiment. Top right: distal end after abrasion. Bottom left: proximal femur before experiment. Bottom right: proximal femur after abrasion. The whole bone is rounded and the surface is smoothed, and not just the salient angles as for digestion (see A.864)

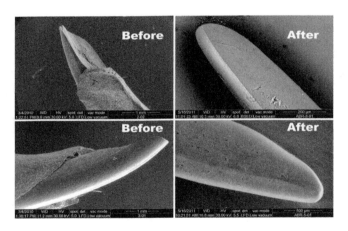

Fig. A.544 Rodent incisor experimentally modern abraded in a tumbler containing gravels and water for 192 h. The incisor tip is rounded, including the enamel, dentine and the mandible edges, providing a truncated tip of the incisor. This is in contrast to the effects of digestive corrosion. Digestion may also produce rounding at the incisor tips (see A.819), but digestion preferentially corrodes the enamel (A.628) and shows traits of chemical corrosion, rather than physical abrasion

Fig. A.547 SEM microphotograph of the effects of abrasion. A fossil ▶ bone from the middle Pleistocene was experimentally abraded for 360 h in fine sands and silt in water (<250 microns grain size). Fine sand showed a higher degree of rounding capacity than coarse sands. See A.551, A.555 and A.559

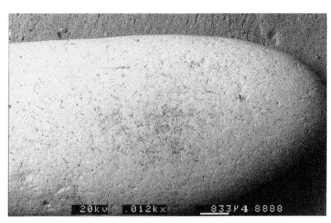

Fig. A.545 SEM microphotograph of the effects of abrasion. A fossil bone from the middle Pleistocene was experimentally abraded for 360 h in a gravel/water mixture. Fossils in gravels were found to be more rounded than weathered bones, and pitting and scratches were present on the fossil bone surfaces. See A.549, A.553 and A.557

Fig. A.546 SEM microphotograph of the effects of abrasion. A fossil bone from the middle Pleistocene was experimentally abraded for 360 h in coarse sands and water (1 mm to 500 microns grain size). Observations before 360 h (72 and 192 h) showed weathered bones were more rounded than fossils. After 360 h, fossil bones were the most rounded, more than weathered, dry or fresh bones. See A.550, A.554 and A.558

Fig. A.548 SEM microphotograph of the effects of abrasion. A fossil bone from the middle Pleistocene was experimentally abraded for 360 h in clay and silts in water (<4 microns grain size). Clay-silts and water have a higher rounding capacity than fine and coarse sands on fossil bones. See A.552, A.556 and A.560

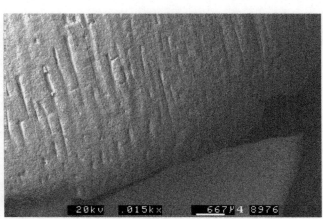

Fig. A.551 SEM microphotograph of the effects of abrasion. A weathered bone (stage 4 weathering) was experimentally abraded for 360 h in fine sands and water. Similar trends were seen for weathered bones in coarse and in fine sands. Initially weathered bones appeared more rounded than fossils until surface flakes peeled off. The final rounding rate in fine sands is slightly higher than in coarse sands. Striations are present as surface histological bone traits. See A.547, A.555 and A.559

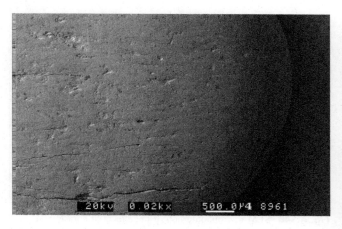

Fig. A.549 SEM microphotograph of the effects of abrasion. A weathered bone (stage 4 weathering) was experimentally abraded for 360 h in gravels and water. Rounding by gravels is less effective on weathered bones than on fossils. Pitting, however, is more intense, and pit size is bigger in weathered bones than in fossils. See A.545, A.553 and A.557

Fig. A.552 SEM microphotograph of the effects of abrasion. A weathered bone (stage 4 weathering) was experimentally abraded for 360 h in clay and water. The surface texture of the bone is rougher than in fine sands, and the edges are more rounded than in fine sand. As seen for sands, weathered bones initially became more rounded than fossil bones, but after the 360 h of the experiment, fossil bones appeared more rounded than weathered bones. See A.548, A.556 and A.560

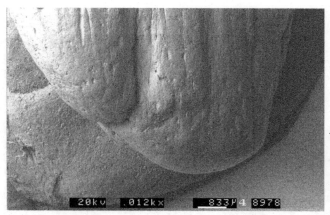

◄ **Fig. A.550** SEM microphotograph of the effects of abrasion. A weathered bone (stage 4 weathering) was experimentally abraded for 360 h in coarse sands and water. Initially, weathered bones appeared more rounded than fossil bones. By the end of the experiment, bone flakes had peeled off completely from the weathered bone surface, and rounding was less effective thereafter than in fossils. See A.546, A.554 and A.558

Fig. A.553 SEM microphotograph of the effects of abrasion. A dry bone was experimentally abraded for 360 h in gravels and water. During most of the experiment, dry bones were observed to have a higher rate of rounding than fresh bones. At the end of the experiment (after 360 h), dry and fresh bones appeared equally rounded. See A.545, A.549 and A.557

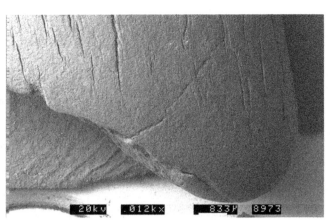

Fig. A.556 SEM microphotograph of the effects of abrasion. A dry bone was experimentally abraded for 360 h in clay and silts (<4 microns grain size) and water. Our results showed similar rounding rates of dry bones abraded by fine sands, clay and silt. Rounding of dry bones is higher than fresh (greasy) bones when abraded by clay and silt. See A.548, A.552 and A.560

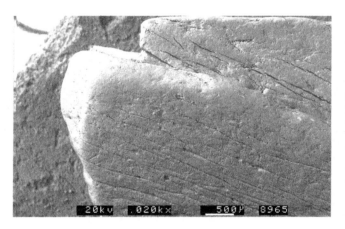

Fig. A.554 SEM microphotograph of the effects of abrasion. A dry bone was experimentally abraded for 360 h in coarse sands (1 mm to 500 microns grain size). Dry bones abraded with coarse sands were altered to the same degree as fresh bones until 192 h into the experiment. At the end of the experiment, edges of dry bones were less rounded than on fresh bones. See A.546, A.550 and A.558

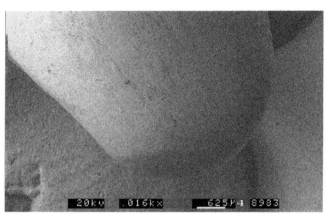

Fig. A.557 SEM microphotograph of the effects of abrasion. A fresh bone was experimentally abraded for 360 h in gravels and water. Fresh bones are less abraded by gravels than any other type of bone during the experiment. After 360 h, however, fresh and dry bones abraded by gravels reached similar degrees of rounding. See A.545, A.549 and A.553

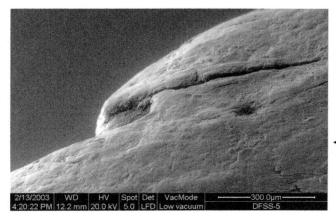

◀ **Fig. A.555** SEM microphotograph of the effects of abrasion. A dry bone was experimentally abraded for 360 h in fine sands (<250 microns grain size) and water. Fine sands produce a higher grade of rounding than coarse sands. The trend seen in other stages of the experiment of dry bone edges being less rounded than fresh bone is more evident here with abrasion by fine sands. See A.547, A.551 and A.559

Fig. A.558 SEM microphotograph of the effects of abrasion. A fresh bone was experimentally abraded for 360 h in coarse sands (1 mm to 500 microns grain size) and water. At the end of the experiment, fresh bones were slightly more rounded than dry bones when abraded by both coarse and fine sands, See A.546, A.550 and A.554

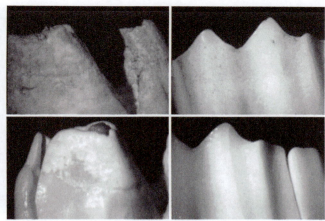

Fig. A.561 SEM microphotograph of the effects of abrasion. Sheep teeth were experimentally abraded for 360 h in gravel and water. Above, before the experiment, bottom, root and crown-edge abraded. Rounding on teeth is less intense than on bones, but tooth roots are comparatively more rounded than enamel (as in A.542 and A.544). These traits could be used as criteria to distinguish abrasion from other agents such as digestion on teeth

Fig. A.559 SEM microphotograph of the effects of abrasion. A fresh bone was experimentally abraded for 360 h in fine sands (<250 microns grain size) and water. All observations of fine sand abrasion showed a higher rate of rounding on fresh bones than on dry bones. See A.547, A.551 and A.555

Fig. A.562 Fossil bone experimentally abraded by gravel and water showing a shiny surface which is described as brightness or polishing. In contrast, bones abraded by sands and silts in water, both modern and fossil, usually acquire a mat surface. See Text Figs. 6.1, 6.2, 6.3, 6.4 and 6.5

◀ **Fig. A.560** SEM microphotograph of the effects of abrasion. A fresh bone was experimentally abraded for 360 h in clay and silts (<4 microns grain size) and water. Rounding is almost absent, with slight abrasion affecting pointed angles. This may probably be due to the greasy surface (periosteum) remaining on fresh bones, for this appears to limit the abrasive capacities of clay. See A.548, A.552 and A.556

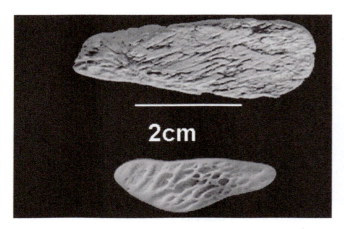

Fig. A.563 Top: corroded bone abraded by clay and silts (<4 microns grain size) in water. Bottom: corroded bone abraded by gravels (>1 mm grain size) in water. This picture was taken after 72 h. None of these bones remained intact after 192 h (compare with Text Fig. 6.2 and A.543)

Fig. A.566 Rounding by wind and fine sand occurs on all sides of small fragments, because they may be rotated by the wind, exposing all surfaces to the wind (see A.341 and A.344). Abrasion on the broken edges is evident, but less intensive or effective than in water and sediment. Wind abrasion, however, produces brighter and more polished surfaces than sediment and water. E. Aguirre's Collection

Fig. A.564 Fossil hominin bones from Sima de los Huesos site. About 25% of transversally broken long bones and 21% of ones with spiral breakage are rounded to a moderate degree with a mat surface in Sima de los Huesos fossil collection. This type of rounding is consistent with clay and silt abrasion, such as may occur in a mudflow (see also A.512)

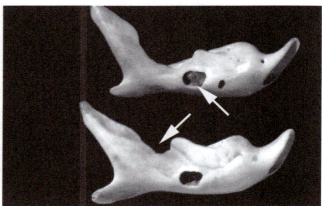

Fig. A.567 Two modern rodent mandibles abraded in water and coarse sands and gravels (>1 mm grain size) in a stone polishing tumbler for 48 h. There are characteristic modifications such as holes at the molar root area of the external side of the mandible and breakage of the ascending ramus (white arrows). The bone surface appears polished and bright (as in A.562)

◄ **Fig. A.565** Rounding by wind and sand from the Sahara desert. The picture shows the opposite side of A.340. This specimen appears heavily rounded, pitted and abraded on one side in contrast to the side facing the ground, shown here, which has sharp edges. This bone was large and stable enough not to be rotated by the wind and so on this side the cancellous tissue looks fresh. Such strong contrast between sides is not frequent in water abrasion. E. Aguirre's Collection

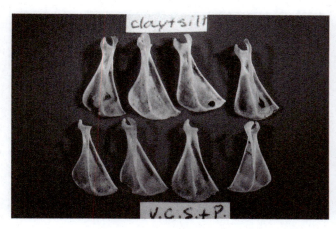

Fig. A.568 Top: modern small mammal scapulae abraded by clay and silts in water after 24 h showing holes in the thinnest part of the blade. Bottom: modern small mammal scapulae abraded in very coarse sand and pebbles (gravels) for 24 h that show brighter surface and rounded edges but no holes on the blade. Compare with Text Figs. 6.2 and 6.4

Fig. A.571 Broken fossil bone showing a heavily rounded spiral break. The fossil was recovered from a fluvial terrace site (A.572). Courtesy of D. Pesquero

Fig. A.569 Rounded fossil bones and teeth from unit 1 at Westbury, transported into the cave system with waterlain sands and gravels. All fossils from this stratigraphic unit show a characteristic rolled aspect suggesting reworking or at least re-sedimentation processes that need further analysis and research. Compare with Text Fig. 6.2

Fig. A.572 Large fossil bone fragment from the same fluvial terrace site as A.571. All sides are rounded suggesting that this large fragment could roll on the riverbed so that all sides were exposed to abrasion. Cemented sediment attached to the bone surface suggests certain calm periods. Courtesy of D. Pesquero

Fig. A.570 Bones from the beach showing relative rounding edges and matt surface that is indicative of sand erosion and typical of sand beach abrasion. Compare with Text Fig. 6.3

Fig. A.573 Monitored modern bones recovered from water streams of different energy. Different degrees of rounding could be established according to the sharpness of the broken edge (A.574). Rounding is higher from right to left South Plate (braided), Lost Creek, ephemeral; Calamus, permanent stream. Photograph: T. Jorstad, Collection of A. K. Behrensmeyer

Fig. A.574 Heavily rounded bone fragment from river rapids of the East Fork river (Wyoming, USA). The edges are more heavily rounded than in A.573. The bone fragment edge is smooth and blunted. Rounding taking place in water may vary with time, by prolonged exposure to water currents, or with water energy. Photograph: T. Jorstad, collection of A. K. Behrensmeyer

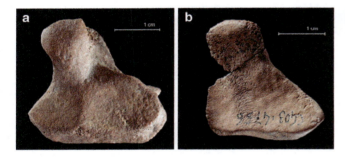

Fig. A.575 Fossil carpals of *Hipparion* from Concud. On the left, a carpal unaffected by abrasion, on the right, a carpal heavily abraded by water and sediment. Abrasion is relatively rare at this lakeshore site. Compare with A.555, A.559 and A.560. Courtesy of D. Pesquero

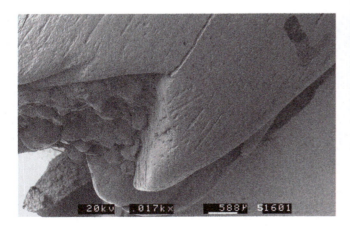

Fig. A.576 SEM microphotograph of a fossil bone fragment from Vanguard Cave (A.118) showing an abraded and rounded broken bone. Fine sand sediment is still attached to the bone, and the surface appears pitted and some "comet like" striations on surface can be distinguished on the surface, suggesting abrasion by wind and sand

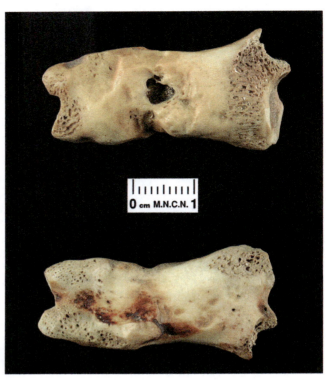

Fig. A.577 Modern bear phalanx. Top: dorsal view. Bottom: ventral view. This specimen was found in a hollow in the floor of a Spanish cave where water was falling. The effects of the water movement were to move the sediment within the hollow so the bone became abraded without ever being moved beyond the limits of the hollow. This bone may show similarities with digestion. High magnification observations at the SEM provide criteria to distinguish abrasion from digestion (see A.349, A.351 and compare with A.744 and following images)

Fig. A.578 Modern bone fragments collected from the Sahara desert showing heavily rounded edges and pitting on the bone surface (white arrow) by sand grains blown across the bone surface by the wind. The heavily pitted surface differs from water abrasion (see A.349, A.351 and compare with A.344, A.346)

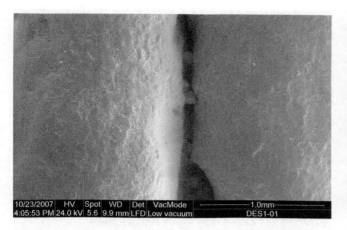

Fig. A.579 SEM microphotograph of a modern bone collected from the Sahara desert. Traits on the surface, such as cracks, cavities and holes, have been rounded. The general aspect is of a rounded, smooth and high brightness surface. The bone is intensively pitted (see A.341 and A.342)

Fig. A.582 SEM microphotograph of a bone fragment from the sand desert in Abu Dhabi showing high degrees of polishing on the surface of the bone due to wind-blown fine sand. The surface has a shiny and almost mirror-like appearance. Fine pitting occurs on the surface of the bone (A.346 and A.580)

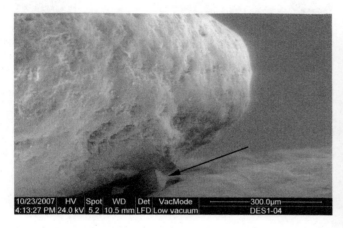

Fig. A.580 SEM microphotograph of a modern bone collected from the Sahara desert. The general aspect is of a rounded and smooth surface that is also heavily pitted. The sediment grain (shown by black arrow) is trapped inside a crack in the bone fragment, and it corresponds in size grade to very fine sand. Similar abrasion in different bones by sand and water, shows a less pitted surface. In contrast to water abrasion, pitting is heavier with wind abrasion (compare with A.351)

Fig. A.583 Rounded fossil fragments from Paşalar, Turkey. Many of the fossils had been exposed to weathering before being transported to the site in a mud flow (Andrews and Ersoy 1990). Although the distance travelled in the mud flow was short, the weathered bones became very rounded while fresh bones were almost unaffected (Text Fig. 6.4). Andrews 1995

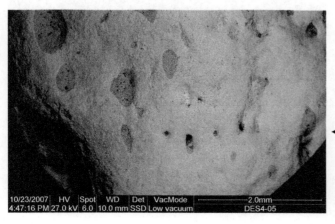

◄ **Fig. A.581** SEM microphotograph of a modern specimen collected from the Sahara desert collection. The general aspect of wind abraded specimens is highly rounded and polished bone surfaces that are smooth to the touch. Observed on the SEM, the bone surface is heavily pitted and uneven. This photograph was taken at BSE mode showing different densities of some areas where sediment has been cemented (A.347) filling holes and bone irregularities

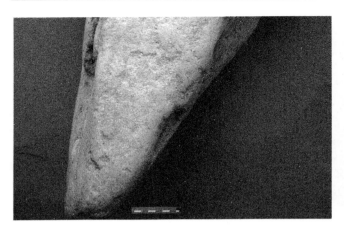

Fig. A.584 SEM microphotograph showing the rounded end of a fossil limb bone from level 3, Westbury. This level has only cave bears (*Ursus deningeri*) preserved, and the age structure of the bears shows that it was a hibernation den for male bears (Andrews and Turner 1992). There is no evidence for reworking of the deposits, and the rounding of the bones was interpreted as due to trampling by the bears and friction of the bone fragment against the sediment (A.585, A.596)

Fig. A.586 SEM microphotograph of a pseudo-tool from Middle Stone Age deposits in Mumbwa Cave. The bone has an abraded and rounded end with featureless surfaces (Barham et al. 2000). This bone can be compared with manufactured bone tools from Late Stone Age deposits from the same site (see A.109). In contrast to this bone, bone tools have shaped pointed end with consistent pattern of macroscopic linear and vertical striations A.587. These are absent on this specimen

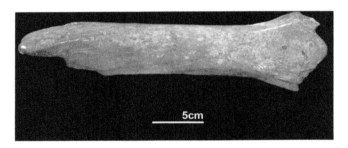

Fig. A.585 Fossil tibia of a rhino from Late Pleistocene deposits in Bacon Hole, a Pleistocene cave site in Wales. The broken end at the left is highly rounded, and it was first considered to indicate use of the bone as a bone tool. After comparison with modern cases, the rounded and polished proximal broken edge is now interpreted as the result of trampling the bone against the sediment (A.584, A.596)

Fig. A.587 SEM microphotograph of a fossil bone from Middle Stone Age deposits in Mumbwa Cave. Although pitting is due to post-depositional damage (e.g., trampling) a consistent pattern of linear marks which are longitudinal to the length of the fragment (shown in this photograph and in A.109) is observed on this bone tool in contrast to A.586

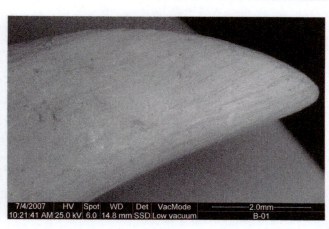

Fig. A.588 Regurgitation of bones by lions, Tanzania. Carnivores ingest hair and bones as well as meat (A.589). Indigestible parts are regurgitated, and some bone splinters are also expelled (white arrows). The black arrow shows a piece of skin. Lions, like most felids, cannot break large bones and they do not usually ingest bones. However, this may exceptionally happen when lions produce regurgitations larger than 15 cm (top right box)

Fig. A.591 SEM microphotograph of a modern large mammal bone fragment regurgitated by hyena. Rounding by digestion (A.595) can be distinguished from abrasion by a characteristic and diagnostic ultra-microscopic cracking (see Chaps. 7 and 8, A.745)

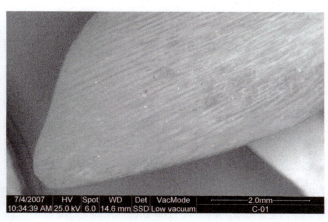

Fig. A.589 Regurgitation of bones by hyenas, Tanzania. Hyenas ingest hair, bones, hoofs and horns. They have strong mandibles that can break and chew almost any hard surface. As a result, regurgitations are frequent and may contain bones. The size of bones is small, up to 10 cm in diameter, because of their capacity to break even large bones (A.588)

Fig. A.592 SEM microphotograph of a modern large mammal bone fragment regurgitated by hyena. Rounding by digestion (A.595) can be distinguished from abrasion by a characteristic and diagnostic ultra-microscopic cracking (see Chaps. 7 and 8, A.744, A.758)

◄ **Fig. A.590** SEM microphotograph of a modern large mammal bone fragment regurgitated by hyena. Rounding by digestion can be distinguished from abrasion by a characteristic and diagnostic ultra-microscopic cracking (see Chaps. 7 and 8, A.744 and A.870)

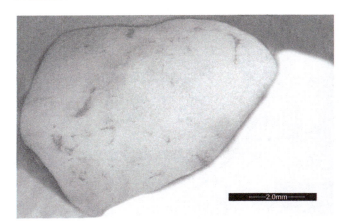

Fig. A.593 SEM microphotograph of a modern large mammal bone fragment regurgitated by hyena. Rounding by digestion (A.610) can be distinguished from abrasion by a characteristic ultra-microscopic bone tissue cracking (see Chaps. 7 and 8, A.758)

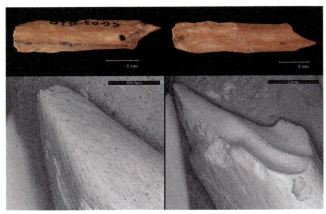

Fig. A.595 Fragments of fossils from Concud showing pointed salient angles. Seen on the SEM (below), the edges are seen to be rounded as a result of digestion after breakage (A.591 and A.592). Courtesy of D. Pesquero

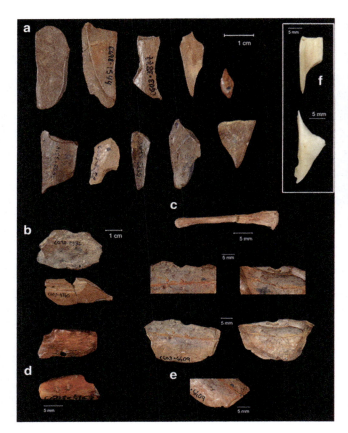

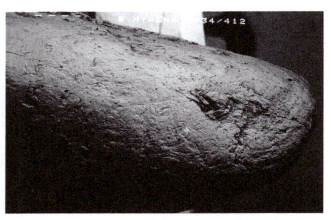

Fig. A.596 Large mammal limb bone collected from a brown hyena den. The bone is rounded at the tip. It was collected from the entry passage to the den, buried in organic-rich soil. One side of the bone is especially highly polished, and the tip is rounded and polished (detail at A.597). Its exact position is unknown, but the distribution of polishing suggests that it is the result of trampling by the hyenas against highly acidic and organic substrate at the den entrance. Photo courtesy of G. Avery. Field width equals 5 mm

Fig. A.594 Fossil bones from Concud, smaller than 2 cm in width. These fossils are similar in shape and size to modern bone regurgitated by hyenas (inside the square on the right). Some of these fragments also have puncture marks resulting from hyena chewing before ingestion. All broken edges are rounded, together with a shiny and polished surface. These traits require close examination with lens or microscope. The rounding is due to the combination of hydrochloric acid and enzyme activity of the gastric juices (see Chap. 8, A.872, A.873 and A.595). Courtesy of D. Pesquero

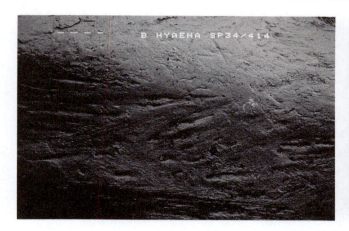

Fig. A.597 Detail of the bone surface of A.596. This type of abrasion and rounding is mixed between organic and inorganic processes. Trampling is by definition considered to be inorganic, as the result of the movement of bone against a rocky/sandy substrate. This particular case at a hyena den has the effects of trampling enhanced by urinated and highly acidic soil produced by hyenas. Photo courtesy of G. Avery. Field width equals 5 mm

Fig. A.600 Rounded fossils from Azokh. The Azokh sediments and the taphonomic study of these fossils show no evidence of abrasion, transport or reworking. The presence of rounded bones, and a certain uniformity of shape, led early workers to consider them as possible bone tools. The lack of specific microwear pattern (A.587), however, suggests that these bones could be trampled like A.585, while microscopic traits (A.601 and A.602) suggest that the rounding is likely due to licking

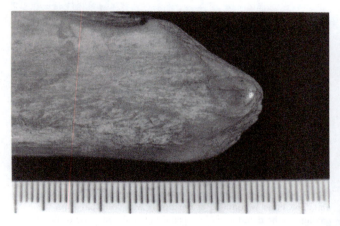

Fig. A.598 Fossil from Bacon Hole, a cave site in Wales, that is interpreted as a hyena den. This is a detail of the specimen photographed in A.585, showing its rounded and shiny end attributed to polishing by trampling in a highly acidic soil. The end of this fossil is rounded in a similar way to the specimen from the hyena den (A.596)

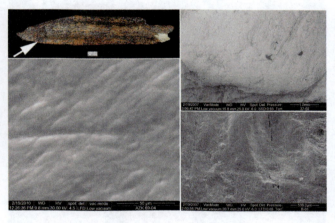

Fig. A.601 The rounded fossils from Azokh Cave (A.600) do not show signs of digestion that could cause rounding (bottom left, compare with A.745). Trampling marks are less abundant on the rounded edges (top right). Tooth grooves or scores (top left white arrow) appear smooth and rounded (bottom right, A.602). The taphonomic study of these fossils show no evidence of abrasion, transport or reworking. The most likely agent to produce this high degree of rounding is licking

◀ **Fig. A.599** Fossil bone from Galeria, Atapuerca with evidence of chipping and licking (A.1017). Chipping has been described by Lyman (1994) as "chewing the edge of a broken long bone; the bone edge is continuously chipped and tooth scoring on the external surface of the bone is frequently associated". He described licking as "chipped edges or ends can produce rounded and polished edges that have the appearance of use wear" (p. 212). Courtesy of I. Cáceres

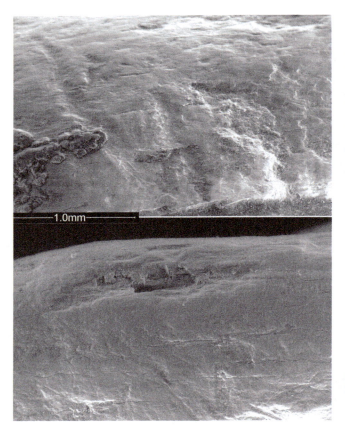

Fig. A.602 SEM microphotograph of the fossil bone surface of rounded fossils shown in A.601 (top right specimen). The bone surfaces of these bones have grooves that appear rounded (top). These fossils have thin transversal scratches along the rounded edges (bottom). Scratches observed at higher magnification also appear rounded (see A.604). Taphonomic studies at this site (Azokh Cave) do not show signs of transport. High magnification of the bone surfaces show smooth surfaces and no signs of abrasion by water (compare A.604 with A.349)

Fig. A.604 SEM microphotograph of a fossil from Azokh. Close up view of the striations of A.602. These striations are narrower than those of Torralba (A.117) observed to be abraded by natural agencies. The interior of these microstriations is irregular (white arrow) and there is a superimposed blurred film-like coat that obscures the images. These features have also been observed on licked fossil bones, suggesting that these bones could have been modified by licking rather than trampling

Fig. A.605 Rounded fossil human phalanx from Sima de los Huesos, Atapuerca. The rounding is probably due to licking or digestion by a carnivore. The distinction between digested and licked bones may be established by analyzing the bone surface at high magnification with SEM images. Digestion, as distinct from rounding is characterized by surface cracking ('torn'-like surface, see Chaps. 7 and 8, A.745). Photo M. Bautista

◀ **Fig. A.603** SEM microphotograph of fossil AZUM-03 (Azokh Cave) showing a peculiar dense and regular microstriation observed on these bones transversally to the length of the fragment. A close up view of these striations is at the top left square, enlarged in A.604. Striations are very thin, much thinner than natural abrasion observed on other fossils (see A.115) or trampled experimental bones. Further observations on modern bones licked by carnivores need to be done

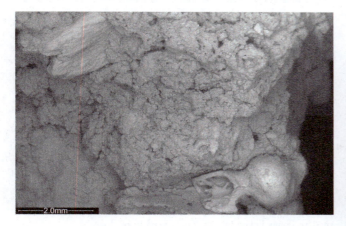

Fig. A.606 Bone fragments have been seen in both modern and fossil carnivore scats. Top: rounded fossil bone fragments in a hyena coprolite from Olduvai. See Chap. 8 and A.607 and A.609. The most reliable source of digested bones is their presence in coprolites, for the fossil record, and pellets or scats for present day bone. Bottom: modern brown hyena dung. (scale bar 3 cm)

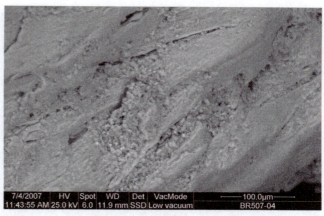

Fig. A.608 SEM microphotograph showing a detail of A.607, a fossil bone from a hyena coprolite from Bois Roche (SW France). Close up view of the large mammal bone fragment recovered from inside the hyena coprolite showing rounded edges produced by digestion. The bone surface is also damaged by cracking, which is only apparent at higher magnifications (see A.625)

Fig. A.609 SEM microphotograph of small fossil bone fragments from Vanguard Cave carnivore coprolite. They have been rounded and polished by digestion (A.606 and A.607)

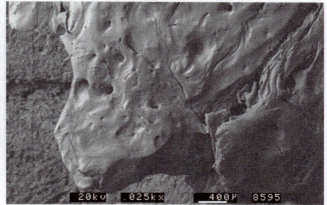

Fig. A.607 SEM microphotograph of fossil mammal bone fragments in an hyena coprolite from Bois Roche (SW France). The site is a late Pleistocene hyena den (69.7 ± 4.1 ka, Villa et al. 2010). The large mammal bone fragment at top left has a rounded edge. At the bottom, a rodent femur (broken while preparing the sample) has been digested on the head of the femur. Digestion has produced rounding on both large and small mammals (A.606 and A.609)

Fig. A.610 SEM microphotograph of a modern experimental bone fragment from a puma scat (*Puma concolor*). The complete fragment appears extremely rounded by digestion (A.611, A.612 and A.869). The cortical layer of bone has almost disappeared, leaving bone histological traits exposed (Haversian system and lamellae). G. Gómez's collection

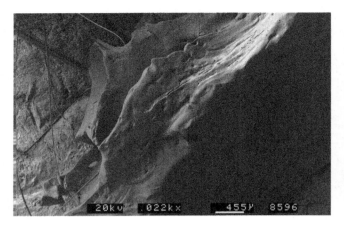

Fig. A.611 SEM microphotograph of a modern experimental bone fragment recovered from *Puma concolor* scats, heavily digested (A.612 and A.869). Broken edges appear rounded by digestion indicating that breakage occurred before digestion (A.613). G. Gómez's collection

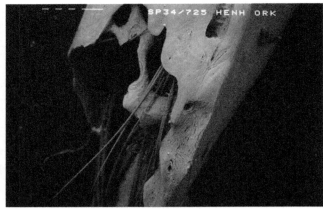

Fig. A.614 SEM microphotograph of the distal end of a modern small mammal tibia. It was broken during feeding and the broken end digested by a hen harrier (*Circus cyaneus* A.746, A.834 and A.857). There is a slight degree of polishing and rounding, but this is restricted for avian predators because digestion only occurs in the stomach. Ingested bones are usually regurgitated from the stomach, sometimes only partly digested, and they do not pass through the rest of the digestive system. Field width equals 1.5 mm

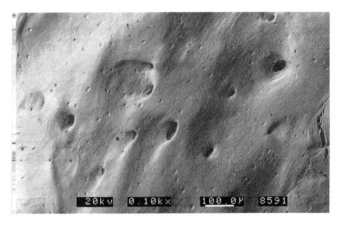

Fig. A.612 SEM microphotograph of a modern experimental bone fragment from a puma scat, *Puma concolor*. The rounding and smooth aspect is produced by extreme degrees of digestion (A.610 and A.869). G. Gómez's collection

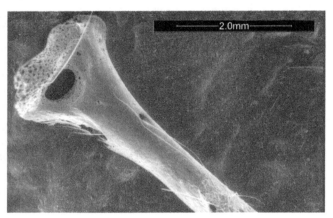

Fig. A.615 SEM microphotograph of a modern experimental small mammal recovered from a Jaguar (*Panthera onca*) scat, fed with rodents in a zoo (A.397). Courtesy of G. Gómez

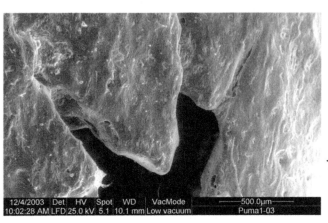

◄ **Fig. A.613** SEM microphotograph of a modern experimental bone fragment recovered from a puma scat *Puma concolor*. Rounding by digestion can be distinguished from abrasion by diagnostic ultra-microscopic cracking (see Chaps. 7 and 8, A.869). G. Gómez's collection

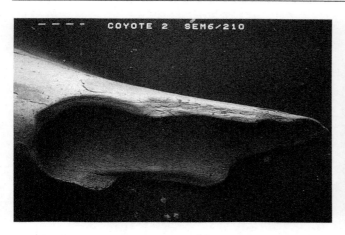

Fig. A.616 SEM microphotograph of a modern small mammal limb bone from a coyote scat (*Canis latrans* A.630, A.832 and A.1023). The bone was broken during ingestion by the coyote, and the rounding of the broken end occurred during digestion. Field width equals 6.5 mm

Fig. A.617 SEM microphotograph of a rodent femur from the pellet of an eagle owl (*Bubo bubo* A.817, A.840, A.846 and A.864). It shows a slight degree of polishing and abrasion on the head of a femur as a result of digestion and penetration of the articular bone on the head of the femur. Field width equals 7.5 mm

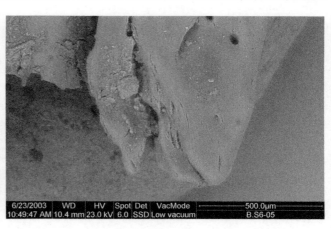

Fig. A.619 SEM microphotograph of a modern amphibian bone fragment from a fox scat (*Vulpes vulpes* A.377). It was recovered from a modern cave (specimen provided by Borja Sanchiz). The end of the bone is strongly rounded as in mammalian bone. (See Pinto Llona and Andrews 1999 for digestion traits and classification in reptiles and amphibians). Compare with A.622, A.758 and A.868

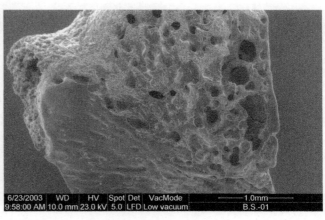

Fig. A.620 SEM microphotograph of a modern amphibian bone fragment from a fox scat (*Vulpes vulpes*). It was recovered from a modern cave (specimen provided by Borja Sanchiz). Surface bone has been lost through digestion A.621, A.622 and A.758. (See Pinto Llona and Andrews 1999 for digestion traits and classification in reptiles and amphibians)

◀ **Fig. A.618** SEM microphotograph of a modern bird bone fragment from the scat of a jackal (*Canis* sp. A.399). The broken end is strongly rounded by digestion

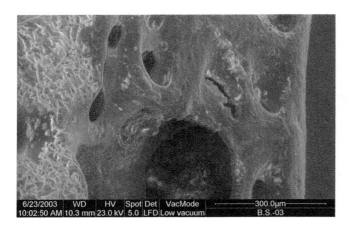

Fig. A.621 SEM microphotograph of a modern amphibian bone fragment from a fox scat (*Vulpes vulpes* A.758 and A.759). It was recovered from a modern cave (specimen provided by Borja Sanchiz). (See Pinto Llona and Andrews 1999 for digestion traits and classification in reptiles and amphibians)

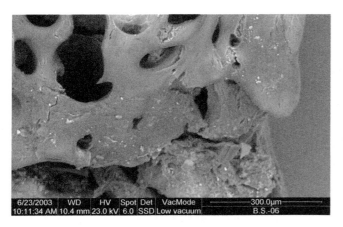

Fig. A.622 SEM microphotograph of a modern amphibian bone fragment from a fox scat (*Vulpes vulpes).* It was recovered from a modern cave (specimen provided by Borja Sanchiz). Surface bone has been lost through digestion (A.619 and A.758). (See Pinto Llona and Andrews 1999 for digestion traits and classification in reptiles and amphibians)

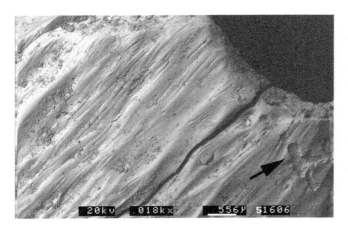

Fig. A.623 SEM microphotograph of a fossil from Vanguard. The fossil bone is strongly rounding by digestion (A.609 and A.867). Root marks are present on the right side (black arrow), and they formed after digestion

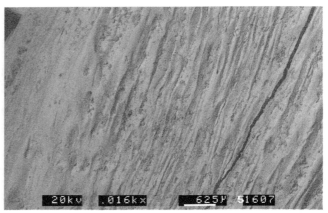

Fig. A.624 SEM microphotograph of a fossil from Vanguard (A.625). Cracking of the bone surface that is characteristic of digestion is present, and this is similar to the digestion in canids and hyenids (A.745 and A.758)

Fig. A.625 SEM microphotograph of a fossil from Vanguard Cave showing a strong rounding aspect (A.624) that is similar to the action of carnivores such as hyenas (A.592). The bone surface appears cracked which is characteristic of digestion, especially in canids and hyenids (A. 745 and A.758)

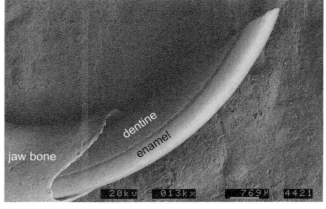

Fig. A.626 SEM microphotograph of an unmodified rodent incisor (see A.544, top left). Enamel covers one side of the dentine, which is in rodents is exposed lingually. This unmodified tooth can be compared with the rodent incisors altered by digestion from the scats and pellets of various predators

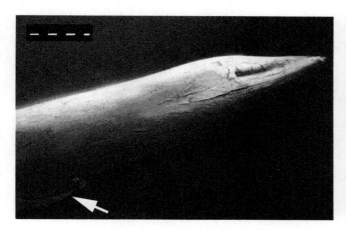

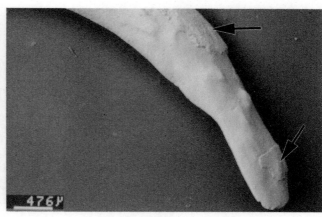

Fig. A.627 SEM microphotograph of a modern rodent incisor digested by an African eagle owl (*Bubo lacteus*). The tip of the incisor shows the dentine to be rounded and the enamel totally removed from the tip. There is a small island of enamel at the bottom left corner of the photograph, indicated by a white arrow (see Text Fig. 8.4). Field width equals 5.3 mm

Fig. A.629 SEM microphotograph of a rodent incisor digested by a polecat (*Mustela putorius*). Most of the enamel has been removed by digestion, but what is left (black arrows) is rounded together with the dentine, having here a wavy aspect. Dentine is usually less affected by digestion (Text Fig. 8.4)

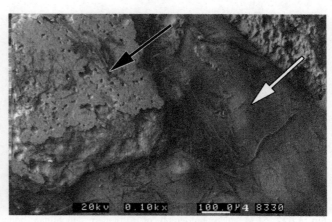

Fig. A.628 SEM microphotograph of a modern rodent upper incisor rounded by digestion. Digestion produces rounding of skeletal tissues, including enamel and dentine, and the tooth is strongly rounded on the tip (seen on the right). Enamel is reduced to small islands on the edge of the incisor (black arrow) indicating a heavy degree of digestion (see Chap. 8 and Andrews 1990 for description of digestion classification in rodents, Text Fig. 8.4)

Fig. A.630 SEM microphotograph of a modern small mammal incisor digested by coyote (*Canis latrans* A.832). Note the wavy and rounded aspect of the dentine surface (white arrow). The enamel also rounded but more characteristically is strongly corroded (black arrow). See also A.616 and A.1023

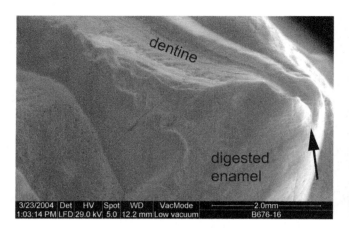

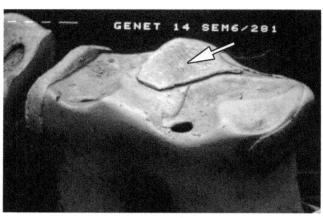

Fig. A.631 SEM microphotograph of a fossil horse incisor from Concud. The salient angles of the enamel are smoothed by digestion (arrow) as also seen on vole molars (see A.841). Abundant evidence of fossil hyena has been found in Concud (see Text Table 2.1), such as coprolites, tooth marks (A.389), and digested small fragments of bone (A.594). Courtesy of D. Pesquero

Fig. A.633 SEM microphotograph of a modern small mammal molar digested by Genet (*Genetta genetta*). Note the strong rounded aspect of the tooth. Digestion preferentially corrodes enamel, unlike the effects of physical abrasion that affects both enamel and dentine. Digestion removes most of the enamel, leaving small islands on the occlusal surface (arrow). The dentine has a wavy and rounded aspect. See also A.768, A.771, A.865 and A.1024. Field width equals 1.7 mm

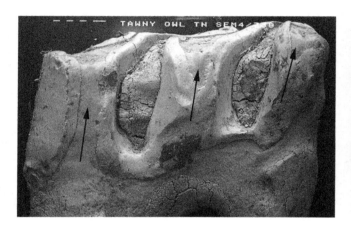

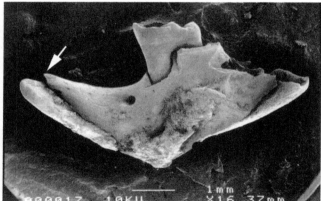

Fig. A.632 SEM microphotograph of a modern rodent molar heavily digested by a tawny owl (*Strix aluco*). The enamel is removed along the salient angles of the crown providing a rounded aspect, as in A.843. The dentine underneath (arrows) remains less affected than enamel, but dentine and enamel are both rounded. See also A.767 and A.833. Field width equals 2.5 mm

Fig. A.634 SEM photograph of a recent rodent mandible. It is heavily rounded by digestion, has affected the tip of the incisor, with the dentine and enamel almost completely removed. The mandibular bone at the edge near the symphysis (white arrow) is also strongly rounded (as in A.850 and A.858), extending down and along the edge of the incisor's alveolar cavity

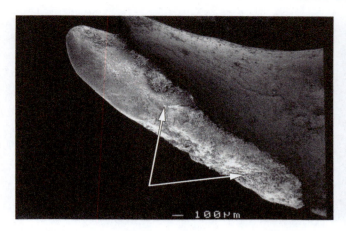

Fig. A.635 SEM microphotograph of A.634, showing the strong rounding of dentine at the tip of the incisor, with a small island of enamel (white arrows). Both rounding and reduction of the enamel are the result of digestion

Fig. A.636 Rounding may also be produced by vegetation. This is a fossil horse incisor from Concud having a whitish surface that has left an irregularly corroded surface on the tooth crown. This has short and irregular root marks on the surface which could have been produced by plants or algae superimposed on another and earlier modification, maybe produced by biofilms (see A.493 and A.502). Courtesy of D. Pesquero

Fig. A.637 SEM microphotograph of a fossil horse incisor from Concud. The smooth (whitish color) surface (white arrow) seen in A.636 is made by root marks, already described at the site (see A.245). The plant that could produce these root marks remains unknown, and root marks are superimposed on a type of bioerosion (black arrow), that was probably produced by a biofilm (an aggregate of microorganisms, see A.493)

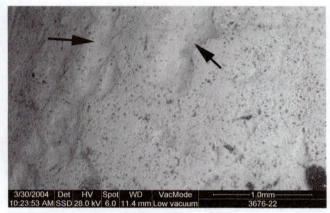

Fig. A.638 SEM microphotograph of a fossil horse incisor from Concud. This photograph shows a less intensive action of the bioerosion that is mildly covering the whole tooth enamel and more restricted areas where root marks (black arrows) acted (see A.245 and A.246)

Part III
Modifications Penetrating Bone Tissue

Chapter 7
Flaking and Cracking

Cracking of the surfaces of bones and teeth is defined as the opening up of splits and cracks penetrating beneath the surface of the bone but without any obvious loss of surface tissue. It is caused by shrinkage of bone, which could be due to loss of water and/or organic matter. On teeth, this shrinkage affects dentine initially, but as the dentine shrinks from loss of collagen and moisture it may be followed by splitting of the enamel. Flaking of the surfaces of bones is defined as loss of surface tissue, exposing underlying bone to varying depths (exfoliation). Flaking is usually associated with cracks in the bone, so that fragments of surface bone are flaked off along the edges of cracks. At this stage, the edges of both cracks and flakes are sharp, but with further flaking the edges become more diffuse or rounded. As discrete flakes increase in size and coalesce, the whole surface of the bone may be lost. Degrees of modification are affected by climate and degree of environmental protection (vegetation, shade or relative humidity).

Agents and Processes

Inorganic processes:

Weathering, A.639, A.644, A.656 and A.663
Fire, A.709 and A.712
Heat (boiling in water), A.690 and A.698
Sediment pressure, A.739
Alkaline environment, A.728, A.730 and A.737

Organic processes:

Digestion, A.744, A.749 and A.771
Root marks, A.775 and A.782

Characteristics

The principal characteristics of flaking and cracking depend on the type of bone affected and the duration of the modifying agent. There are few differences in the morphology of flaking and cracking produced by different taphonomic agents, but there are differences between different classes of vertebrates. Only mammals and amphibians will be considered here: reptiles and birds have patterns that are probably close to those of amphibians, but they have not yet been fully studied. For mammals, distinction is made between large (>5 kg) and small mammals (<5 kg). The extent of flaking and cracking is categorized into stages in the literature (Behrensmeyer 1978), and as neither has yet been quantified we will follow this practice here, first for cracking, since this may precede flaking in some instances, and then flaking.

Inorganic Flaking and Cracking

Cracking of Surface Bone

The formation of cracks in fossil bone is very common, but the agents and processes responsible are often not apparent. The effects of heating by boiling or by fire both produce deep cracking of bone (A.693). Boiling in particular has an overall effect on the surface of the bone that produces cracking first, followed by flaking of the bone surface (A.698 and A.702) (see below and Chap. 5). Direct exposure to fire produces deeper cracking, as there is rapid loss of organic matter and drying out of the bone, but it does not necessarily lead to flaking of the bone surface (A.711). The degree and extent of cracking depends on the duration and

© Springer Science+Business Media Dordrecht 2016
Yolanda Fernández-Jalvo and Peter Andrews, *Atlas of Taphonomic Identifications: 1001+ Images of Fossil and Recent Mammal Bone Modification*, Vertebrate Paleobiology and Paleoanthropology, DOI 10.1007/978-94-017-7432-1_7

temperature of the fire (Shipman et al. 1984; Cáceres 2002; Cáceres et al. 2002).

Cracking also forms in the presence of highly alkaline or acidic conditions in the environment, and as a result of digestion. Alkalinity appears to affect the collagen in the bone, denaturing it and shrinking the bone all over so that extensive cracking occurs (A.726 and A.734). Acidity dissolves the mineral content of bone so that the bone becomes soft and pliable with the loss of the supporting mineral framework. The bone does not shrink to any great extent, however, so that it does not form cracks except at the microscopic level. The cracking from alkalinity initially affects bone, tooth root and dentine, and enamel is not affected. With continued exposure, however, or extreme alkaline conditions, the whole bone is heavily affected. Small mammals are affected in the same way as large mammals. Cracking during digestion appears to be mainly the result of high acidity.

Bones in highly humid conditions, such as in closed environments in caves or lakeshores, may also exhibit cracking to a limited degree (A.717 and A.722). The feature that characterizes cracks by humidity is that the sides of the crack appear elevated and contorted, either both sides or one side only, and this feature is very specific to these highly humid or damp conditions. Cracking may occur along the linear marks produced by plant roots, but little is known about which types of plant produce root marks. It may be that when cracking is present, this may not only identify it as a root mark but may also be characteristic of particular plant/fungal-bacterial associations. At present, it is not certain if this cracking is caused by soil processes, micro-organisms associated with the growing roots, or the micro-chemical environment generated by the roots.

Probably the most common cause of cracking of bone is weathering. By this we mean exposure of bones on the surface of the ground to elements of the weather (sun, wind, relative humidity, temperature and rain). We have identified other forms of modification that are sometimes referred to as 'weathering' as corrosion (Chap. 8): corrosion penetrates deeply into bone on exposed edges, exposing the underlying bone, but weathering progresses through a number of surface stages as defined by Behrensmeyer (1978) for the tropical, semi-arid environment at Amboseli in Kenya. The first stage is cracking of the bone as the organic matters decay and the bone shrinks. This is Behrensmeyer's stage 1 weathering (Table 7.1) as applied to large mammal bone, and the cracking generally follows the fiber structure of the bone. The same is seen in small mammals, although the cracking is generally finer in scale (Andrews 1990). Cracking is followed by flaking of the bone surface and gradual loss of bone tissue. The two sets of stages, for large and small mammals, are shown in Table 7.1 with the time scale at which the different stages occur. Stage 1, for instance, develops in <1 to 3 years on large mammals in the tropical environment that Behrensmeyer (1978) studied, while on small mammals in the temperate environment that Andrews (1990) studied, stage 1 cracking develops in 1 to 2 years. Once the bone is penetrated by the cracking, break-down of bone tissue proceeds rapidly, as Behrensmeyer (1978) noted. She showed that mammals less than 100 kg appear to weather more rapidly than larger mammals in tropical environments, and we have found that small mammal bones in temperate conditions have essentially disintegrated after 5–7 years.

Large mammal weathering has also been studied in temperate environments in Wales A.657 and in Spain A.658 (Andrews and Armour-Chelu 1998). The cool temperate environment at Neuadd in Wales (52° N) has been monitored over the past 30 years, and after this time period the greatest degree of weathering on the least sheltered specimens is only stage 2. In most cases, bones were buried, protected by vegetation (providing shade and humidity) or dispersed long before they had reached this stage, and it was only by removing bones from their place of deposition and exposing them on platforms where they could not be buried or removed by scavengers that higher degrees of weathering could be achieved. In this situation, bones showed weathering changes as follows:

- Temperate stage 1 weathering, 4–10 years at Neuadd (surface cracking), at which time bones in the tropics were all at stages 3 and 4.
- Temperate stage 2 weathering, 10–15 years at Neuadd (incipient flaking of bone surfaces), while bones exposed on the surface in tropical environments had been destroyed after 15 years.
- Temperate stage 3 weathering, not reached at Neuadd after 30 years exposure, latitude 52° North.
- Temperate stage 3 weathering, 15–30 years at Riofrio, latitude 41° North.
- Bones that were buried or were covered by vegetation were still at stage 0 after 30 years.

Table 7.1 Weathering stages for large and small mammals

Stage	Behrensmeyer large mammal categories	Years	Andrews small mammal categories	Years
0	No modification, bone still greasy, marrow present, skin or other tissue may still be present	0–1	No modification, bones may still be connected together	0–2
1	Cracking parallet to fibre structure, articular surfaces may have mosaic cracking	0–3	Slight cracking parallel to fibre structure, chipping of enamel and cracking of dentine	0–5
2	Concentric flaking associated with cracks and loss of most of outer bone, crack edges angular	2–6	More extensive cracking but little flaking; of enamel and deep dentine cracks extending into the enamel leading to loss of parts of the crown	3–5
3	Bone surface with patches of rough compact bone resulting in fibrous structure extending to cover entire surface, penetration up to 1.5 mm, crack edges rounded	4–15	Deepening cracks in both bone and dentine of teeth; cracks deep in enamel	4–5
4	Bone surface is coarsely fibrous and rough, splinters Falling away, cracks open and have rounded edges	6–15	Deep cracks and some loss of segments of bone; sections of enamel broken away	4–5
5	Bone falling apart in situ with splinters of bone around the remaining bone core which is fragile and easily broken	6–15	Bone split apart and disintegrating, teeth fragmented	5–7

Categories taken from Behrensmeyer (1978) and Andrews (1990)

Weathered bones from Neuadd show cracking that only penetrates a few microns into the bone. This shallow penetration has also been observed in Riofrio in Central Spain, a more open and drier environment where bones may reach weathering stage 3. Riofrio is 11° latitude south of Neuadd, and this is likely to have influenced this result. Weathering is thus slower and more superficial in temperate and wet climates than has been observed near the equator. There is some indication that bones exposed in tropical rain forest (Virunga National Park 0.5° latitude), on sites that are permanently or mostly protected by vegetation, also remained unweathered after 10 years. This may be related to protection from direct UV sun exposure and low temperature variation, already suggested by Tappen (1994), but see below.

Another long term study investigated large mammal weathering in a subtropical desert environment at Jebel Barakah (A.640) (Abu Dhabi, 24° latitude, Andrews and Whybrow 2005). This shows similar stages to the tropical pattern, not surprising since insolation is strong in both areas, but the timing of the weathering stages is different. Figure 7.1 compares the Abu Dhabi weathering pattern, based on the bones from a single individual of camel, with the much larger Amboseli sample and shows that after two years exposure, most of the Abu Dhabi bones were still at stage 0 while all of the Amboseli bones were at stages 1 and 2. Similarly after 4 to 8 years exposure, the Abu Dhabi bones were at stages 0 and 1 in approximate ratio of 3:1, but all the Amboseli bones were at stages 3 and 4. The weathering profile in Wales described above is similar to the Abu Dhabi one thus far, but at 10 to 15 years, the latter had a distribution of stages 1 to 3 (Fig. 7.1) while the Welsh bones were still mostly at stage 1 and only a few at stage 2, and the Amboseli bones at stages 3 to 5. Furthermore, we have observed that the effects of weathering on bones from temperate environments does not alter the organic matrix (Fernández-Jalvo et al. 2010) as significantly as on specimens from the Behrensmeyer collection from Amboseli (Trueman et al. 2004), where weathering was related to destruction of the organic matrix of bone (collagen), thus facilitating dissolution and remineralization. The onset of weathering is clearly a complex issue depending on both the local climate and the type of bone, and much more data are needed before the weathering morphologies can be related to environment and especially to time (see Lyman and Fox 1989).

Many fossils are found with cracks running through the bone, which may or may not reflect pre-fossilization weathering. These may result in the breakage of the bones, in which case the fragments may either remain in place held together by the sediment (see Chap. 9) or may become separated by subsequent sediment movement. In other cases, the cracks may be infilled with calcite and the fragments separated by the cracks fused by the calcite. As noted earlier, fossil bone with deep cracks may appear intact in the sediment but may break up during excavation or screening.

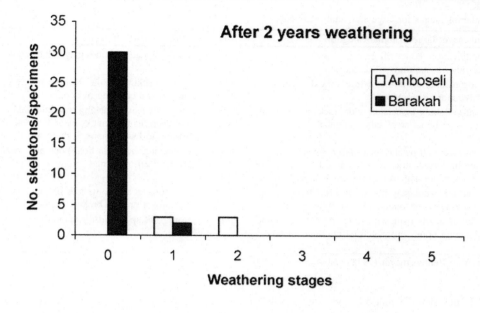

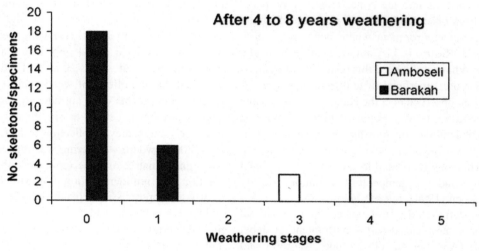

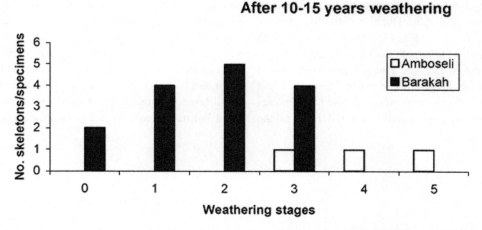

Fig. 7.1 Weathering stages of bones from a subtropical environment in Abu Dhabi are compared with the tropical weathering profile described by Behrensmeyer (1978). The weathering stages are shown on the horizontal axis, and the numbers of skeletons (Amboseli) or skeletal elements (Abu Dhabi) present in each after three periods of weathering: after 2 years, after 4–8 years, and after 10–15 years

Flaking of Surface Bone

Weathering often proceeds on exposed bones from cracking to flakes peeling from the cracked edges of the surface of the bones (A.641). Much of the outer surface bone may still be present at stage 2 weathering (Behrensmeyer 1978), but most of it has gone by stage 3. In other words the flaking of the surface is so advanced that little remains of the surface bone. It is common for the same bone to have two grades of weathering if they are in a stable position, one on the exposed side and another on the protected side, and it is also possible for different ends of a single bone to be at different weathering stages, for example if one end is buried and the other end exposed. At stage 4 weathering, segments of the bone have started to peel off so that the bone has begun to fall apart. The final stage of this for large mammals is the complete disintegration of the bone in stage 5.

Flaking of surface bone is not a general phenomenon in small mammal limb bones, and based on our small sample it appears that bones subject to taphonomic processes pass straight from cracked surfaces to splitting apart and finally disintegration. Rodent femora exposed for 4–5 years show extensive cracking but little or no flaking (Table 7.1). Rodent molars develop cracking in the dentine, and as the cracks expand the cracking spreads into the enamel at stage 2. On the other hand, cranial elements may show extensive flaking, with large flakes becoming detached from the bodies of rodent mandibles exposed to weathering, with similar flaking from the rodent crania.

Other processes such as boiling may cause flaking directly on the bone. The difference compared to flaking produced by weathering is that boiled bones show flaking of the outer cortical layer has the edges of the flakes curled up (A.706) (looking like mud-cracks in drying soil (A.707), Fernández-Jalvo and Marin-Monfort 2008). Flaking has also been observed on bones exposed to highly alkaline environments, experimentally reproduced at extreme pH (~ 14) to accelerate the process and mainly observed on fossils from limestone caves and from alkaline volcanic sediments (e.g., Olduvai, Tanzania). This modification has been named desquamation (Fernández-Jalvo 1992; Fernández-Jalvo et al. 1998), to avoid confusion with weathering and it is characterized by bone surfaces obtaining superficial exfoliation without previous cracking or splitting.

Organic Cracking

Microscopic cracking occurs on digested bones and along the inner surfaces of some types of root marks. Both cases contribute additional evidence to distinguish these modifications. Digestion that produces rounding is distinguishable from abrasion by a characteristic torn flaked surface (A.745 and A.750). The surface of the bone has the appearance of being torn apart, with fibers separated and with longitudinal as well as lateral displacement of surface tissue. The digested bone may be both rounded (see Chap. 6) and cracked, and this characteristic microscopic "torn" cracked surface is an important identifying feature that distinguishes digestion from fluvial abrasion. The presence of microscopic cracks in the interior of the root marks may be caused by different type of plants, although experimental work is currently in progress to substantiate this. These microscopic cracks are one of the identifying characteristics of root marks.

Atlas Figures

A.639–A.785

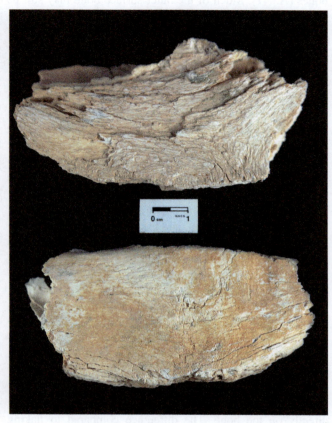

Fig. A.639 Fragmentary recent camel mandible from Jebel Barakah. There is heavy flaking of the surface of the buccal side (top), at stage 3 weathering (Behrensmeyer 1978), and a lesser degree of flaking on the lingual side (bottom), which was the down side when the specimen was collected, stage 2 weathering (as in A.640). The mandible was at the entrance to a hyena den and was being transported down slope from the den. Compare with A.651 and A.688

Fig. A.640 Fragment of recent camel mandible from Jebel Barakah. It was exposed for an unknown number of years. The buccal side (upper figure) shows extensive flaking on the surface: advanced stage 3 weathering (Behrensmeyer 1978). The lingual side (lower figure) has a lesser degree of flaking: incipient stage 2 weathering (as in A.639)

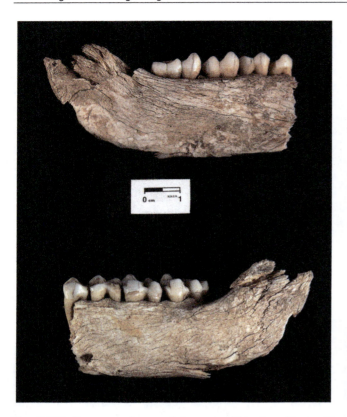

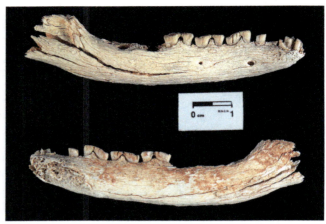

Fig. A.643 Recent mandible of fox from Qatar at a high stage of weathering with teeth flaked and cracked. Weathering is very similar on both sides, as in A.651

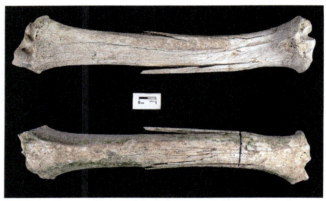

Fig. A.641 Fragment of recent baboon mandible from Kenya rift valley (A.645). It has incipient flaking and deep cracks on the surface of both buccal and lingual sides (advanced stage 2 weathering). (Behrensmeyer 1978)

Fig. A.644 Recent radius of sheep collected from, Barranco de las Ovejas, Huesca (Pyrenees, Spain). It was exposed for an unknown length of time, with cracking and flaking and splitting of bone segments: stage 3 weathering. See A.658

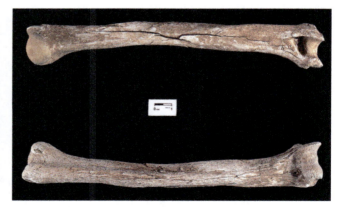

Fig. A.645 Recent humerus of baboon from the Kenya rift valley. It was exposed for an unknown period in a tropical environment with extensive flaking and the bones splitting apart: stage 2 weathering. Compare with A.653 and A.654

◄ **Fig. A.642** Recent hyrax mandible *Procavia* sp. from Kenya near Kajiado with weathering cracks on both sides: advanced stage 1 (Behrensmeyer 1978), see also A.655 and A.689

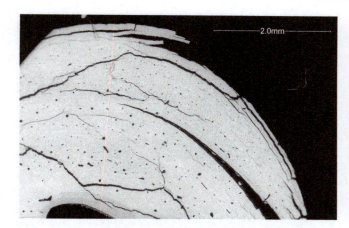

Fig. A.646 SEM microphotograph of a sectioned bone of a recent baboon humerus (A.645). It has deep weathering cracks penetrating through the cortical bone. Cracks and flaking producing splinters of bone indicate stage 2 weathering as identified by surface modifications (Behrensmeyer 1978)

Fig. A.648 Fossils uncovered on the ground at Langebaanweg by seasonal rains and wind (A.999 and A.1078). Fossils are then exposed to modern weathering, which produces breakage, with the bones splitting into many fragments. This is the final stage 5 of weathering as defined by Behrensmeyer (1978). Fossil bones become weathered in a manner difficult to distinguish from pre-burial weathering

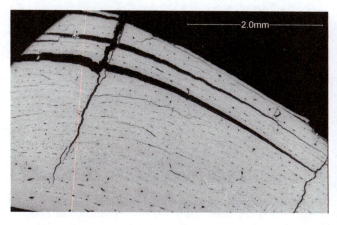

Fig. A.647 SEM microphotograph of a sectioned bone of a recent baboon femur (A.649). This shows the depth to which weathering reaches through the cortical bone. Cracks and flaking producing splinters of bone are more superficial than seen in the previous section indicating stage 2 as identified on surface (Behrensmeyer 1978)

Fig. A.649 Recent femur of baboon from the Kenya rift valley. It was exposed for an unknown period in a tropical environment with extensive flaking and some splitting longitudinally along shaft: stage 2 weathering (see A.647)

Fig. A.650 Recent vertebra of camel from a desert environment in ▶ Qatar. It was exposed for an unknown number of years and heavily weathered on all sides: stage 3 or 4 weathering. Weathering is very similar on both sides, as in A.643 and A.651

Fig. A.651 Recent astragalus of a camel from Qatar. It was exposed for an unknown length of time and is extremely weathered and also with rounded edges. Weathering is strong on both sides of the bone in contrast to A.639 and A.640

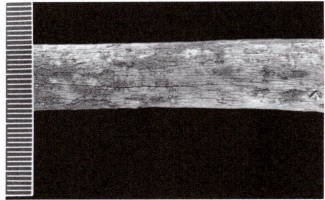

Fig. A.654 Recent bone fragment from Qatar. It was exposed for an unknown length of time and is heavily weathered with splitting and cracking on both sides of the bone fragment. Weathering stage 2/3, as in A.641, A.652 and A.653

Fig. A.652 Recent bone fragments from Qatar. It was exposed for an unknown length of time, is extremely weathered and also broken by weathering. Note, however, differential stage of weathering between one and the other side of the fragment. Weathering stage 2, as in A.640, A.653 and A.654

Fig. A.655 Monitored recent limb bone from Neuadd exposed for eight years, with early stage 1 cracking and flaking of the bone surface, as in A.642, A.657 and A.669. The bone was exposed on the surface in a cool temperate climate at latitude 52° North

◄ **Fig. A.653** Recent bone fragment from Qatar. It was exposed for an unknown length of time and is heavily weathered, with splitting and cracking on both sides of the bone fragment. Weathering stage 2, as in A.639, A.652 and A.654

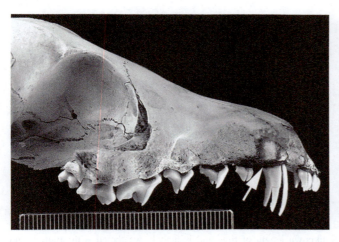

Fig. A.656 Skull of a modern fox. The bone is still greasy and only sutures appear detached (stage 0 of weathering) The teeth show strong cracking and splitting, and part of the gum remains between the tooth-maxilla (white arrow) protecting this area from further modifications (see A.943, A.944 and A.945)

Fig. A.658 Barranco de las Ovejas (the Sheep Gully, Huesca Pyrenees, Spain). Carcasses are abandoned by the local shepherds and scavenged by Lammergeier or Bearded Vulture, *Gypaetus barbatus*. The radius at the bottom of the photo is at stage 3 weathering (A.644) and the mandible in advanced stage 2. Linear marks are present on the mandible (square, see A.202) and puncture marks (white arrow, see A.404) on the scapula, all made by the vultures' beaks

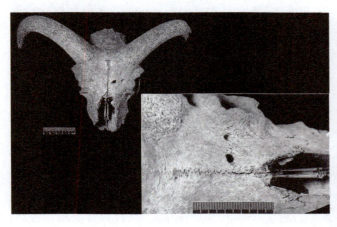

Fig. A.657 Monitored sheep skull from Neuadd. There is fine flaking of the surface bone, but after being exposed for more than 18 years, it is still at weathering stage 1, as in A.642, A.655 and A.669. The skull was exposed on the surface of the ground with no shade or vegetation in a cool temperate climate at latitude 52° North

Fig. A.659 Skull of zebra from Tswalu Kalahari Reserve (South ▶ Africa). The tooth enamel is cracked by weathering, as in A.656. The carcass was at a hyena feeding place and it was no longer than a few months at that location

Fig. A.660 Modern long bone from Riofrío. It is weathered at stage 1 on surface (as in A.655) and broken during breakage experiments. The fresh appearance of the interior of the bone shows that weathering has not penetrated deep into the bone

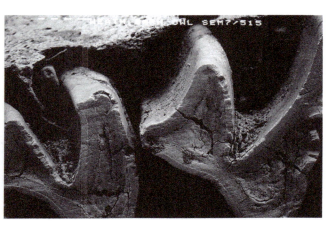

Fig. A.663 SEM microphotograph detail of A.662. The cracked occlusal surface of modern vole molars has been exposed for 18 months in the cool temperate climate of Neuadd. Cracks in the dentine are beginning to split the enamel. Field width equals 2.3 mm

Fig. A.661 Two bones found close each other at Riofrío. They are at different stages of weathering: the long bone (top) shows advanced stage 2–3 and the pelvis (bottom) has incipient stage 1. Different weathering stages are also seen in A.658

Fig. A.664 SEM microphotograph detail of A.663 showing the occlusal surface of a modern monitored vole molar exposed for 18 months in the cool temperate climate of Neuadd. Both dentine and enamel are cracked. Field width equals 1.2 mm

Fig. A.662 SEM microphotograph of modern rodent teeth from Neuadd (A.663, A.664, A.665, and A.666). They are at early stage 1 weathering, with cracks in the dentine starting to penetrate into the enamel: 18 months exposure in the cool temperate climate (Andrews 1990). Field width equals 4.8 mm

Fig. A.665 SEM microphotograph of part of the occlusal surface seen in A.663 of a modern monitored vole tooth exposed for 18 months in the cool temperate climate of Neuadd. There is slight chipping of enamel edges, and the dentine shows moderate cracking. Field width equals 397 microns

Fig. A.666 SEM microphotograph detail of part of the occlusal surface of A.662, a modern monitored vole molar. It was exposed for 18 months in the cool temperate climate of Neuadd. The enamel is slightly detached from the dentine. Field width equals 1.2 mm

Fig. A.669 SEM microphotograph of modern monitored rodent femur with very early stage 1 weathering cracks (A.670). It was exposed for 18 months in the cool temperate climate of Neuadd. The bone surface and femur head show slight cracking parallel to fiber arrangement. Field width equals 3 mm

Fig. A.667 SEM microphotograph of modern monitored vole maxillary suture, detached after exposure for 18 months in the cool temperate climate of Neuadd (see A.662). Field width equals 5.4 mm

Fig. A.670 SEM microphotograph of a close up view of A.669: the monitored femur head has a network of fine cracks after exposure for 18 months in the cool temperate climate of Neuadd. Field width equals 1.2 mm

Fig. A.668 SEM microphotograph of an upper incisor of modern monitored vole exposed for 18 months in the cool temperate climate of Neuadd. The enamel is cracked longitudinally along the length of the incisor. Compare with A.680. Field width equals 5.2 mm

Fig. A.671 SEM microphotograph of a monitored femur of modern small mammal from Neuadd. It was exposed for 18 months. The bone surface and femur head show no cracking: weathering stage 0, as in A.673. Field width equals 3.1 mm

Fig. A.672 SEM microphotograph of the occlusal surface of a modern monitored vole molar exposed for 29 months in the cool temperate climate of Neuadd. The body of the mandible has stage 1 weathering, the molar enamel is cracked and the dentine cracking is extensive, leading to the loss of tissue (compare with A.662). Field width equals 4.3 mm

Fig. A.673 SEM microphotograph of a distal end of humerus of a modern monitored small mammal after 29 months exposed on the ground in the cool temperate climate of Neuadd. Almost no modification can be observed, apart from roots covering the bone surface: weathering stage 0, as in A.671. Field width equals 2.4 mm

Fig. A.674 SEM microphotograph of the proximal end of a femur from Neuadd. The modern monitored small mammal was exposed for 29 months in the cool temperate climate of Neuadd. Cracks are evident on the bone surface and there is some flaking: stage 2 weathering (as in A.678). Field width equals 1.2 mm

Fig. A.675 SEM microphotograph of a modern monitored rodent femur exposed for 29 months in the cool temperate climate of Neuadd. The femur head has mosaic cracking from weathering: stage 1 weathering (as in A.669). Field width equals 2.6 mm

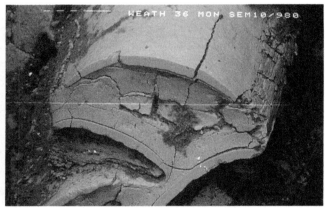

Fig. A.676 SEM microphotograph of part of an occlusal surface of modern monitored vole molar from Neuadd. It was exposed for 36 months in the cool temperate climate of Neuadd, showing more intense cracking of the dentine than in A.672, passing laterally into the enamel and along the height of the crown. Field width equals 1.2 mm

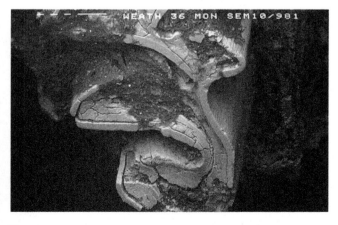

Fig. A.677 SEM microphotograph of part of the occlusal surface of a modern monitored vole molar (see A.676). It was exposed for 36 months in the cool temperate climate of Neuadd. There is extensive chipping of the enamel, and the dentine is more cracked and detached from the enamel than in previous stages. Field width equals 1.1 mm

Fig. A.678 SEM microphotograph of the ascending ramus of a modern monitored rodent mandible from Neuadd. It was exposed in the temperate environment of Neuadd for 36 months. The bone has deep splits and loss of tissue within and between the splits: stage 2/3 weathering (see A.679). Field width equals 5.8 mm

Fig. A.681 SEM microphotograph of part of the incisor of A.680. Enamel and dentine show deep splitting along the length of the incisor after 5 years of exposure in the cool temperate climate of Neuadd: stage 3 weathering

Fig. A.679 SEM microphotograph of the ascending ramus (detail of A.678) of a modern monitored rodent mandible from Neuadd. It was exposed in a temperate environment of Neuadd for 36 months. The bone has deep splits and loss of tissue within and between the splits: stage 2/3 weathering (compare with A.688). Field width equals 2.9 mm

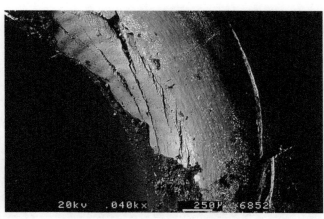

Fig. A.682 SEM microphotograph of part of the incisor of A.680. There is deep splitting in the dentine and deep cracks and splitting on the enamel of the incisor after 5 years exposure in the cool temperate climate of Neuadd: stage 3 weathering

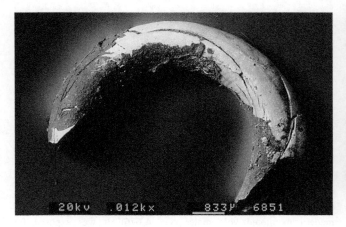

Fig. A.680 SEM microphotograph of a monitored rodent incisor from Neuadd (A.681 and A.682). It was exposed for 5 years in the cool temperate climate of Neuadd, with deep splitting running along the length of the incisor: stage 3 weathering

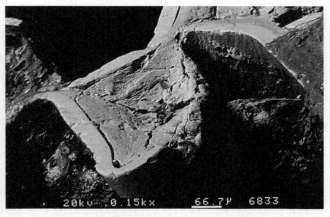

Fig. A.683 SEM microphotograph of a modern monitored vole molar from Neuadd. The molar was exposed for 5 years in the cool temperate climate of Neuadd. Both enamel and dentine are extensively chipped and cracked, with collapse of dentine, which is characteristic of stage 3 weathered teeth (compare with A.672 and A.662)

Fig. A.684 SEM microphotograph of fossil vole molar from Gran Dolina (TD10, Atapuerca). It has extensively chipped and cracked enamel with collapse of dentine, similar to the experimental weathered specimen seen in A.683. This is equivalent to stage 3 weathering. Field width equals 428 microns

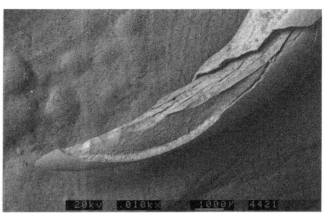

Fig. A.687 SEM microphotograph of a fossil small mammal lower incisor from Vanguard Cave. Damage on the dentine has produced deep splitting, and deep cracks and flaking on the enamel. Weathering could cause this as previously seen on experimental specimens (A.682)

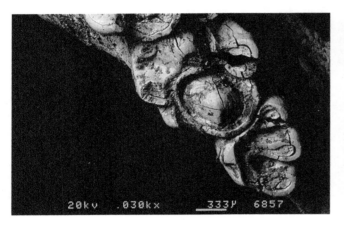

Fig. A.685 SEM microphotograph of modern monitored murid molars from Neuadd. The specimen was exposed for 5 years in the cool temperate climate of Neuadd, and it is extensively cracked through both the dentine and the enamel of the molars. There is little comparative data on murid teeth, which have different molar morphology and enamel structure from vole teeth, but this could be provisionally categorized as stage 2 weathering (compare with A.677)

Fig. A.688 SEM microphotograph of a modern monitored vole mandible exposed in the temperate environment of Neuadd for 5 years. Deep splitting and some loss of deep segments or "flakes" between splits is present, indicating stage 3 weathering, as in A.639 and A.640 (compare with A.679)

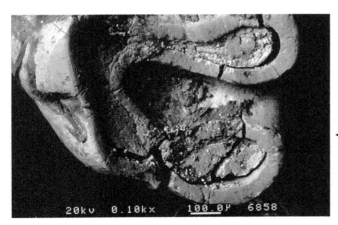

◀ **Fig. A.686** SEM microphotograph detail of A.685, a modern monitored murid molar from Neuadd. The specimen was exposed for 5 years in the cool temperate climate of Neuadd, showing enamel and dentine extensively cracked. The dentine is detached from the enamel. There is little comparative data on murid teeth, which have different molar morphology and enamel structure from vole teeth, but this could be provisionally categorized as stage 2 weathering

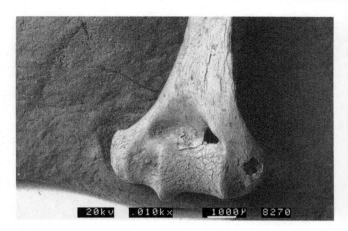

Fig. A.689 SEM microphotograph of a modern distal rodent humerus. It was exposed for an unknown number of years on open savanna near Olduvai Gorge. The bone surface has deep splitting and flaking on the diaphysis and mosaic cracking on the articulation: stage 1 weathering (as in Figs. A.669 and A.642)

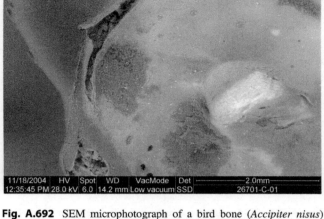

Fig. A.692 SEM microphotograph of a bird bone (*Accipiter nisus*) from the taxidermic collection of the MNCN after boiling. Incipient cracks have been formed on the bone surface (see A.705). The dark grayish color is fat still adhering to the bone

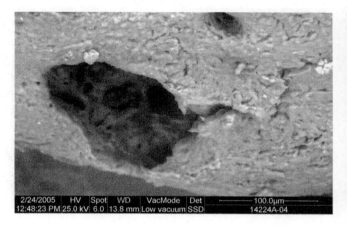

Fig. A.690 SEM microphotograph of a small mammalian carnivore bone (*Genetta genetta*) from the MNCN taxidermic collection (A.691). The bone surface appears flaked and cracked after boiling in water (100 °C) for several hours, to detach the meat for taxidermic preparation (compare with A.702)

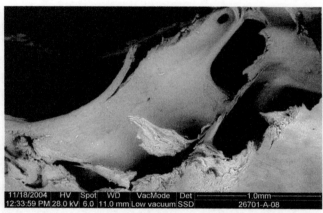

Fig. A.693 SEM microphotograph of deep cracks on a bird bone (*Accipiter nisus*) from the taxidermic collection of the MNCN after boiling (A.692)

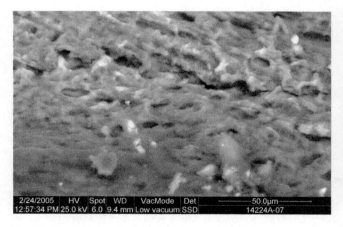

Fig. A.691 SEM microphotograph detail of A.690 of a small mammalian carnivore bone (*Genetta genetta*) from the MNCN taxidermic collection. The texture obtained after boiling has a characteristic curly flaked surface, as in A.694 (Fernández-Jalvo and Marin-Monfort 2008)

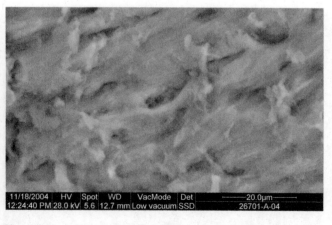

Fig. A.694 SEM microphotograph of a bird bone (*Accipiter nisus* A.691) from the MNCN taxidermic collection after boiling. The bone surface has acquired a characteristic curly flaked surface. (Fernández-Jalvo and Marin-Monfort 2008)

Fig. A.695 SEM microphotograph of a bird bone (*Accipiter nisus*) from the MNCN taxidermic collection after boiling. The bone surface has increased in porosity, and fat (grayish darker patches) is embedded into the microstructure (compare with A.694) (Fernández-Jalvo and Marin-Monfort 2008)

Fig. A.696 SEM microphotograph of a bird bone (*Aquila adalberti*) from the MNCN taxidermic collection after boiling (detail of A.695). Grayish darker spots are fat embedded into the bone microstructure (Fernández-Jalvo and Marin-Monfort 2008)

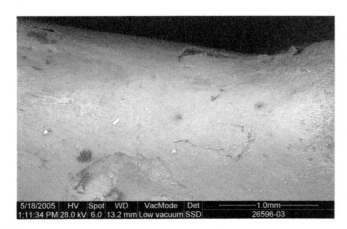

Fig. A.697 SEM microphotograph of a bird bone (*Aquila adalberti*) from the MNCN taxidermic collection after boiling. The surface appears superficially flaked, as in A.698 and A.705

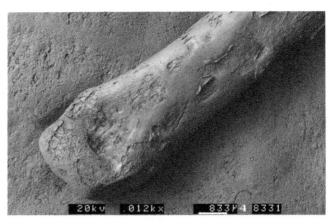

Fig. A.698 SEM microphotograph of a modern rabbit bone (*Oryctolagus cuniculus*) boiled in water. Damage is stronger on this specimen than on the bird bone from the MNCN taxidermic collection (A.697). It has a heavily flaked curled up surface. Both were boiled, but the difference in modification was either caused by a longer exposure to boiling in this specimen or to different bone physiology in birds and mammals. Specimen provided by S. Parfitt

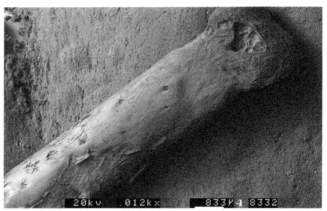

Fig. A.699 SEM microphotograph of a modern rabbit bone (*Oryctolagus cuniculus*) boiled in water. Damage is already evident at low magnification, showing a flaked curled up surface (A.698). Specimen provided by S. Parfitt

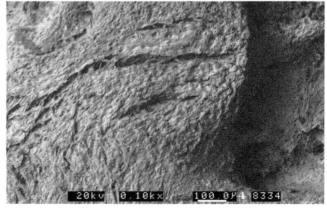

Fig. A.700 SEM microphotograph detail of A.699 of a modern rabbit bone (*Oryctolagus cuniculus*) boiled in water. Cracks penetrate the bone surface. Specimen provided by S. Parfitt

Fig. A.701 SEM microphotograph of a modern rabbit bone (*Orycto-lagus cuniculus*) boiled in water. It has superficial cracks and a flaky surface, as in A.700, A.702 and A.703. Specimen provided by S. Parfitt

Fig. A.702 SEM microphotograph of a modern rabbit bone (*Orycto-lagus cuniculus*) boiled in water. Flakes of bone have come off from the surface, as in A.700, A.701 and A.703. Specimen provided by S. Parfitt

Fig. A.703 SEM microphotograph of a modern rabbit bone (*Orycto-lagus cuniculus*) boiled in water. Flakes of bone have come off from the surface, as in A.700, A.701 and A.702. Specimen provided by S. Parfitt

Fig. A.704 SEM microphotograph of a modern rabbit bone (*Orycto-lagus cuniculus*). After boiling the bone becomes more fragile, cracked and broken, as in A.692. Specimen provided by S. Parfitt

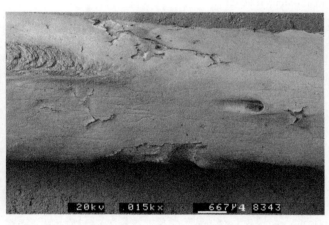

Fig. A.705 SEM microphotograph of a modern rabbit bone (*Orycto-lagus cuniculus*). After boiling the bone becomes more fragile and the surface develops a flaky texture, as in A.697, A.698 and A.708. Specimen provided by S. Parfitt

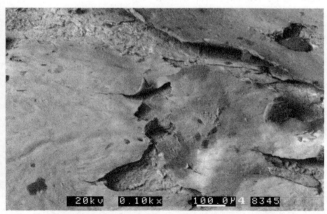

Fig. A.706 SEM microphotograph of a modern rabbit bone (*Orycto-lagus cuniculus*). After boiling, thin bone flakes curl-up, looking like mud cracks in drying soil as A.707 (Fernández-Jalvo and Marin-Monfort 2008). Specimen provided by S. Parfitt

Fig. A.707 Mud cracks at Langebaanweg (compare with A.706 and A.708). Courtesy of A. Louchart and J. Carrier

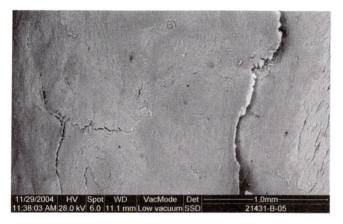

Fig. A.708 SEM microphotograph of a modern bone from the MNCN taxidermic collection. Detail of bone cracking caused by boiling after lengthy natural maceration. Thin flakes of bone curl-up from the bone surface, as in A.705, A.706 and A.698

Fig. A.709 Experimental burnt bones (metapodials of pig). Both show heavy cracking on the articular surfaces produced by exposure to the fire. On the left calcined bone (category 5 according to Text Fig. 5.2 and A.522) exposed directly to the flames and on the right metapodial burnt to grade 4 (grayish) located on fire embers but not directly exposed to the flames

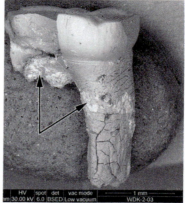

Fig. A.710 Fossil rodent molar from Wonderwerk Cave (South Africa) cracked. The black staining on the color image is due to both burning and manganese deposition. The SEM image on the left shows localized manganese deposition which has higher electron emission (lighter color, A.513) indicated by the black arrows. The rest of the black staining has no manganese and is due to burning, which has also produced cracking on the root and enamel (see A.511 and Text Fig. 5.3)

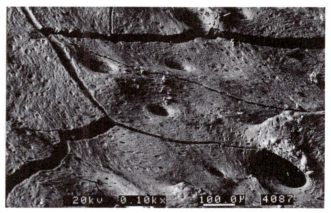

Fig. A.711 SEM microphotograph of modern experimentally burnt bone, detail of cracking of the bone surface by exposure of bone to fire. Detail of fissures, as in A.710 and A.712

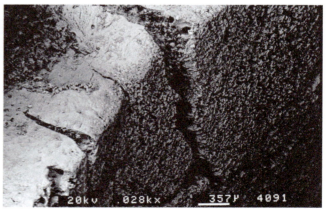

Fig. A.712 SEM microphotograph of modern experimentally burnt bone, detail of cracking due to exposure to fire shows that it penetrates deep into the bone. The sides of the cracks are separated and indicating fast dehydration. These cracks differ from those caused by boiling (compare with A.702) or by changes in humidity (compare with A.717 and A.722) because they are not distorted and are at the same level (similar to weathering cracks)

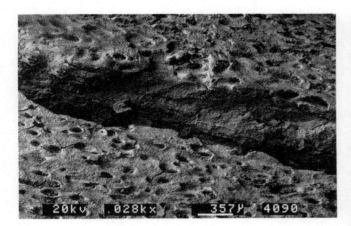

Fig. A.713 SEM microphotograph of modern experimentally burnt bone, detail of the sides of a crack produced by direct exposure to flames. Note the edges are at the same height, and they do not curl up or appear warped (elevated or contorted). Edges are separated and even (compare with A.702 and A.706). This also occurs in bones cracked by weathering

Fig. A.714 SEM microphotograph of modern experimentally burnt bone, detail of cracking on articular surfaces. Mosaic cracking due to exposure to fire is present on articular surfaces (see Figs. A.709, A.710 and A.711)

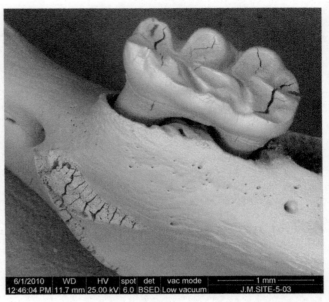

Fig. A.715 SEM microphotograph of rodent mandible from Forat de la Conqueta (Neolithic site at Lerida, Spain). Small mammals are associated with cremated human remains. Cracking is present on the body of the mandible (lower left) and on the dentine exposed on the occlusal surface of the molar, as in A.710 and A.716. The dentine cracking extends laterally into the enamel similar to the effects of weathering (A.685). Specimen provided by J. Martínez

Fig. A.716 SEM microphotograph of a rodent molar from Forat de la Conqueta (Neolithic site at Lerida, Spain). Small mammals are associated with cremated human remains, and the rodent bones and teeth have cracked surfaces suggesting they have been exposed to fire, as A.710 and A.715. The edges of fire-produced cracks on bone surfaces dental cementum do not curl up or appear warped. Specimen provided by J. Martínez

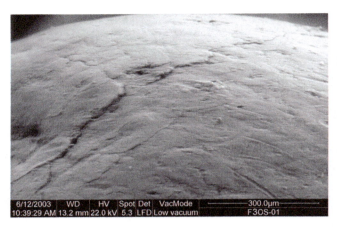

Fig. A.717 SEM microphotograph of fresh bone fragment after exposure to water during several weeks immersed in water. One side of the crack is elevated (curl up, but not as much as A.706) and contorted or warped, as in A.722

Fig. A.720 Fossil equid third phalanx from Senèze showing cracking by humidity characterized by curled up edges, as in A.718, A.719 and A.721

Fig. A.718 Fossil equid sesamoid from Senèze. The crack was produced by high humidity, and it is characterized by curled up edges along the crack, as in A.719, A.720 and A.721. The site is a paleo-lakeshore environment, and most of these fossils were immersed in water for long periods

Fig. A.721 Fossil equid mandible from Senèze showing cracking by humidity characterized by curled up edges, as in A.718, A.719 and A.720

Fig. A.719 Fossil equid third phalanx from Senèze showing cracking by humidity characterized by curled up edges, as in A.718, A.720 and A.721

Fig. A.722 SEM microphotograph showing cracking of fossil bone from Gough's Cave probably caused by changes in humidity of the cave environment, see A.717

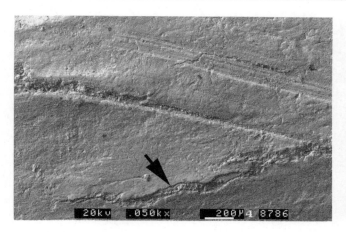

Fig. A.723 SEM microphotograph showing cracking of fossil bone (black arrow) from Gough's Cave probably caused by changes in humidity of the cave environment, see A.717 and A.722

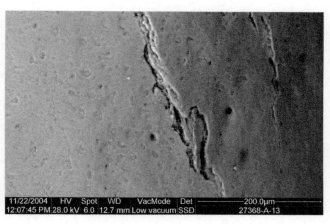

Fig. A.726 SEM microphotograph, detail of A.725, of a bird bone (*Athene noctua*) after taxidermic preparation with KOH. It has cracks with warped (elevated and contorted, A.724 and A.728) edges (Fernández-Jalvo and Marin-Monfort 2008)

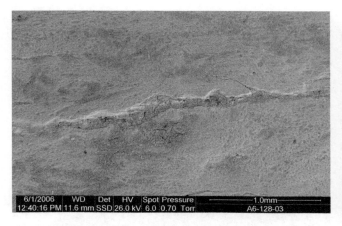

Fig. A.724 SEM microphotograph of a fossil bone from Gough's Cave. Notice the sides of the crack are warped (elevated and contorted), probably caused by alkaline environments and the high humidity of the cave environment, see A.717 and A.722

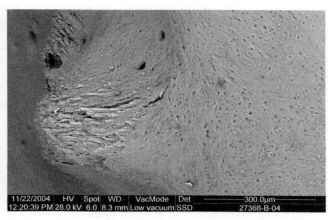

Fig. A.727 SEM microphotograph of modern bird bone, detail of cracking concentrated on bony protuberances. This resulted from preparing skeletons in alkaline solutions (KOH) A.726

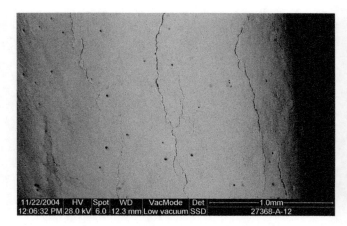

Fig. A.725 SEM microphotograph of a bird bone (*Athene noctua*) after taxidermic preparation with KOH. It has cracks with warped (elevated and contorted) edges (Fernández-Jalvo and Marin-Monfort 2008) not as much as A.708

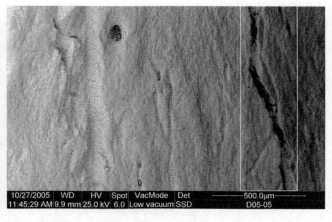

Fig. A.728 SEM microphotograph of a fossil from Azokh cave. It is penetrated by a deep crack, the edges of which are warped (elevated and contorted). This trait is characteristic of environments with high humidity (A.717 and A.722). Similarities with KOH preparation may also indicate the presence of alkaline environments in caves (Fernández-Jalvo and Marin-Monfort 2008)

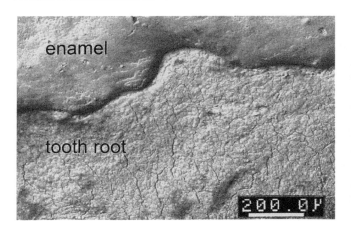

Fig. A.729 SEM microphotograph of a human tooth experimentally exposed to extreme alkaline conditions with KOH (pH ∼ 14). Cracking may affect dentine, roots and bone, but the enamel remains unaffected (see also A.732)

Fig. A.732 SEM microphotograph of a modern rodent incisor after experimental exposure to alkaline conditions (KOH, pH ∼ 14) for 4 min. Intense flaking affects the dentine, but the enamel remains intact, as in A.729

Fig. A.730 Cracking of a modern rodent mandible by experimental exposure to alkaline conditions (KOH, pH ∼ 14). Note the heavily cracked bone surface also referred to as desquamation to distinguish it from exfoliation by weathering (Fernández-Jalvo et al. 1998). Compare with A.688

Fig. A.733 SEM microphotograph of a detail of modern rodent incisor, specimen A.732, after 4 min exposed to caustic soda (KOH, pH ∼ 14). The dentine surface shows a mosaic pattern. Examining it in detail, the dentine and the sides of the cracks are warped (elevated and contorted)

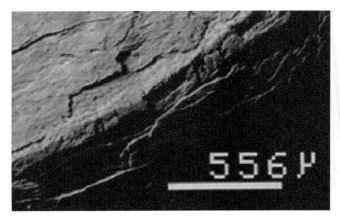

Fig. A.731 SEM microphotograph of a modern rodent mandible, showing cracking of a rodent mandible by experimental exposure to alkaline conditions (KOH, pH ∼ 14). Detail of A.730. Note the sides of the cracks are warped (elevated and contorted)

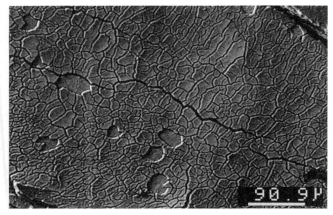

Fig. A.734 SEM microphotograph of a modern rodent, detail of incisor A.733 showing mosaic cracking on the dentine as result of 4 min exposed to caustic soda (KOH, pH ∼ 14)

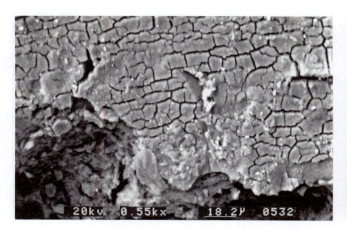

Fig. A.735 SEM microphotograph of a modern rodent incisor from Swildon's Hole. Mosaic cracking on the dentine is similar to experimental bones exposed to strongly alkaline conditions (see A.734). Highly alkaline soil characterizes Swildon's Hole congruent with this strong cracking seen on the dentine surface

Fig. A.738 SEM microphotograph of a modern small mammal long bone. It was exposed to weathering for 4 years, and this has produced deep cracks and splitting, together with superficial flaking or exfoliation. Compare with A.679 and A.688

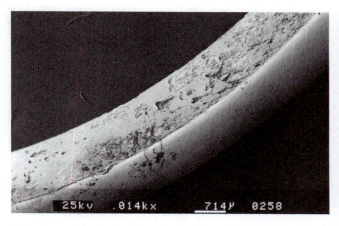

Fig. A.736 SEM microphotograph at a lower magnification than A.735 of a modern rodent incisor from Swildon's Hole. Enamel is unaffected but dentine appears strongly flaked and corroded by the alkaline environment

Fig. A.739 Fossils from Langebaanweg excavation area (A.999). They have exfoliation, flaking and cracking before burial. Bones preserved at this site likely accumulated in lagoons and river bank environments, and the wet environment has sometimes destroyed bone structure and resulted in the fragmentation of the fossils (A.958). Fine sedimentation also preserved most fossils (A.1078)

◀ **Fig. A.737** SEM microphotograph detail of A.736 of a modern rodent incisor from Swildon's Hole. The dentine surface appears flaked and pieces of superficial dentine have been chipped off. The sides of the cracks are warped (elevated and contorted) as shown in the squares

Fig. A.740 Long bone of Sivathere from Langebaanweg. Bone flakes are detached from the bone surface (A.739 and A.959). Bones preserved under wet conditions that has sometimes destroyed bone structure and resulted in the fragmentation and deformation of the fossils. Fine sedimentation also preserved most fossils

Fig. A.742 Detail of a specimen from Langebaanweg preserved under damp conditions (lagoon or river bank). Cracks on the surface appear to have warped edges (elevated and contorted). This type of crack also forms under the influence of high humidity (A.717 and A.721). This is in contrast to cracks formed by weathering or fire for which the edges are separated and even (A.711)

Fig. A.741 Mandible of Sivathere from Langebaanweg. The enamel is cracked and flaked by exposure to wet environments. Mandibles and long bones are also deformed (A.957 and A.1078)

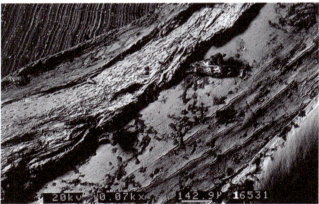

Fig. A.743 SEM microphotograph of a modern murid mandible exposed in a peat-bog (pH 4.1) for 5 years from Neuadd 33. This is a detail of the dentine of the incisor showing a cracked and flaked surface. Cracks are similar to those observed in bones under high humidity (cracked edges uneven, elevated and contorted, A.717 and A.721) but only affecting the dentine

Fig. A.744 SEM microphotograph of a modern bone from hyena regurgitation. The bone surface appears cracked due to digestion. Digested bones may acquire a rounded shape (A.609 and A.616) and be similar to abraded (A.543) or licked bones (A.599). What is different from these, however, is this distinctive "torn" like cracked surface seen at the microscopic level which distinguishes digestion from abrasion and licking

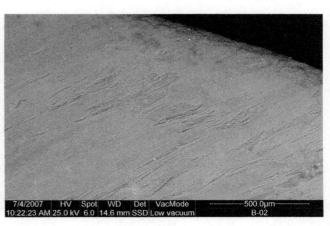

Fig. A.747 SEM microphotograph of a modern bone fragment partially digested by hyena (*Crocuta crocuta*). The edge of the bone is rounded by digestion and the bone surface is cracked. The presence of cracking is diagnostic of rounding produced by digestion (A.745) compared with rounding due to abrasion (A.549)

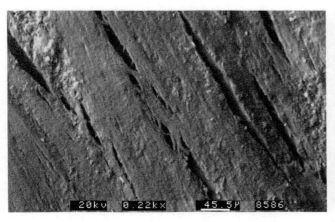

Fig. A.745 SEM microphotograph of modern bone partially digested by hyena. This is a detail of previous, showing a close up view of the "torn" like cracking observed as characteristic of digestion and exposure to acidic environments. Digested bones are also rounded and polished (A.589). Water or wind abrasion (see A.111 and A.341), however, do not show this characteristic cracking, which is the result of chemical action

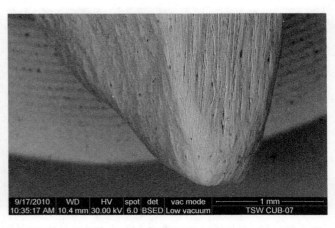

Fig. A.748 SEM microphotograph of a modern bone fragment recovered from a brown hyena (*Hyaena brunnea*) scat near the den. The bone surface appears cracked as well as having the edges rounded by digestion (see rounded abraded bones in Text Fig. 6.2, and rounded digested bones A.610 and A.614). See close up view at A.749

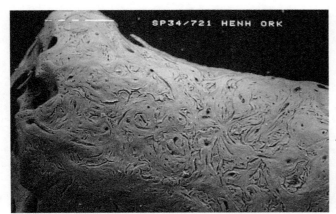

◀ **Fig. A.746** SEM microphotograph of an articular surface of a modern rodent humerus heavily digested by hen harrier, *Circus cyaneus* A.614, A.834 and A.857). The bone surface is cracked as well as rounded. Cracking of the articular bone surface has a different arrangement of cracking to that observed on long bone shafts, reflecting differences in the internal bone structure (see A.760). Field width equals 1.1 mm

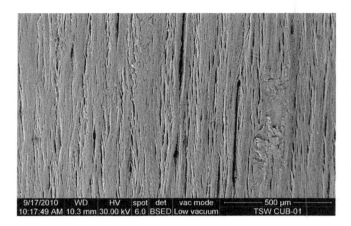

Fig. A.749 SEM microphotograph of a modern bone fragment recovered from a brown hyena scat (*Hyaena brunnea*). This shows a detail of A.748. The bone surface is cracked by digestion. Cracks are microscopic, but they are characteristic and diagnostic of digestion showing torn-like cracking

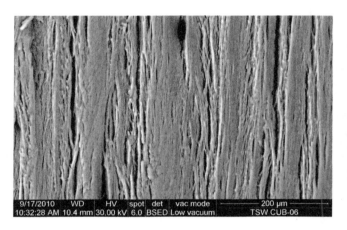

Fig. A.750 SEM microphotograph of a modern bone fragment recovered from a brown hyena scat (*Hyaena brunnea*). This shows a detail of A.748 and A.749. The surface of the bone is cracked by hyena digestion, showing the torn-like cracking

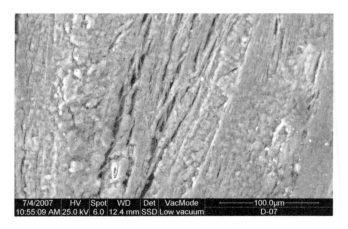

Fig. A.751 SEM microphotograph of a modern long bone fragment partially digested by spotted hyena (*Crocuta crocuta*). The cracked bone surface has torn-like cracking (A.757), which is diagnostic of digestion and similar to corrosion by acidic environment (A.889)

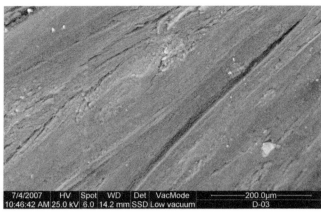

Fig. A.752 SEM microphotograph of a modern long bone fragment partially digested by spotted hyena (*Crocuta crocuta*). This shows the cracked surface which is diagnostic of digestion (A.744)

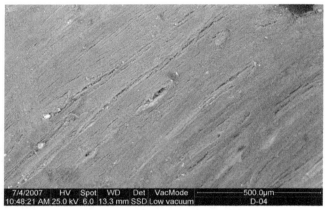

Fig. A.753 SEM microphotograph of a modern long bone fragment partially digested by spotted hyena (*Crocuta crocuta*). This shows the cracked surface which is diagnostic of digestion (A.749) and similar to corrosion by acidic environment (A.889)

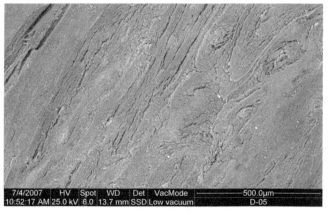

Fig. A.754 SEM microphotograph of a modern long bone fragment partially digested by spotted hyena (*Crocuta crocuta*). This shows the cracked surface which is diagnostic of digestion (A.749) and similar to corrosion by acidic environment (A.889). Cracking by digestion adapts to the bone collagen arrangement. Histological traits (canals, osteons and lamellae) may emerge when digestion is heavy, as is the case with digestion by hyenas

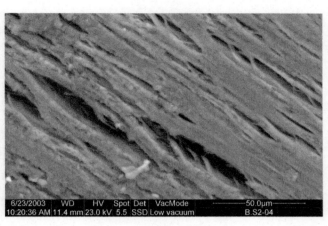

Fig. A.755 SEM microphotograph of a modern long bone fragment partially digested by spotted hyena (*Crocuta crocuta*). This shows the cracked surface which is diagnostic of digestion (A.749) and similar to corrosion by acidic environment (A.889)

Fig. A.758 SEM microphotograph of a modern amphibian metapodial digested by fox (*Vulpes vulpes*). This is a detail of the tissue and superficial cracking in A.756 with a 'torn'-like cracking

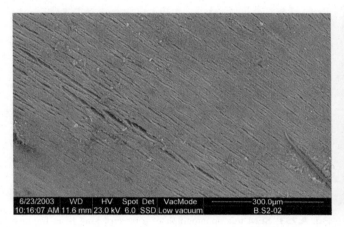

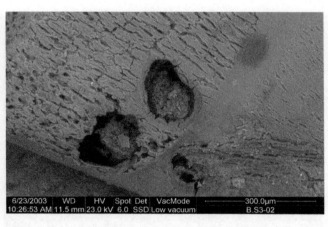

Fig. A.756 SEM microphotograph of a modern amphibian metapodial digested by fox (*Vulpes vulpes*). The characteristic cracked surface that is a diagnostic trait of digestion on mammalian bone (A.745) can also be distinguished in small vertebrate bones (see Pinto Llona and Andrews 1999 for further detail of digestion in amphibians). Characteristics of digestion on the diaphysis of long bones are similar to those of large and small mammals. Specimen provided by B. Sanchiz

Fig. A.759 SEM microphotograph of a modern amphibian phalanx digested by fox (*Vulpes vulpes*). There is deep cracking at the metaphysis (near the articular end, A.760). The different cracking arrangement, compared to rodent bone (see A.746) depends on differences in the bone tissues and histology (see further description of digestion on amphibian bones in Pinto Llona and Andrews 1999)

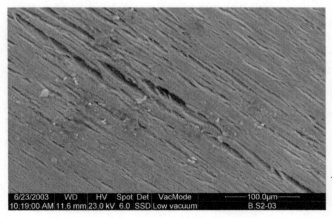

◄ **Fig. A.757** SEM microphotograph of a modern amphibian metapodial digested by fox (*Vulpes vulpes*). This is a close-up view of A.756 showing the tissue and superficial cracking

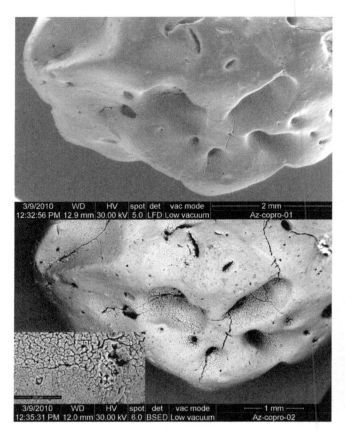

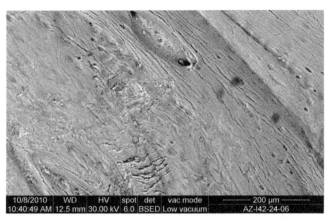

Fig. A.762 SEM microphotograph of a fossil bone fragment from Azokh Cave. The edge of the large mammal bone is heavily rounded and cracked by digestion. The bone surface is also heavily cracked (see A.756)

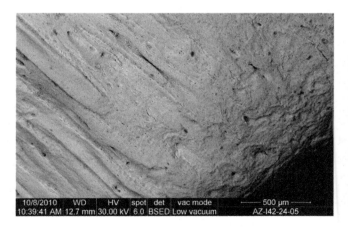

Fig. A.760 SEM microphotograph of a heavily digested fossil bone recovered from a coprolite at Azokh Cave. The fossil has been photographed with SE detectors (above) and at BSE mode (at the bottom). The fragment is highly rounded by digestion, but cracking cannot be distinguished with the SE detector. The image of the fossil at BSE mode shows a form of mosaic cracking, as in A.746 and A.759 (small box bottom left) that suggests this tiny fragment is part of an articular area

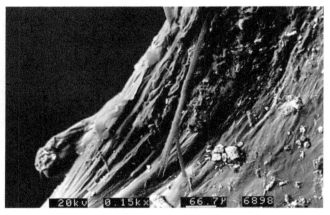

Fig. A.763 SEM microphotograph of a modern rodent mandible fragment digested by a crocodile (A.871). Digestion is so extreme that what remains after digestion has a fibrous texture, rather than flaking or cracking of the bone surface. This is due to the destruction of the bone mineral component, hydroxyapatite, by crocodile gastric juices that are basically composed of HCl (A.765). Specimen provided by Dan Fisher

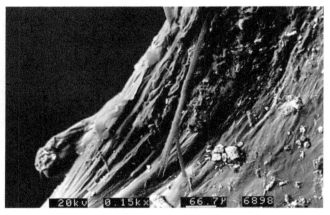

Fig. A.761 SEM microphotograph of a fossil bone fragment from Azokh Cave. The edge of the large mammal bone is heavily rounded and cracked by digestion (see A.747)

Fig. A.764 SEM microphotograph of a modern rodent femur digested by crocodile (A.871). Crocodile gastric juices are virtually HCl, with almost complete absence of enzymes. Collagen fibbers resist digestion by crocodiles, but the mineral phase of the bone is destroyed. Specimen provided by Dan Fisher

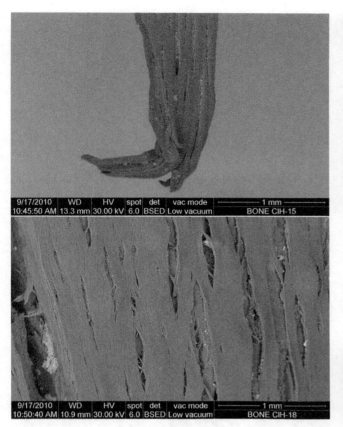

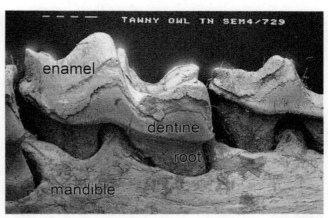

Fig. A.767 SEM microphotograph of a modern soricid teeth digested by a Tawny owl (*Strix aluco*). The enamel is cracked and flaked, as well as the dentine, the tooth root and the mandibular bone surfaces. Enamel has been lost upwards from the alveolar junction towards the top of the crown, for the enamel-dentine junction is the area most easily penetrated by digestion. Compare this manner of digestive corrosion with physical abrasion shown in A.542. Field width equals 2.5 mm

Fig. A.765 SEM microphotographs of a modern bone splinter of large mammal treated with HCl at 10% for a few hours. Top: note the edge of the bone does not show any rounding, but the bone surface is cracked, shown in detail in the lower picture (bottom). This experiment shows that cracking is formed by acid even in the absence of enzyme activity, and it is consistent with observations on bones digested by crocodiles, where there is cracking but no rounding. Enzymes also cause cracking (A.766). Cracking on the bone surface is thus produced both by HCl and by enzymes

Fig. A.768 SEM microphotograph of a modern soricid mandible digested by a genet *Genetta genetta* (A.633 and A.865). Both enamel and bone are flaked and cracked. Enamel has been lost from both the alveolar border and from the tips of cusps, where the enamel dentine junction is exposed on worn teeth. Field width equals 3.5 mm

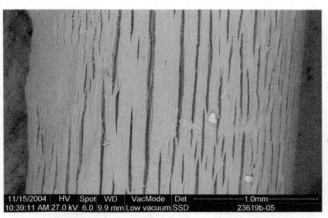

◀ **Fig. A.766** SEM microphotograph of a modern bird bone (*Anas crecca*) from the MNCN taxidermic collection after skeleton preparation using enzymes (Neutrase). This is a bacterial protease produced by a selected strain of *Bacillus amyloliquefaciens*. Cracking is present on this specimen overexposed to enzymes. Similarities between digestion and enzymatic activity (A.873 and A.874) have been shown by Denys et al. (1995)

Fig. A.769 SEM microphotograph of fossil murid maxillary bone and upper molar recovered from a coprolite from Vanguard Cave. The coprolite may be that of a wolf (A.609). The tooth and bone surface are heavily digested (A.852). Both enamel and bone are flaked and cracked. Enamel has been lost from both the alveolar border and from the tips of cusps, where the enamel dentine junction is exposed on worn teeth

Fig. A.772 Modern metapodials (top three) and a proximal end of radius (bottom) of fallow deer taken from Riofrío. Breaks and cracks are transverse. The second metapodial from the top is half broken (see A.773). These bones were collected from open plains in an area where deer herds cross path ways (see A.1063). Experiments are in progress (A.1012), but further work is needed to test this pattern of cracking and breakage

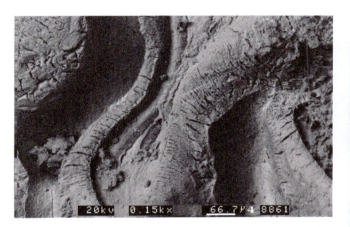

Fig. A.770 SEM microphotograph of fossil murid maxillary bone and upper molar recovered from a coprolite from Vanguard Cave. Detail of A.769, showing the heavily cracked enamel surface due to heavy digestion

Fig. A.773 Modern metapodial of fallow deer (second metapodial from the top A.772) taken from Riofrío path way shown in A.1063. Top: lateral view of the metapodial showing the top of the bone is moderately weathered. Below, detail of transverse breakage and cracking

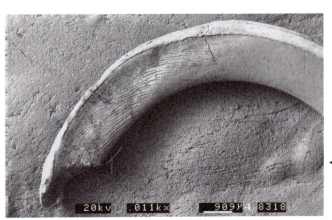

◄ **Fig. A.771** SEM microphotograph of a modern small mammal incisor digested by a genet *Genetta genetta* (A.633 and A.768). Dentine cracking and reduction of the enamel is restricted to the part of the tooth extending beyond the alveolar cavity of the mandible. This indicates that the tooth was still in place in the mandible during digestion

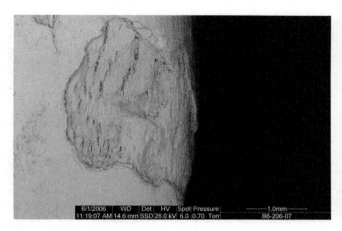

Fig. A.774 SEM microphotograph of a fossil bone fragment from Gorham's Cave damaged by root marks. The mark has a sinuous shape and cracked surface at the interior (see A.228 and A.429). Cracking is not always present in root marks, but when a sinuous linear mark bears cracks inside, this is a diagnostic criterion to identify it as a root

Fig. A.777 SEM microphotograph of a fossil specimen from Gorham's Cave. The puncture root mark has cracking in the interior of the root mark, see A.776. The outer cortical layer of the bone has been corroded by the root, leaving histological traits such as Havers canals and osteons (arrows) exposed

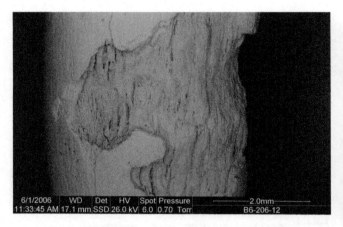

Fig. A.775 SEM microphotograph of a mammal fossil bone from Gorham's Cave, showing characteristic cracks in the interior of the root mark (see A.228 and A.429)

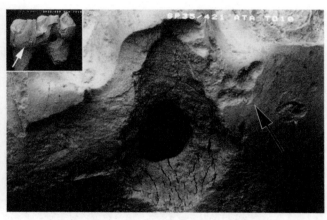

Fig. A.778 SEM microphotograph showing a perforation made by a root on a fossil murid tooth from Atapuerca, Gran Dolina (TD10, A.423). Note the internal cracking in the interior of the root mark perforation, as in A.426, A.427 and A.428. Arrows show sinuous trajectories also made by roots. Field width equals 1.2 mm, and field width of the small box top left equals 2.8 mm

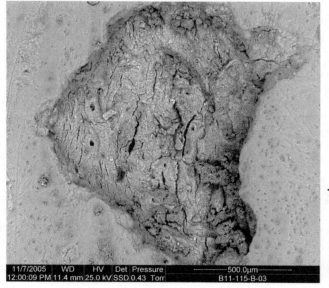

◀ **Fig. A.776** SEM microphotograph, a fossil specimen from Gorham's Cave. The puncture root mark has cracking in the interior (A.243 and A.425). Roots may have a sinuous trajectory or penetrate into the bone. The type of plant that makes perforations in contrast to branched or sinuous trajectories is unknown. It is possible that perforations may be formed by different parts of the root. Further investigations are needed to establish these differences and identify the types of plant that form root marks

Fig. A.779 SEM microphotograph of root marks on a fossil rodent bone from Atapuerca, Gran Dolina (TD11). Extensive cracking is present in the interior of the corroded area made by the root (see A.242, A.244 and A.424, A.430). Field width equals 1.1 mm

Fig. A.781 SEM microphotograph of a fossil bone fragment from Atapuerca, Galeria showing a perforation made by roots. Note the internal cracking in the interior of the root marks. These cracks, however, are not always present, but when this happens, this is diagnostic of root marks (see A.883)

Fig. A.780 SEM microphotograph of a small mammal fossilized long bone from Atapuerca, Galeria. There are very extensive perforations made by roots, so much so that the surface of the fossil is destroyed. Note the internal cracking in the interior of the root marks (A.423 and A.429)

Fig. A.782 SEM microphotograph of a fossil small mammal fragment with a root mark from Atapuerca and internal microcracking inside the mark (A.242 and A.244)

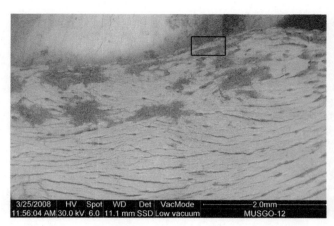

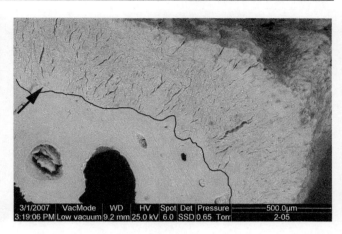

Fig. A.783 SEM microphotograph of a modern collected bone from a neonate fallow deer from Riofrío seen in A.257. This bone was exposed on surface for one year. Moss growing on the bone surface produced a form of cracking that may be confused with weathering, but the immature tissue of the individual may increase the cracked aspect. Square at the top right is enlarged in A.784

Fig. A.785 SEM microphotograph of laboratory cultured fungi (*Chaetomium*) growing for 5 years on modern phalanx. This is a transverse section of the bone. The PCL (periosteal cortical layer) has been cracked and shows a distinctive crumbly irregular textured fringe. Wedl tunneling may also be distinguished (see A.282, A.928 and A.940)

◀ **Fig. A.784** SEM microphotograph of a modern collected bone from a neonate fallow deer from Riofrío, detail A.783. Some of the moss is still present and covering the bone. Some moss was removed from the bone surface to see marks left on the bone. Root marking on the surface was distinguished by the naked eye by reddish staining (see A.256 and A.257). Observed on the SEM the apparent root marks are remains of moss penetrating into the bone and flaking the bone surface

Chapter 8
Corrosion and Digestion

Corrosion and digestion are similar modifications but produced by different agents. Corrosion is generally perceived as due to inorganic processes, although they may be mixed with biological processes in some cases. Digestion on the other hand is entirely due to organic processes. There are some differences in the types of modification produced, and guidelines for distinguishing different processes will be emphasized here, although it must be pointed out that this distinction is not always clear in the absence of information on context.

Corrosion

The term corrosion has been used for many different forms of modification, but here it is defined as surface modifications arising out of chemical attack due to either biological or geochemical action. This is intended to imply that for corrosion to occur on bones, they must be exposed to moist, chemically reactive conditions and removed in some degree from direct contact with the air (thus distinguishing corrosion from weathering). This can arise from burial in soil, being covered by dense ground vegetation so that the bones are in intimate contact with the ground, or long-term immersion in still or stagnant water, which degrades bone surfaces either through biological attack, e.g., from algae, or through water acidity. Bone microstructure may be preserved in fossils many millions of years old, and this is currently being investigated by Zheng and Schweitzer (2012). There is a special case of corrosion (A.786) that occurs in the protected environment of caves, where biogenic acids (e.g., bat guano (A.788) or hyena urine and droppings in their dens A.789), the high humidity and protection from subaerial environments and weathering produces conditions analogous to burial in active soils. This form of modification is currently being studied at the cave site of Azokh cave (rich in bat guano) and was also observed at Kent's Cavern site (hyena den).

Agents and Processes

Inorganic processes:

> Immersion in water A.809
> Cave corrosion A.786

Organic processes:

> Soil corrosion A.902
> Gastric juices A.870
> Fungi and algae A.893 and A.891
> Moss and lichen A.900, A.894
> Bacteria A.906

Characteristics

The main characteristic of corrosion is loss of tissue through chemical action, the loss being unsystematic in the sense that it is not concentrated in striations or patterns of grooves. There is clearly some overlap with striations produced by root action, where the striations are produced by chemical solution of bone surface by growing roots (see Chap. 3). Corrosion from most of the agents described above has only been investigated as surface effects on bone, except for some pioneer work by Behrensmeyer and colleagues (Trueman et al. 2004, and see Fernández-Jalvo et al. 2010). This will be mentioned briefly, but most of the following descriptions are based on the more comprehensive work done on surface corrosion.

> Corrosion damage will be described under three headings:

> Location and distribution of corrosion
> Depth reached by corrosion
> Degree of corrosion

© Springer Science+Business Media Dordrecht 2016
Yolanda Fernández-Jalvo and Peter Andrews, *Atlas of Taphonomic Identifications: 1001+ Images of Fossil and Recent Mammal Bone Modification*, Vertebrate Paleobiology and Paleoanthropology, DOI 10.1007/978-94-017-7432-1_8

Location of Corrosion

A common characteristic of corrosion caused by soil biological activity, immersion in water and cave corrosion is that specimens are affected on all exposed surfaces. The agents causing the corrosion are usually ubiquitous in these mediums, so that all parts of the bones or teeth in such environments are more or less equally affected. This is one of the main ways corrosion can be distinguished from digestion, for in the latter the modifications are generally localized to specific parts of the skeletal elements (Andrews 1990). The one exception to this general result is seen on bones continuously covered by low dense vegetation cover. In this instance the bones are exposed to two different taphonomic agents, with those parts of the bone in contact with the soil being modified by one and the parts of the bone exposed to the air to the other. This produces localized corrosion, usually on parts of the bone in contact with the soil (Fig. 8.1), and this type of damage can be confused with other types of modification. The loss of tissue from the extremities of the bones shown in Fig. 8.1 could be mistaken for carnivore chewing,

although in this case the bones came from an individual that was regularly monitored from time of death and loss of bone tissue was observed as gradually accumulating over time. The bones remained in contact with wet soil (pH 6.0) beneath dense vegetation for 18 years and were observed to become progressively corroded over this period of time. The differential corrosion comes about because the upper parts of the bone are exposed to air with high humidity, and the under parts where the bone is in contact with the soil are exposed to a substrate that is more or less permanently damp and weakly acid. The permanent cover of dense ground vegetation prevented the soil from drying out, and this could be confirmed in the examples above where the humidity above and below the canopy level of the ground vegetation was monitored (A.902 and A.903). Humidity at ground level beneath the vegetation remained high even when the external humidity was low. Where this combination is not met and the soil dries out, this type of corrosion is less likely to occur.

Localized corrosion also occurs directly from plant action, for example from growth of moss and lichen. Moss produces a very superficial surface corrosion (A.898), and it

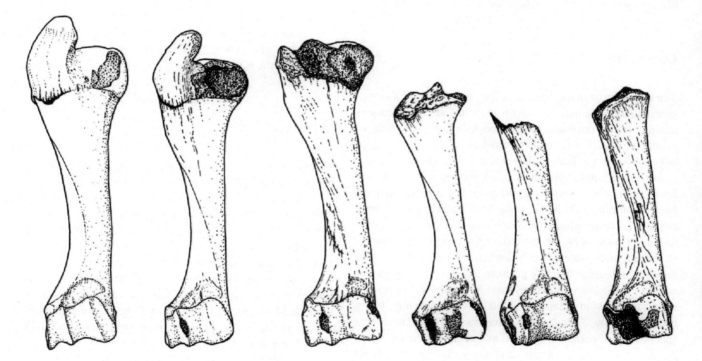

Fig. 8.1 Bone corrosion mimicking carnivore chewing, showing the sequence of corrosion on sheep humeri from Neuadd, Wales. The four drawings on the left are all the same bone from a monitored skeleton (Neuadd 12) spanning 2, 6, 11 and 18 years exposure in a wet drainage ditch with heavy tree cover. Corrosion started where the ends of the bone were in contact with the wet soil (pH 6.0) and progressed through the epiphyseal lines. The two bones on the right are from a separate individual of unknown age from Neuadd 12 and show the extreme loss of articular bone after extended corrosion

is always restricted to those parts of the bone exposed to light. There is some linearity in the corrosion, and this is likely to be the result of angles of the bone that cut off light to lower parts of the bone. Corrosion caused by lichen has a similar distribution to that caused by moss, for the same reason, but the penetration beneath the surface of bone is greater (A.895). The pattern of damage is similar to that produced by water immersion (see below), but since lichen requires light for the symbiotic algae to grow, the damage is usually restricted to one side only of the bone.

Corrosion covering all surfaces of bone occurs as a result of immersion in water. The surfaces of the bones lose patches of surface tissue, giving a mottled appearance, but they do not show any flaking or cracking such as is seen with weathering. Damage is greatest along epiphyseal lines, where there is penetration into underlying bone tissue, but the diaphyses are also affected along their lengths. No modification of dental enamel is apparent unless the water is highly acidic. Similar modifications occur on both large and small mammals, and with different degrees of acidity. In highly acid environments, for example in one of the Dartmoor mires with pH 3.5, the high acidity produces a more penetrating type of corrosion affecting all parts of the bones (A.790). The high acidity has penetrated the bone and dissolved parts of the interior. A similarly extreme modification is seen in bones affected by hyena urine and other organic acids, as well as bones heavily modified by roots. Another type of modification is seen with cave corrosion where so much of the interior of the bone has been removed that the outer surfaces have started to collapse into the interior (A.354).

Depth of Corrosion

Penetration of corrosion below the surface is often superficial, even in the case of heavily corroded outer surfaces (A.903), only passing a few microns into the compact bone (Fernández-Jalvo et al. 2010). It appears that immersion in water produces only surface effects, with dispersed damage to these surfaces. In conditions of high acidity, penetration is deeper, producing a lattice work of pits passing into the interior of the bone. Penetration is even deeper in the case of cave corrosion, and this may be associated with highly alkaline conditions. It is likely that greater depths of penetration occur with increasing time, but we have no data on this.

With regard to corrosion produced by microorganisms, depth of corrosion is variable. Bacteria may affect only the outer cortical lamellae or be restricted to the outer and inner cortical layers. Bacteria may also penetrate through the entire compact bone (A.922) and/or through nutrient foramina (A.908). The distribution of the corrosion, together with the arrangement of bacterial patches (around histological features or diagenetic fissures, and/or affecting periosteal and endosteal-medullar layers, but not medial layers of cortical bone) may indicate whether the bacteria come from within the animal's body or from the soil outside, i.e., endogenous or exogenous. Bacterial colonies arranged around canals of Havers may not distinguish endogenous bacteria, as exogenous bacteria may penetrate into the histological system through them. So far, we are beginning to distinguish a particular bacterial pattern (a "flower like" shape) around canals of Havers in specimens that decayed naturally (A.475 and A.927). However, more extensive studies and observations are needed to find the appropriate pattern to distinguish exogenous from endogenous bacteria, if that is possible.

Fungi have been found penetrating a few microns into the cortical bone (A.941), but more frequently they are limited to the outer cortical lamellae penetrating through tunnels made by fungi (Wedl tunneling) and also providing corrosion on the outer cortical lamellae (A.940). Apparent re-mineralization of incipient Wedl tunneling (Fig. 8.2) has produced a brighter color (i.e., higher electron density) on a sample from Azokh Cave analyzed by chemical element mapping using energy dispersive spectrometry (EDS). This shows the tunneling to be a weak Silica deposit (double head arrows, Fig. 8.2). The transition from bone to sediment (see solid line in Fig. 8.2) is not always straightforward in terms of chemical mapping. Silica mapping (top row, center) shows a relative high concentration on the outer bone edge (solid line). The Calcium and Phosphate mapping show that the edge of the bone can be better distinguished (dashed line at the top SEM photo). Iron is almost absent.

Superficial corrosion of teeth may occur in calm water as a result of modification by a microbial biofilm, which only penetrates a few microns into the enamel. It contrasts with bacterial action, which does not corrode enamel and which may penetrate completely through the bone or the dentine. Sometimes the action of bacteria maintains and preserves the microstructure of bones after earlier corrosive damage. In some cases, the areas affected by bacteria are the only remains that survived, possibly due to a denser structure or other protective qualities of the re-deposited bone.

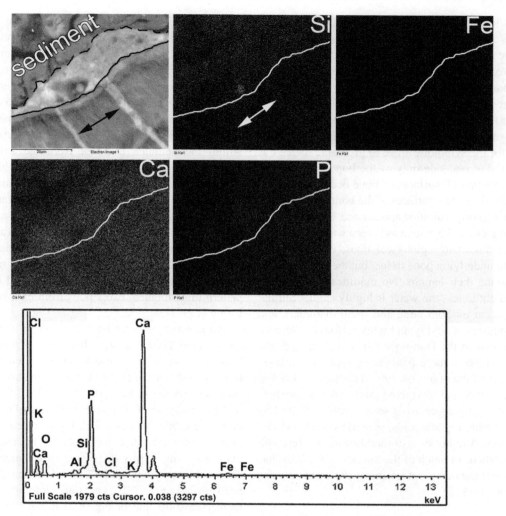

Fig. 8.2 Chemical element mapping (EDS) of basic chemical components (silica: Si, iron: Fe, calcium: Ca, phosphorous: P) of a heavily diagenetically altered fossil bone from Azokh Cave. Transition from bone to sediment is shown as a solid line at the top left SEM microphotograph, and the dashed line indicates an intermediate transitional area (lighter area) between sediment and fossil bone. Double arrow shows Wedl tunneling, probably made by fungi. Clockwise from the right of the SEM microphotograph shows Silicon, Iron, Calcium and Phosphorus, and at the bottom, is the spectrum of the element composition of the fossil, with calcium phosphate of bone and silica and iron uptaken during diagenesis

Degree of Corrosion

There is little information on degrees of corrosion, for example the depth and extent in relation to time, as few long term studies have been done. In the Neuadd sample (see Chap. 2) there is some evidence for this relationship. For example, Fig. 8.1 shows the progressive corrosion over a period of 18 years of a sheep humerus (see above), and these specimens and others under study (Andrews, in prep.) may provide a measure of how corrosion increases with time, but the data are not yet analyzed and it is evident that both degree and rate are highly correlated with environmental parameters such as climate, substrate and aspect.

Bacteria may corrode bones at very different rates and degrees, and the intensity is not related to the depth of corrosion (A.913/A.914 vs. A.473/A.475). Bacteria may corrode intensively as well as lightly, affecting small areas or extended all over the bone, on the surface of the bone or penetrating deep into its section. In contrast, fungi corrode superficially and lightly. Superimposed colonies of bacteria (A.472) was firstly noticed by T.G. Bromage (pers.com). This together with restriction of bacterial attack to inner and outer cortical layers may suggest intermittent environmental changes, which alternately favor or inhibit bacterial growth (Fernández-Jalvo et al. 2010).

Digestion

Digestion has two discernible effects on bones and teeth: modifications to the surface of the bone and chemical modification to the internal structure. Although the modifications differ, they are both the product of digestion, being the result of two processes operating during digestion: high levels of acidity in predators' stomachs and digestive enzymes. The former produces acid etching of bone surfaces, and digestive enzymes break down the organic constituents of bone. Vertebrate digestion varies among predators, with some species having higher acidity levels in their digestive tracts than others (Grimm and Whitehouse 1963). Both acidity and digestive enzymes require time to operate, and so the degrees of modification are related to the time bones are exposed to the digestive processes. Differences are also related to ways of ingestion. In contrast to nocturnal raptors that ingest prey whole, mammalian carnivores chew their prey and diurnal birds tear it apart using beak and claws. Breakage before digestion increases the effects of digestion, allowing digestive juices to penetrate into medullary cavities and increasing bone areas exposed to gastric acids. In addition, the degrees of modification may depend on the state of hunger of the predator, such that a hungry predator may retain a meal in its stomach for longer or until everything is digested. This explains both the variation observed in degrees of modification and the fact that some micromammal individuals are completely digested and never recovered from the pellets or scats of the predator (Raczynski and Ruprecht 1974; Dodson 1973; Dodson and Wexlar 1979; Lowe 1980; Yalden and Yalden 1985). Finally, many avian predators regurgitate less digestible parts of their prey, such as hair, feathers, bones, and soil (in cases where earthworms are eaten). Mammalian carnivores, reptiles and birds such as ospreys regurgitate infrequently or never, so that their ingested prey passes all the way through their digestive system (Andrews and Evans 1983; Andrews 1990).

Agents and Processes

Digestion by nocturnal raptors
Digestion by diurnal raptors
Digestion by mammalian carnivores
Digestion by reptiles

Characteristics

The principal characteristic of digestion is the surface etching of bone surfaces and dental enamel by stomach acids. Less easily observed are the chemical changes to bone caused by digestive enzymes. Within these two subdivisions, modifications will be described according to major body part, teeth or postcrania. Descriptions are based as follows.

Location and distribution of digestion
Depth reached by digestion
Degree of digestion

Location of Digestion

Digestion may be localized or may extend over the whole surface of bones. This distinction tends to be correlated with type of predator, with all-over digestion occurring in large mammal predators such as hyenas and large reptiles like crocodiles (Fisher 1981a, b). In these the bone is comminuted and the fragments rounded by digestion. In the case of teeth, digestion by crocodiles may completely remove the enamel because of the high acidity of crocodile digestive juices, leaving rounded fragments of dentine (A.871). In contrast to this, avian predators that regurgitate the bones of their prey produce localized digestive modifications, which are concentrated on epiphyses and articular surfaces. The effects are loss of tissue, particularly at the ends of bone, broken ends, processes and articulations (A.864 and A.865). Destruction may be so great as to completely destroy the articular ends of bones, and this applies particularly to small mammals that can be ingested whole and all parts of the skeleton subjected to digestion. Birds and amphibians ingested whole by predators have similar tissue loss, particularly with penetration of the articular end.

The effects of digestion are seen less commonly in large mammals as their bones are too large to be ingested whole by even the largest predators. Hyenas ingest and regurgitate many bone fragments, and sometimes small bone splinters and small mammal bones are recovered from their scats (A.606). When they are ingested they are broken beforehand by the predators' teeth (or beaks in the case of avian predators), and the articular ends may be broken beyond recognition before digestion. An exception to this on occasion is digestion of larger bones by vultures (A.870), but there is little information on this type of modification.

Degrees of Digestion

Four categories of digestion of small mammal postcranial bones are currently recognized (Table 8.1). These categories are based on presence of digestion of the femur head, as this

Table 8.1 Digestion categories of limb bones as represented by femur heads

Digestion category	Femur heads
Light <20%	Barn owl, long-eared owl, short-eared owl, Verreaux eagle owl, snowy owl
Moderate 20–35%	Spotted eagle owl, great grey owl, tawny owl, European eagle owl
Heavy 30–40%	European eagle owl
Extreme 50–80%	Little owl, kestrel, peregrine
	Hen harrier, canids, felids, viverrids, mustelids

Fig. 8.3 Schematic drawing of four of the five stages of digestion of ▶
arvicolid molars (from Fernández-Jalvo and Andrews 1992). White
enamel, grey dentine, black empty space (collapsed dentine). The first
stage could not be shown here because the modification is confined to
slight stippling of the surface enamel. Extremely light pitting of the
surface enamel (not shown). **a** Light: loss of enamel from the occlusal
corners of the salient angles of the molars so that they appear rounded.
Digestion does not penetrate to the dentine. **b, c** Moderate: enamel is
removed along half or superficially most of the salient angles, exposing
the dentine and leaving a smooth edge. **d** Heavy: all edges of the teeth
are heavily rounded and enamel extensively removed from the entire
lengths of the salient angles. **e** Extreme: loss of most of the enamel with
rounding of the dentine, leading to collapse of the crown from within.
The shape of the molars becomes almost unrecognizable

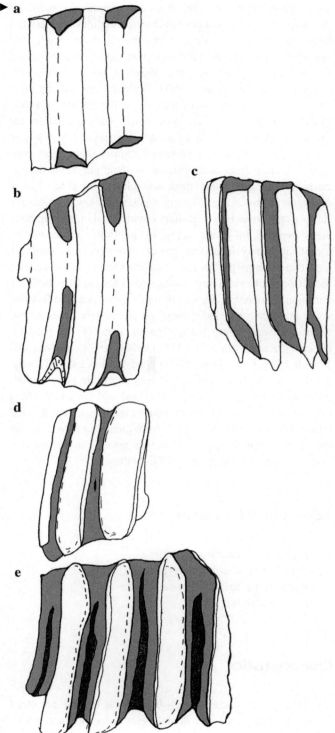

is one of the most commonly preserved elements in small
mammal assemblages (Andrews 1990). Digestion on the
distal humerus articulation, which is also commonly preserved in the fossil record, is comparable to that of the
femur, and only the one element was used for this purpose. It
is categorized as light, moderate, heavy and extreme, and the
different degrees of digestion for postcranial bones and the
predators that produce them are shown in Table 8.1.

Digestion may also produce rounding of exposed ends of
bones without penetration of tissues (see Chap. 6), polishing
of articular bone (see Chap. 6), cracking of limb bone shafts
(see Sect. 7) and breakage of bone (see Chap. 9). On teeth,
the effects of digestion are primarily loss of enamel, starting
by thinning, which may expose the prism structure of the
subsurface enamel, with loss of tissue at exposed places like
tips of incisors (A.819), the enamel-dentine junction
(A.846), alveolar margin, or salient angles of teeth (A.841);
at its most extreme, all enamel may be lost. This can happen
with only minor modification of the dentine, which, because
it is less heavily mineralized than enamel, is little affected by
the acidity in the predators' stomachs.

Five categories of digestion of arvicolid molars are currently recognized (Andrews 1990; Fernández-Jalvo and
Andrews 1992; Williams 2001). This was first established
for this rodent taxon, but other taxa with different tooth
morphologies are affected differently and these should be
considered separately (see below). Arvicolid molars have a
characteristic morphology, often high crowned, with sharply
V-shaped salient angles and deep valleys in between that are
sometimes filled with cementum. The salient angles are
exposed to digestion on three sides of each angle (Fig. 8.3)
as well as from both root and occlusal ends, and enamel is
removed along the angles starting at both ends, exposing the
dentine beneath. The first category of modification is
extremely light digestion only seen as a slight pitting of the
surface of the enamel but with no penetration of the salient
angles and no rounding of the occlusal and root ends (not
shown here, described by Williams 2001). Light digestion
(Fig. 8.3) has some removal of surface enamel in addition to
surface pitting, but this is limited to the top of the crown and

little dentine is exposed (A.842). Moderate digestion extends
along half or superficially most of the height of the salient
angles but only exposes dentine to a limited degree (A.841).
Heavy digestion and extreme digestion show a greater
degree of dentine exposure, with the collapse of the tooth
structure in the extreme case (A.844) (Fig. 8.3).

Table 8.2 Digestion categories (Andrews 1990): comparison of molar and incisor digestion between arvicolids and soricids

Digestion stage	Predator arvicolids	Category, soricids	Molar digestion	Incisor digestion
Absent or very light digestion <10%	1	0	Barn owl	Barn owl
Light digestion Molars % 0–3, Incisors % 8–13	1	0	Long-eared owl, short eared owl, verreaux eagle owl	Short-eared owl, snowy owl
Moderate digestion Molars % 4–6, Incisors % 20–30 (tips only)	2	0	Snowy owl, spotted eagle owl, great grey owl	Long eared owl, verreaux eagle owl, great grey owl
Heavy digestion Molars % 18–22 Incisors % 50–70	3	1	European eagle owl, tawny owl, viverrids	European eagle owl, tawny owl, spotted eagle owl, little owl
Extreme digestion Molars % 50–70 Incisors % 60–80	4	2	Little owl, kestrel, peregrine, mustelids	Kestrel, peregrine
Extreme digestion Molars % 50–100 incisors % 100 (dentine corroded)	5	3	Hen harrier, buzzard, red kite, canids, felids	Hen harrier, buzzard, canids, felids, viverrids, mustelids

It is likely that vultures also produce extreme digestion on small mammals, but we have no data on this. Small mammals digested by crocodiles have been analyzed, but the effects are extremely destructive, characterized by a lack of digestive enzymes and excess of hydrochloric acids of gastric juices. This mechanism is characteristic of crocodiles and produces extreme loss of enamel (practically eaten away) and milder effects (although still heavy) on the dentine and bone (Fisher 1981b; Andrews and Fernández-Jalvo 1998). Some predators that have been studied in detail are shown in Tables 8.1 and 8.2, now under revision the degrees of modification they produce on the bones of their prey.

Several distinct morphologies of tooth types must be distinguished, as these have different resistance to digestion; rodent incisors; soricid and carnivore canines; microtine molars; murine molars; and soricid molars. The scheme established for arvicolid rodents is not directly applicable to other small mammal taxa. A preliminary classification of modification stages to take account of this variation in small mammals molars is shown in Table 8.2, but more work is needed to clarify differential effects of digestion on different tooth types. For example, both murines and soricid appear to be at least one category lower than arvicolids, so that the same predators produce a lesser degree of digestion on them

with time than they do on arvicolids (Table 8.2). The reason for this is that the strongly angled teeth of arvicolids make them particularly susceptible to digestive action, and having dentine exposed on the occlusal and root surfaces provides a weakened area, through which gastric juices may more easily penetrate. As a result, the occlusal ends and the alveolar junction with the roots are the regions most affected. In contrast, murines and soricids have the occlusal area covered by enamel before wear, and this provides a more homogeneous structure with no weakened areas that facilitate gastric juice penetration, except for the alveolar junction (A.856). Data available on non-microtine small mammal prey are shown in Table 8.2, but these data require checking in future work.

Digestion categories have also been set up for rodent incisors (Table 8.2). These have been shown schematically in Fig. 8.4, which shows the four categories involving loss of enamel (Fernández-Jalvo and Andrews 1992). A fifth category is now also recognized (Williams 2001) for extremely light digestion consisting of light pitting of the enamel surface, with no loss of enamel. Light digestion, the top two incisors shown in Fig. 8.4 (A.819), has loss of enamel at the tip of the incisor but no other modification other than the light pitting seen above. Moderate digestion also tends to

Fig. 8.4 Five stages of digestion on rodent incisors (from Fernández- ▶
Jalvo and Andrews 1992). White = enamel, grey = dentine, black =
empty space (collapsed dentine). Extremely light: light pitting on the
surface of the enamel but no enamel loss (not shown). **a, b** Light: slight to
moderate pitting and loss of enamel from the tips of incisors. Sometimes,
light digestion shows a higher degree of pitting and loss of enamel
concentrated at the tips of incisors as shown in (**b**). **c** Moderate: the whole
surface of the enamel is affected with enamel removed extensively from
the tip. The dentine is modified with a wavy surface. **d** Heavy: digestion
removes most of the enamel from the tooth and leaves a wavy outline on
the dentine, which may become cracked in a way similar to that produced
by weathering. **e, f** Extreme: the enamel is now mostly missing, and the
dentine is cracked and with an uneven surface; (**f**) sometimes dentine is
removed to an extent greater than enamel, and in this case the enamel
collapses in on itself, lacking the support of the dentine. The shape of the
incisors becomes almost unrecognizable

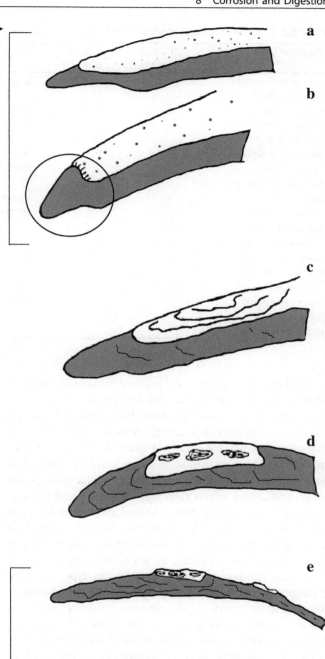

have greatest modification at the tips of the incisor (A.829)
and along the rest of the crown, with limited areas of
removal of the enamel and wavy dentine. Heavy digestion
has extensive enamel loss all along the crown (A.833)
(Fig. 8.4) but with islands of enamel remaining (A.834).
Extreme digestion has removed almost all of the enamel,
with tiny islands of enamel only remaining (A.836), and at
its most extreme the whole structure of the tooth collapses
(A.838 and A.839).

We have found that the single most useful element in
identifying predator damage to small mammals is the rodent
incisor. Most rodents have superficially similar incisors, in
contrast to their molars which differ greatly in size, shape,
degree of hypsodonty or bunodonty, and enamel thickness,
all of which affect the impact of digestion and which make it
difficult to set up a consistent scheme of modification by
which predators can be identified. We therefore recommend
the rodent incisor as the most useful anatomical element to
identify and calibrate this taphonomic agent and strongly
recommend that these should be collected during excavation.

Depth of Digestion

Little work has been done on depth of penetration of digestion into bone. Even in the case of heavy digestion of large
mammal bones by hyenas, it appears that penetration by
digestion is very low and only affects the outer 30 microns
(A.875) (Fig. 8.5) (Dauphin et al. 1988, 2003).

Effects of Enzymes and Acidity

The question whether the modification of animal bones in the digestive system of predators is due to acidity or digestive enzymes is beyond the scope of this book, but we will make a brief comment here. There are large differences in the degree of damage to ingested prey remains in different predator species, and it appears that there is an approximate relationship between the level of acidity in the predators' stomachs and the degree of damage. This is clearly not the whole story, however, partly because the relationship between acidity level and degree of damage has not yet been adequately quantified. Experimental work separating the effects of acids (HCl) and enzymes (proteases) has allowed us to understand better the effects of digestion by predators (Denys et al.1995, 1997; Fernández-Jalvo et al. 2014). Acid by itself causes dissolution of the mineral constituents of bone, with the enamel wholly removed. When teeth are treated experimentally by HCl, the mineral component of the teeth may be lost completely, leaving a soft and malleable tooth. This condition is rarely seen in predator assemblages with the exception of those subject to crocodile digestion (crocodiles have almost no enzymes in their gastric juices (Diefenbach 1975; Fisher

1981a, b). Where enamel remains in place, it is extensively cracked by HCl treatment, and this differentiates it from predator assemblages, as cracking is not seen in digested enamel. The dentine also shows corrosion, but not rounding, and this distinguishes it again from abrasion. The effects of enzymes by themselves are closer to the condition in predator digestion, with penetration of bone surfaces and removal of thin articular bone to expose cavities beneath (see Denys et al. 1995). In the experimental work on these modifications, the combined effect of acid and enzyme was not tried on teeth, but only on bone. Comparison between experimental and digested long bones, especially distal ends, showed highly similar results. Experiments on dental tissues produced an effect similar to that of predator digestion described in the previous section (Fernández-Jalvo et al. 2014).

Mineral Composition

Chemical changes are also observed on bones that have been digested. Figure 8.5 shows an increase in aluminum in the area affected by digestion, in contrast to the intact bone. On

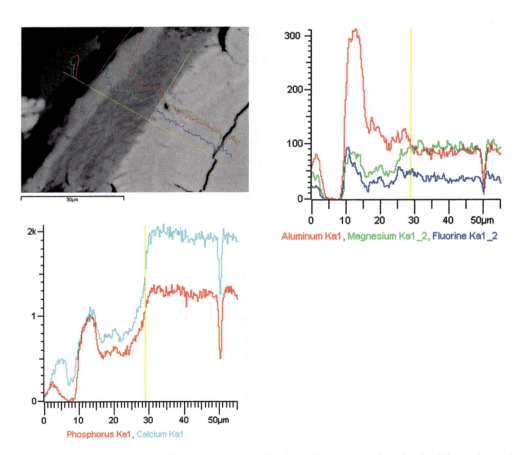

Fig. 8.5 Left, sectioned bone showing the greyish fringe that penetrates 30 microns below the surface. On the right are the results of the chemical analysis: top right showing the element content of aluminium (red), magnesium (green) and fluorine (blue); bottom left is the element content of phosphorus (red) and calcium (blue)

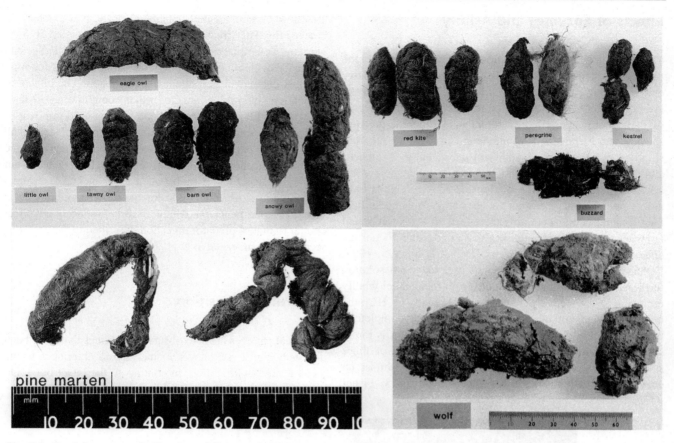

Fig. 8.6 Top left: a range of pellets from different owls. Top right: a range of pellets from diurnal birds of prey. Bottom left: two scats of *Pine martem*. Bottom right: different scats produced by wolf

the contrary, magnesium, and especially calcium and phosphorous, are depleted, while fluorine shows variable behavior. Phosphorus and calcium content of small mammal bones is comparatively lower when they are preserved in predator pellets compared with fresh bone (Fig. 8.4, Fernández-Jalvo et al. 2002). Both increase with fossilization but they also vary from site to site (Fig. 8.6, Denys et al. 1997). Mineral changes due to digestion seem only to affect the most external layer from recently digested bones (Fig. 8.5), and chemical analyses have only been done on a few modern bones. However, it has been observed that digestion increases bone porosity (Fernández-Jalvo et al. 2002), and this may facilitate chemical exchange between the soil and the bone. Diagenetic modification in fossil bones may be extreme, and changes between levels within a site may provide evidence of environmental change, but it is not known at present the extent to which earlier taphonomic

modifications such as digestion may affect the diagenetic modification. It has been shown, for example, that fossil bones from Olduvai Bed I show evidence of digestion, that the digested bone is chemically altered with respect to fresh bone, that the bones from different levels at Olduvai have differences in diagenetic modification, and that these differences may be due to the fluctuating lake level during the early Pleistocene (Denys et al. 1996). What is not known, however, is whether one or another of these stages of modification affects or interacts with other stages. There is clearly some potential for the identification of predators based on the chemical composition of the bones of their prey (Dauphin et al. 2003), and a more exhaustive work on changes in chemical composition of bones from digestion by different predators is needed to provide a statistical basis for understanding these variations.

Atlas Figures

A.786–A.950

Fig. A.786 SEM microphotograph of a modern bat humerus. This is part of an articulated skeleton collected from the surface of the upper chamber of Wookey Hole. The entire surface of the bone is evenly corroded by cave corrosion (as in A.793). Field width equals 5.7 mm

Fig. A.787 SEM microphotograph of a modern bat humerus. This is a detail of the cave corrosion evenly affecting the whole bone surface. Another type of corrosion from the same site can be seen in A.354. In that case, corrosion has locally penetrated the bone and tissue removed from within, so that remaining areas of cortex are collapsing into the interior of the bone. Field width equals 2.3 mm

Fig. A.789 Recent bovid bone from Tanzania. It was buried in the ▶ floor of a spotted hyaena den (*Crocuta crocuta*) and corroded by urine and organic acids from the hyaena droppings (A.1057). The whole surface is degraded with large scale pits and a few patches of unaltered cortical bone (A.1018)

Fig. A.788 Cave corrosion by bat guano at Azokh Unit II (A.1075). For many thousands of years, bats have been dwelling in an area previously occupied by cave bears. The guano produced by the bats has formed thick layers. The cave is seasonally humid and, in the past, the connection to the open air was almost sealed, increasing the relative humidity. Water highly enriched in phosphates from the guano filtered throughout the sediment. The result of these conditions is a heavy corrosion that has seriously damaged the Azokh fossils. The fossil at top left is strongly corroded, as its other side, shown on the lower figure. The 'fossil' at top right has just the silhouette of the bone remaining and it cannot be recovered from the sediment. The yellowish clasts in the sediment were originally limestone blocks that are today corroded and soft. (See other cave soil corrosion types by bat guano from the same site (A.1079) and from other caves (A.354 and A.786)

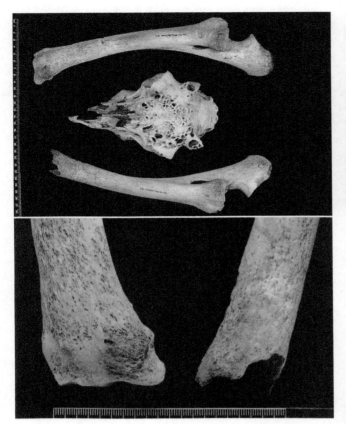

Fig. A.792 SEM microphotograph of a modern monitored rodent bone from Neuadd 33. It was buried in acid soil (pH 4.1) for 5 years, and much of the surface bone has been destroyed. Some patches of cortex are less damaged and have been preserved. This damage is similar to the corrosion seen on A.787 from Wookey Hole cave

Fig. A.790 Modern monitored sheep bones from a mire in a moorland setting at Fox Tor, Dartmoor, Devon. The water is strongly acidic, pH 3.5. There is corrosion of the ends of the bones where they were resting in the acidic water, with solution of the mineral component of the bones. On the thin bone of the frontal and nasal bones of the skull there has been deep penetration of the bone. Below is a close up of the ends of the two radioulnae, which were partially buried in the mire, showing the acid corrosion of the ends (compare with A.789). A.J. Sutcliffe collection

Fig. A.793 SEM microphotograph of a modern monitored rodent bone from Neuadd 33. It was buried in acid soil (pH 4.1) for 5 years, and much of the surface bone has been destroyed (see A.796). This damage is similar to the corrosion seen on A.786 from Wookey Hole cave. This bone is from a juvenile individual and this may also increase the acid etching effects (see further details in Andrews 1990)

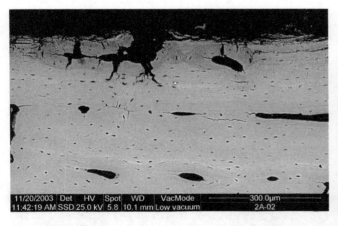

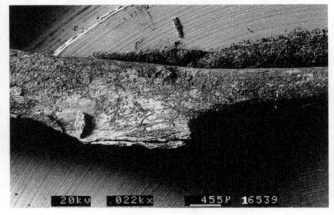

Fig. A.791 SEM microphotograph of cut section of modern monitored horse femur (ND10) from Neuadd 2. The bone was exposed for 13 years under thick ground vegetation in moorland vegetation (soil pH 4.0). Soil corrosion has heavily damaged the bone surface, but also lichen has penetrated into the bone (see A.276). It is apparent, however, that the corrosion is superficial, penetrating less than 100 microns into the compact bone

Fig. A.794 SEM microphotograph of a modern monitored rodent bone from Neuadd 33. It was buried in acid soil (pH 4.1) for 5 years, and much of the surface bone has been destroyed (see A.796)

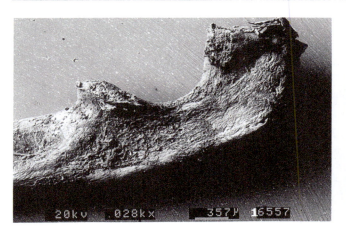

Fig. A.795 SEM microphotograph of a modern monitored rodent bone from Neuadd 33. It was buried in acid soil (pH 4.1) for 5 years, and much of the surface bone has been destroyed. The articular edges appear corroded (see A.796 and A.903)

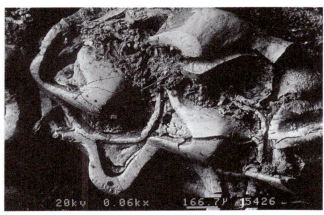

Fig. A.798 SEM microphotograph of a modern monitored rodent bone from Neuadd 33. It was buried in acid soil (pH 4.1) for 5 years, and much of the surface bone has been destroyed. No apparent etching is observed on the molar tooth enamel except for light pitting. Cracking is the main damage affecting the enamel (see A.796)

Fig. A.796 SEM microphotograph of a modern monitored rodent bone from Neuadd 33. It was buried in acid soil (pH 4.1) for 5 years, and much of the surface bone has been destroyed, as in A.795. The tooth enamel on both the incisor and molars has not been affected

Fig. A.799 Modern monitored vertebra of an adult sheep from Neuadd 10/4. Partial skeleton buried for 14 years in a manure heap exposed to the wet Welsh weather. The vertebral processes show certain corrosion on salient angles, possibly increased by bacterial attack seen affecting this specimen (see A.299)

Fig. A.797 SEM microphotograph of a modern monitored rodent bone from Neuadd 33. It was buried in acid soil (pH 4.1) for 5 years, and much of the surface bone has been destroyed. Close up view of previous A.796, showing the etched and cracked surface of the mandible

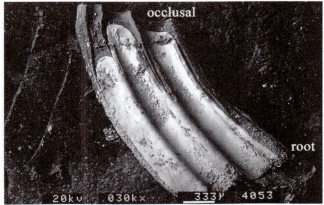

Fig. A.800 SEM microphotograph of a modern vole molar experimentally exposed to acid (HCl), for 3 min. This shows extensive splitting of the enamel. (pH 1). Enamel has been lost at both the occlusal and root extremities of the tooth. (See A.765, A.802 and A.872)

Fig. A.801 SEM microphotograph of a modern vole molar experimentally exposed to acid (HCl), for 3 min. This shows extensive splitting of the enamel. (pH 1). The enamel appears intensively damaged, similar to the effects of digestion, but the enamel is wholly removed from the occlusal end of the tooth, and the dentine also shows corrosion, but not rounding, both of which distinguish this from digestion. (See A.765, A.802 and A.872)

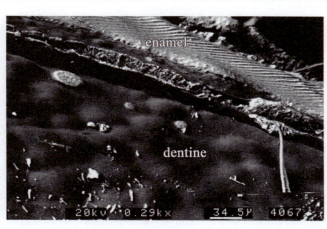

Fig. A.804 SEM microphotograph of the incisor experimentally exposed to HCl for 3 min. The end of the tooth was still inside the incisor alveolus, and therefore protected by the mandible, so that it has a small island of enamel remaining. Compare it with A.732, A.733 and A.734 of incisor exposed to pH ~ 14

Fig. A.802 SEM microphotograph of a modern vole incisor experimentally exposed to acid (HCl), for 3 min. The enamel has been completely removed from the incisor. (See A.765, A.800 and A.872)

Fig. A.805 Surface corrosion on fossil bones has been observed at Senèze (A.353, A.807 and A.808). Corroded surfaces of articular ends could be mistaken for scooping out by carnivores (A.902). The main difference is that salient angles are the bone parts most affected by corrosion, where they are in contact with the ground, while carnivores may alter any part of the bone and may leave tooth marks (absent in this specimen and in A.902)

◄ **Fig. A.803** SEM microphotograph of a vole incisor experimentally exposed to acid (HCl), for 3 min. The tip of the incisor has no enamel left (see A.800 and A.872), and the remaining dentine is curled around. Compare it with A.732 of incisor exposed to pH ~ 14

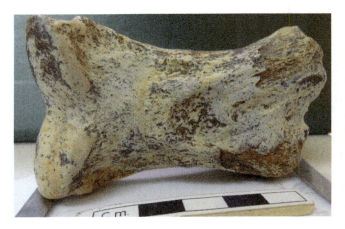

Fig. A.806 Corrosion shown on only one side suggests the bone was most of the time lying in one position. Observations made at Neuadd (experiments and monitoring) and Senèze (see this picture and A.807, A.808 and A.902) indicate that decay processes in a permanent humid substrate or under calm water, combined with corrosion by vegetation (moss or algae), cause what has been traditionally named, "soil corrosion or soil acidity" (see Fernández-Jalvo et al. 2010)

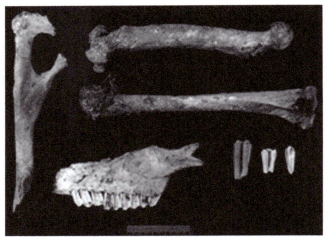

Fig. A.809 Modern monitored sheep bones from Neuadd 24. This skeleton was immersed in water for 12 years at pH 6.0. The surfaces of the bones are degraded, not by acidity, for the pH is close to neutral, but by decay and corrosion by plant and micro-organisms, with patches of tissue loss, and penetration of the bone along the epiphyses. The teeth are not affected. See A.538 and A.799

Fig. A.807 Fossil phalanx of *Eucladoceros* from Senèze. The site was a volcanic crater filled with water, and the acidity of the volcanic sediment has produced strong corrosion on some of the fossils. On the other hand, slow decay of the bodies under permanent water may have allowed the preservation of complete or partially complete skeletons. Modification of the bones may have been enhanced by aquatic vegetation (possibly algae, see A.265 and A.436)

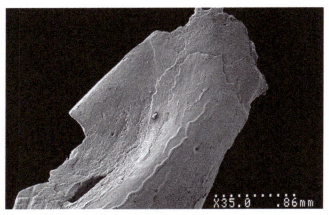

Fig. A.810 SEM microphotograph of the ascending ramus of a modern rodent mandible experimentally exposed for 2 years to sodium bicarbonate ($NaHCO_3$). The bone surface suffered superficial exfoliation or desquamation. Similar desquamation is observed in experimental mandibles exposed to NaOH (pH $\sim$ 14) A.730 and A.731 but with the modification occurring in a shorter time (4 min)

◀ **Fig. A.808** Soil corrosion affecting a fossil fragment of *Eucladoceros* right femur. At first glance, the surface could be mistaken for carnivore scooping out of limb bones (Sutcliffe 1970; Haynes 1980, 1983; Binford 1981). However, punctures of different sizes in the cancellous tissue (white arrows) and the chemical damage observed on the cancellous tissue, suggest that the punctures were likely to have been made by stones in contact with the cancellous tissue. See A.353

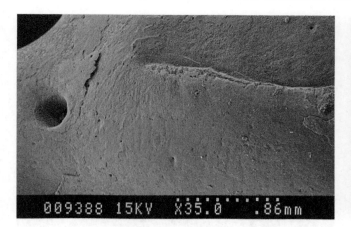

Fig. A.811 SEM microphotograph of the mandibular body of a modern rodent exposed for 2 years to sodium bicarbonate (NaHCO₃). It shows superficial exfoliation or desquamation. Faster and more destructive results were obtained in experimental mandibles exposed to KOH (pH ∼ 14) A.730 and A.731. This type of desquamation has been observed on fossil bones from cave environments and open air sites that are highly alkaline, like Olduvai (the carbonatite sediments have high pH ∼ 10)

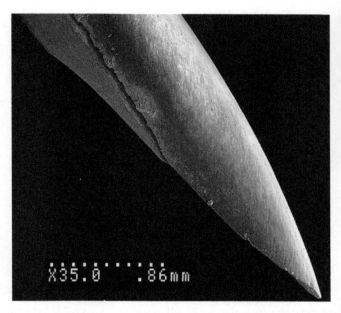

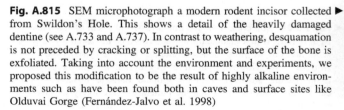

Fig. A.812 SEM microphotograph of the tip of a modern rodent incisor. It was exposed to sodium bicarbonate for 2 years. The enamel is intact, see A.626, although it is detached from the dentine, and the dentine is slightly cracked

Fig. A.815 SEM microphotograph a modern rodent incisor collected ▶ from Swildon's Hole. This shows a detail of the heavily damaged dentine (see A.733 and A.737). In contrast to weathering, desquamation is not preceded by cracking or splitting, but the surface of the bone is exfoliated. Taking into account the environment and experiments, we proposed this modification to be the result of highly alkaline environments such as have been found both in caves and surface sites like Olduvai Gorge (Fernández-Jalvo et al. 1998)

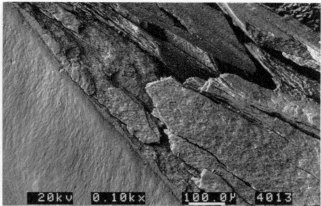

Fig. A.813 SEM microphotograph of a modern rodent incisor exposed to KOH (pH ∼ 14) for 4 min. This shows a detail of A.732. The dentine is heavily damaged, but the enamel is not affected. The dentine has the appearance of being dried out suddenly

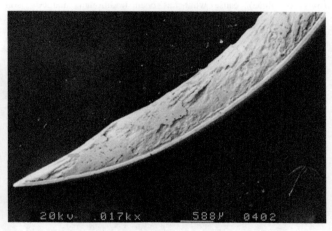

Fig. A.814 SEM microphotograph of a modern rodent incisor collected from Swildon's Hole. It has been heavily damaged on the dentine by the high alkalinity of the cave environment. The enamel is intact and unaltered. See A.732 and A.736

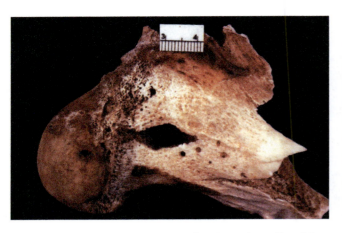

Fig. A.816 Proximal end of humerus of modern zebra collected from Amboseli. The bone surface shows perforations (as in A.870), rounding and signs of corrosion. This could be a result of partial gastric digestion in spite of the large size of the bone fragment. It is possible that the bone was ingested and regurgitated, note the broken edge of the bone fragment is pointed but rounded (see A.594). Basically, gastric juices are a combination of hydrochloric acid and enzymes. Photo T. Jorstad, Collection of A.K. Behrnesmeyer

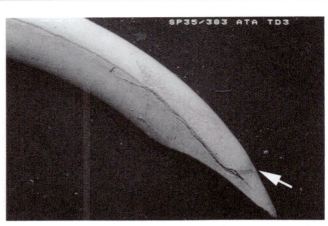

Fig. A.819 SEM microphotograph of a fossil incisor from Atapuerca TD3. This has light digestion on the incisor tip which is seen as a reduction in thickness of the enamel. The dentine is exposed at the tip of the incisor after the removal of the enamel by digestion (white arrow), and the surviving enamel surface is lightly pitted (see A.626 for a non-digested incisor). Field width equals 7 mm

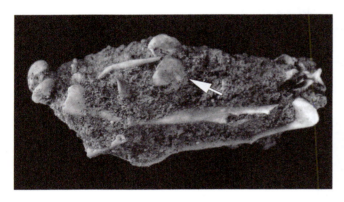

Fig. A.817 Fossil pellet from Ibex Cave in Gibraltar. The taphonomic study of the preserved fossil bones inside the pellet indicates that it was produced by European eagle owl, and this pellet was regurgitated by this raptor (*Bubo bubo*). The arrow points to postcranial bones visibly digested. See A.606

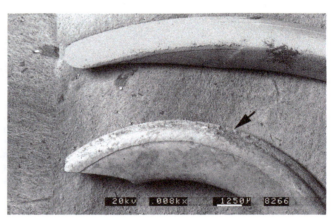

Fig. A.820 SEM microphotograph of modern teeth, detail of light digestion. Enamel has not been lost from the tip, but on the tip of the incisor on the upper specimen (on the left) the enamel surface has been deeply pitted. The lower incisor has had the middle part of the enamel more strongly pitted. The arrow points to the area most heavily digested on the bottom incisor. See Text Fig. 8.4

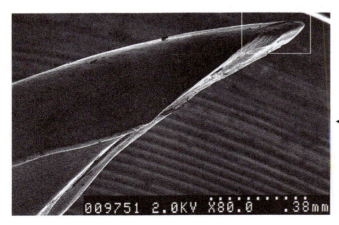

◄ **Fig. A.818** SEM microphotograph of an incisor from a barn owl pellet (*Tyto alba*), showing very light digestion. This grade of digestion, lighter than the minimum grade categorized by Andrews (1990) was established by Williams (2001) in his study on barn owls. Digestion is only evident at the very tip as a fading of the polished enamel surface of the enamel. Very light digestion is seen on light microscopes as an opaque enamel surface. (See A.626 for a non-digested incisor and Text Fig. 8.4)

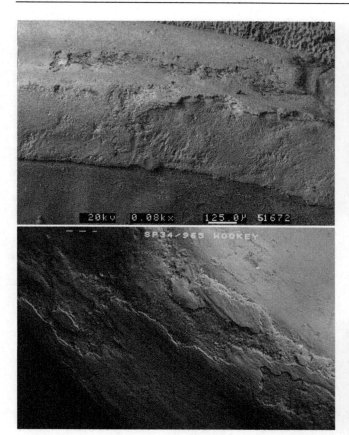

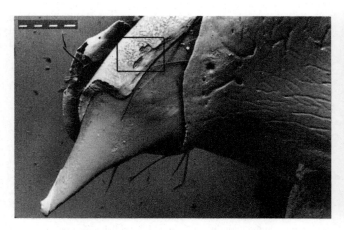

Fig. A.821 SEM microphotograph of modern rodent incisors with the enamel lightly digested by barn owls *Tyto alba*. See Text Fig. 8.4. Top: modern tooth from an owl pellet. Parts of the surface enamel has been digested but not through to the dentine. The dentine on the lower part of this image has not been affected by digestion. Bottom: tooth from Wookey Hole digested by barn owl. This case also it is only the most superficial layer of enamel that is corroded. Field width equals 716 microns

Fig. A.823 SEM microphotograph of a lightly digested rodent incisor. This is a detail of the incisor from a pellet of long eared owl *Asio otus*, and it is shown boxed in A.822. While the tip of the tooth has lost all of its enamel, there is also damage further along the crown, where corrosion affects the enamel superficially (see A.830 and A.831). Field width equals 400 microns

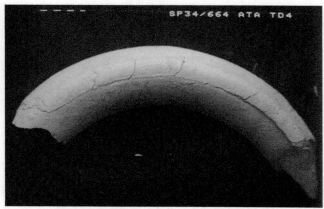

Fig. A.824 SEM microphotograph of a fossil rodent incisor from Atapuerca TD4. It is lightly digested with enamel totally removed from the tip. The grade of digestion is similar to the modern specimen shown in A.822 and similar to the fossil specimen shown A.825. Field width equals 5.8 mm

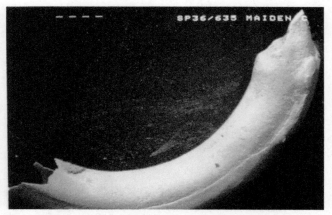

Fig. A.822 SEM microphotograph of a modern rodent incisor from a long eared owl *Asio otus* pellet. This is classed as light digestion concentrated at the tip of the incisor, and it often occurs when the incisor is still retained in the jaws during the whole period of digestion. The square is enlarged at A.823. See A.824 and A.845. Field width equals 2.8 mm

Fig. A.825 SEM microphotograph of a rodent incisor from Maiden Castle, an Iron Age hill fort in the UK. It has light digestion concentrated on the tip of the incisor, with enamel totally removed from the tip (see Text Fig. 8.4). This is light digestion similar to that seen on incisors from long eared owl pellets (A.822) and Atapuerca TD4 (A.824). Field width equals 4.6 mm

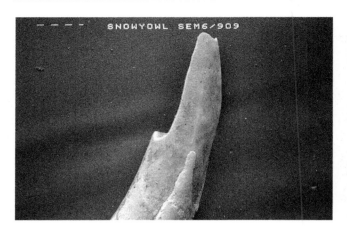

Fig. A.826 SEM microphotograph showing digestion of a modern rodent incisor from the pellet of a snowy owl (*Nyctea scandiaca*). Digestion is concentrated on the tip of the incisor where the enamel is totally removed (see Text Fig. 8.4). The wavy outline of dentine indicates that this is moderate digestion, as in A.828 and A.829. Field width equals 7.1 mm

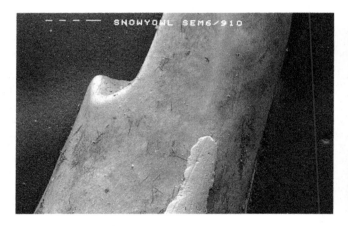

Fig. A.827 SEM microphotograph of moderate digestion on a modern rodent incisor from a snowy owl (*Nyctea scandiaca*) pellet. Digestion is concentrated at the tip of the incisor where enamel is totally removed: detail of A.826. The dentine may be slightly digested, producing a wavy outline. The digested edge of the enamel and the wavy outline of dentine shows that this is moderate digestion. Field width equals 2.1 mm

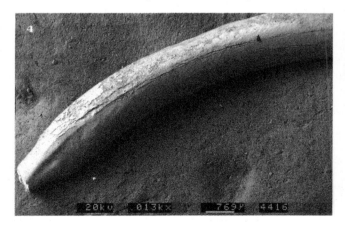

Fig. A.828 SEM microphotograph of a fossil rodent incisor from Vanguard Cave. Moderate digestion affects the enamel superficially showing slight pitting. The dentine may be slightly digested on the tip, producing a wavy outline, as in A.826 and A.829

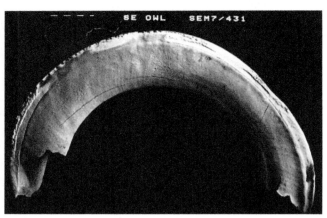

Fig. A.829 SEM microphotograph of a modern rodent incisor from the pellet of short eared owl (*Asio flammeus*). The tooth is moderately digested. The surface of the enamel is more intensively affected than is seen with light digestion, and the dentine is also modified with an evident wavy surface. The prism structure of the enamel is beautifully exposed at higher magnification in A.830. Field width equals 7.6 mm

Fig. A.830 SEM microphotograph of a modern rodent incisor from the pellet of short eared owl (*Asio flammeus*). This shows a detail of A.829. The prism structure of the enamel is beautifully exposed at higher magnification. Field width equals 983 microns

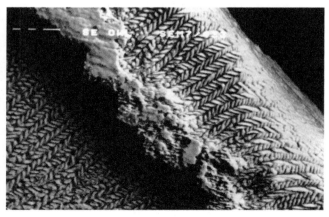

Fig. A.831 SEM microphotograph of a modern rodent incisor from the pellet of short eared owl (*Asio flammeus*). Detail of the digested enamel of A.830, exposing the prism structure of the enamel. Field width equals 2 mm

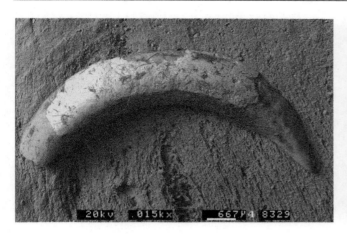

Fig. A.832 SEM microphotograph of a rodent modern incisor from a coyote scat, *Canis latrans*. The incisor is heavily digested. Enamel has been removed entirely from the tip of the incisor and in places along the rest of the tooth, with deep pitting elsewhere on the enamel (see Text Fig. 8.4, A.833 and A.834)

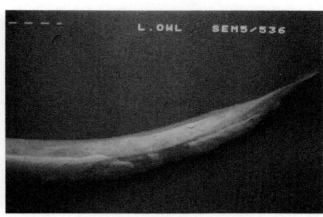

Fig. A.835 SEM microphotograph of a modern rodent incisor from the pellet of a little owl, *Athene noctua*. The incisor is extremely digested with tooth enamel restricted to islands along the crown (see Text Fig. 8.4, A.836, A.838 and A.839). Field width equals 6.5 mm

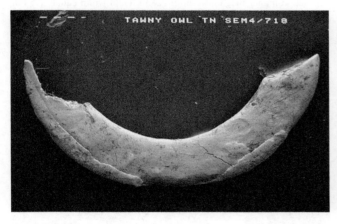

Fig. A.833 SEM microphotograph of a modern rodent incisor from the pellet of a tawny owl, *Strix aluco*. Heavy digestion has removed much of the enamel and exposed the dentine along the length of the tooth except for islands of enamel not fully removed (see Text Fig. 8.4, A.832 and A.834). Field width equals 7.1 mm

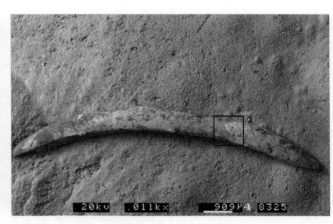

Fig. A.836 SEM microphotograph of modern lower incisor of a rodent from the pellet of a common kestrel *Falco tinnunculus*. The incisor has been extemely digested and is reduced to small islands on the surface of the dentine. (See Text Fig. 8.4, A.835, A.838 and A.839)

Fig. A.834 SEM microphotograph of a modern rodent incisor from the pellet of a hen harrier, *Circus cyaneus*. This shows heavy digestion of the incisor with the removal of much of the enamel and exposing the dentine along the length of the tooth except for islands of enamel not fully removed (see Text Fig. 8.4, A.832 and A.833). Field width equals 4.8 mm

Fig. A.837 SEM microphotograph, detail of A.836 of modern lower incisor of a rodent from the pellet of a common kestrel *Falco tinnunculus*. The incisor has been extremely digested and here can be seen the wavy aspect of the dentine and the edge of the small island of almost entirely eroded enamel

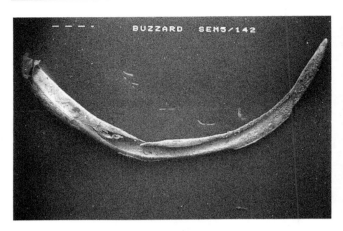

Fig. A.838 SEM microphotograph of the lower incisor of a modern rodent from the pellet of a common buzzard, *Buteo buteo*. This is extreme digestion, and the main trait at this stage of digestion is the loss of the recognizable shape of the incisor, almost complete absence of enamel and collapse of the structure in on itself (Text Fig. 8.4, A.835, A.836 and A.839). Field width equals 11.3 mm

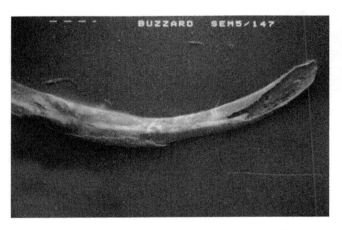

Fig. A.839 SEM microphotograph of the lower incisor of a modern rodent from the pellet of a common buzzard, *Buteo buteo*. This is extreme digestion, and the main trait at this stage of digestion is the loss of the recognizable shape of the incisor, almost complete absence of enamel and collapse of the structure in on itself. (See Text Fig. 8.4, A.835, A.836 and A.838). Field width equals 9.4 mm

Fig. A.841 SEM microphotograph of the molar of a modern lemming from a snowy owl pellet (*Nyctea scandiaca*). Top: general view (Field width equals 4.2 mm). This tooth is moderately digested as the enamel has been removed at the occlusal ends of the salient angles and along the edges of the enamel-dentine junctions of the salient angles. (Note that lemmings lack enamel along the salient angles on the lateral columns). Bottom: SEM microphotograph detail of the picture above (square), showing the characteristic smoothened and rounded dentine edge of the salient angles of the tooth typical of digestion (Text Fig. 8.3, A.845). Field width equals 1.4 mm

◀ **Fig. A.840** SEM microphotograph of a modern arvicolid molar digested by an eagle owl *Bubo bubo*. Light digestion can be seen on the top third of the three salient angles of the crown. The lower two thirds of the crowns have the slight pitting that is characteristic of extremely light digestion. No modification is observed in murid, cricetid or soricid teeth when exposed to this low degree of digestion. (Text Fig. 8.3, A.842 and A.843). Field width equals 4.9 mm

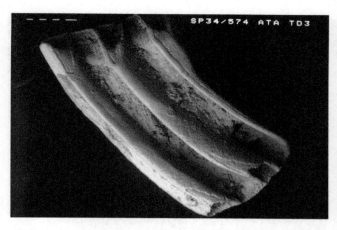

Fig. A.842 SEM microphotograph of the molar of a fossil lemming from Gran Dolina (TD3, Atapuerca). This is lightly digested and the enamel has been removed along the top edges of the salient angles. (Text Fig. 8.3, A.840 and A.843). Field width equals 3 mm

Fig. A.843 SEM microphotograph of a modern arvicolid molar from the pellet of a European eagle owl, *Bubo bubo*. Light digestion is present, with removal of enamel along half the length of the salient angles and at the occlusal surface. It should be noted that with the teeth still in place in the mandible, the lower half of the crowns would be protected from digestion, and this could give a misleading impression of the extent of digestion (Text Fig. 8.3, A.840 and A.842). Field width equals 2.1 mm

Fig. A.845 SEM microphotograph of a modern arvicolid mandible ▶ from a little owl (*Athene noctua*) pellet. The molars are moderately digested, the ascending ramus is missing, and the inferior border of the mandible is broken, exposing the incisor and molar roots. In contrast, the incisor is only lightly digested (Text Fig. 8.3, A.841). Field width equals 7 mm

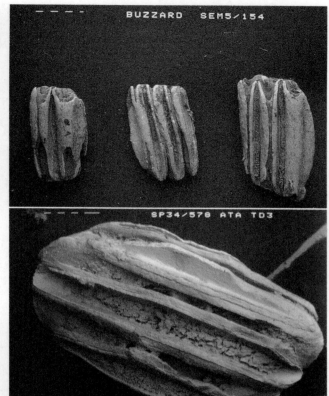

Fig. A.844 SEM microphotographs. Top: three modern arvicolid molars (field width equals 7.5 mm) from separate pellets of a common buzzard, *Buteo buteo*. Digestion on the molars ranges from moderate digestion on the left and on the right (see Text Fig. 8.3 and A.841) and heavy digestion in the middle. The pattern overall for buzzards is heavy digestion. Bottom a fossil molar from Gran Dolina (TD3, Atapuerca) showing extreme digestion on a fossil tooth, with removal of enamel and dentine along the length of the salient columns of the tooth (Text Fig. 8.3). Field width equals 2 mm

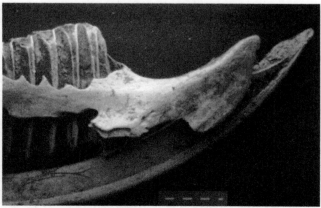

Fig. A.846 SEM microphotograph of a modern murid molar from an eagle owl pellet *Bubo bubo*. The teeth are moderately digested (A.847 and A.855) with the effects of digestion concentrated along the enamel dentine junctions, both at the alveolar margin and where dentine is exposed by wear. At this stage of digestion on murid teeth, arvicolid teeth would be heavily digested because the sharp salient angles of their molars are more exposed to digestive action. Soricid teeth would be lightly to moderately digested. Field width equals 3.1 mm

Fig. A.849 SEM microphotograph of a modern cricetid molar by a mongoose (*Cynictus penicillata*). Although the salient angles of the low crowned cricetid molars are less prominent than that of arvicolid molars, digestion can still be seen to have penetrated the length of the crown down to the root (white arrow). This is a heavy degree of digestion (A.848). Compare the effects of digestion displayed here with physical abrasion (see A.542)

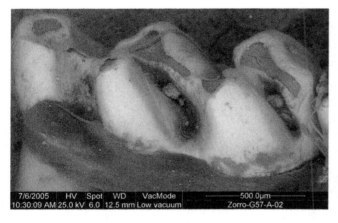

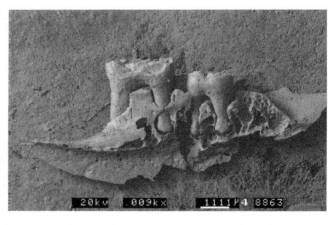

Fig. A.847 SEM microphotograph. Murid and cricetid molars that are digested show enamel partially removed along the edges of the wear facets and at the enamel-dentine junction. Dentine is not affected. This is moderate digestion (A.846 and A.855), and arvicolid teeth in the same fox scat would be heavily digested because the sharp salient angles of their molars are more exposed to digestive action

Fig. A.850 SEM microphotograph of a fossil murid mandible recovered from a coprolite in Vanguard Cave. Mandibular bone is very heavily digested with no remains of the inferior border and the bone surface is holed and corroded by gastric juices. The coprolite from Vanguard could possibly have been produced by wolf. For digestion of the tooth enamel, see A.851

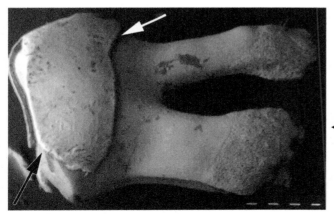

◀ **Fig. A.848** SEM microphotograph or modern rodent molar from the scat collected from a captive margay (*Felis wiedii*) in London Zoo. Digestion is heavy (A.849) and the penetration from the enamel-dentine junction is particularly evident, indicated by the white arrow. The roots and dentine of the occlusal surface are unaffected, but the enamel junction on the wear facet (black arrow) has been heavily damaged. Field width equals 3.5 mm

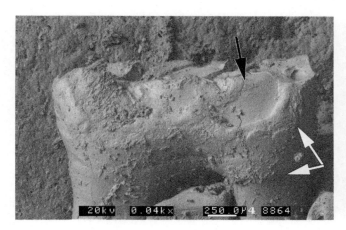

Fig. A.851 SEM microphotograph of the fossil murid molar from the mandible A.850 recovered from a coprolite in Vanguard Cave. In a first examination, the tooth appears less damaged than the jaw (as in A.852 and A.854), with signs of light digestion at the bottom of the crown (white arrows) with little loss of enamel and no damage at the enamel junction and lateral side. But the occlusal surface shows great enamel loss (black arrow)

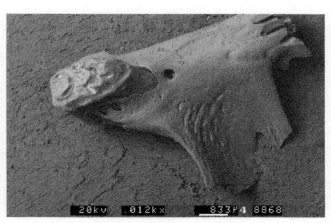

Fig. A.854 SEM microphotograph of a fossil murid maxillary fragment and upper M3 recovered from a possible wolf coprolite in Vanguard Cave. The molar is moderately digested (A.855), but the zygomatic arch is heavily digested (as in A.852), showing the edge (bottom side of the figure) perforated and thinned. The equivalent digestion stage of arivicolid molars from the same scat would be moderate to heavy since they are more vulnerable to digestion

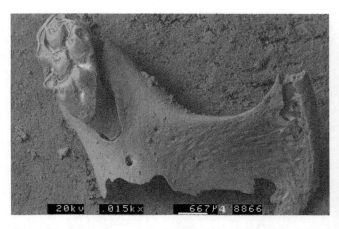

Fig. A.852 SEM microphotograph of a fossil murid maxilla recovered from a possible wolf coprolite in Vanguard Cave. The bone of the zygomatic arch is penetrated and heavily corroded by digestion, and the broken edge (bottom side of the figure) has been thinned. The tooth appears less damaged (A.853) than the jaw (as in A.850 and A.854)

Fig. A.855 SEM microphotograph of a fossil murid maxillary fragment. Detail of A.854 showing the upper M3 recovered from a possible wolf coprolite in Vanguard Cave. The molar is moderately digested with local enamel removal (black arrows)

◀ **Fig. A.853** SEM microphotograph of a close up view of a fossil murid molar recovered from a coprolite in Vanguard Cave (A.852). The buccal and distal surfaces of the molar have some loss of enamel. Considering that murid teeth have a more robust structure, and that digestion is less evident than on arvicolid teeth, digestion observed here can be considered to be moderate. Extreme digestion occurs when enamel is reduced to small islands on the dentine

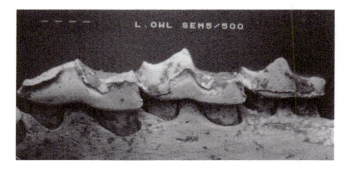

Fig. A.856 SEM microphotograph of a modern soricid mandible from a little owl pellet, *Athene noctua*. This is heavy digestion (milder than A.857, A.858 and A.859), with much of the enamel removed from the enamel-dentine junction. Soricid teeth, as well as murid molars, are more resistant to digestion than arvicolid molars, with no equivalent damage at the initial light-moderate degrees of digestion categorized by Andrews (1990) for arvicolids (see text for explanation of digestion categories in soricids, murids and arvicolids). Field width equals 4.3 mm

Fig. A.857 SEM microphotograph of modern soricid mandible recovered from a Hen harrier *Circus cyaneus* pellet. The tooth has extreme digestion (A.858), and the bone, dentine, tooth root and enamel are cracked and flaked. The intensive grade of digestion by this diurnal bird of prey has produced the collapse of the enamel. The dentine also is etched, indicating extreme digestion in soricid teeth (see text for explanation of digestion categories in soricids, murids and arvicolids). Field width equals 1.2 mm

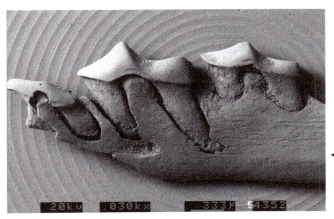

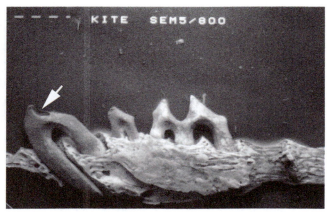

Fig. A.859 SEM microphotograph of a modern soricid mandible from the pellet of a red kite *Milvus milvus*. This is the most extreme degree of digestion (compare with A.856, A.857 and A.858). Enamel is entirely missing, the dentine is smoothed and collapsed (arrow), and the inferior border of the mandible is missing. At this stage of digestion of a murid or soricid molar, arvicolid molars would have been totally destroyed. Field width equals 6.9 mm

Fig. A.860 Fossil incisor of horse (*Hipparion*) from Concud showing the rounded edge of the enamel probably caused by digestion (see also A.631). Hyenas were taphonomic agents active in the site, with tooth prints on bones and heavily digested bone splinters (A.594). This incisor has also been affected by plant roots (see A.636) and previously by a type of unknown agent that we interpret as produced by a biofilm of microorganisms (see A.493) Courtesy of D. Pesquero

◄ **Fig. A.858** SEM microphotograph of a modern soricid mandible digested to an extreme degree (A.857). Enamel is entirely missing and the dentine is smoothed and etched, and it has entirely collapsed on the tooth at far left of the mandible. (See text for explanation of digestion categories in soricids, murids and arvicolids.)

Fig. A.861 SEM microphotograph of a modern rodent femur head from a barn owl (*Tyto alba*) pellet showing a very light degree of digestion (A.862 and A.863). The cancellous tissue is exposed, with some areas of cortex still remaining (white arrow). Field width equals 984 microns

Fig. A.864 SEM microphotograph of the proximal femur of a modern arvicolid rodent from the pellet of a European eagle owl, *Bubo bubo*. Both the head and greater trochanter are penetrated by digestion. This is classed as moderate digestion (compare with A.862 and A.867). Field width equals 7.8 mm

Fig. A.862 SEM microphotograph showing the proximal end of a femur of a modern arvicolid rodent lightly digested (A.861) by short eared owl, *Asio flammeus*. The cancellous tissue is completely exposed with no remains of bone cortex. Field width equals 3.2 mm

Fig. A.865 SEM microphotograph of the modern distal humerus of a rodent from a genet scat *Genetta genetta*. There is loss of tissue due to heavy digestion which has penetrated the articular bone. Digestion by mammalian carnivores may also produce rounding of broken edges of bone fragment and on the edges of cancellous tissue as seen here. This is a characteristic trait distinguishing between digestion and soil corrosion. (see A.786 and A.810). Field width equals 8.3 mm

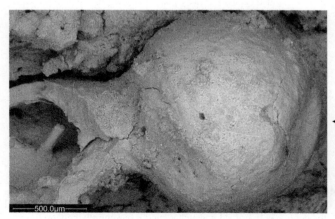

◄ **Fig. A.863** SEM microphotograph of a proximal femur of a fossil rodent recovered from a hyena coprolite from Bois Roche (France, see A.607). Small mammal remains can resist the strong digestion of hyena (and crocodiles A.874), and those recovered from scats or coprolites show a relative low grade of digestion. This may be that they are protected to some degree by hair, skin or other large mammal bones during digestion. If small mammals are not protected, they dissolve completely

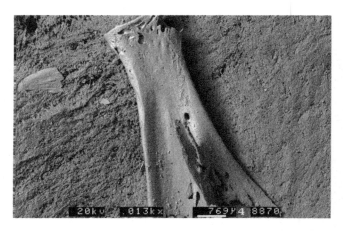

Fig. A.866 SEM microphotograph of a fossil rodent scapula recovered from a carnivore scat, probably wolf, from Vanguard cave. It has very heavy digestion (A.865) at the epiphysis and loss of tissue on the scapular blade. (Fernández-Jalvo and Andrews 2000)

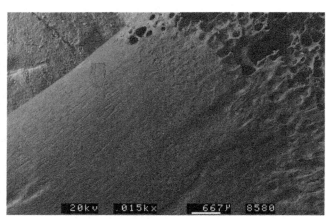

Fig. A.869 SEM microphotograph of a modern bone fragment extremely digested (A.867) by puma, *Puma concolor*. G. Gómez collection

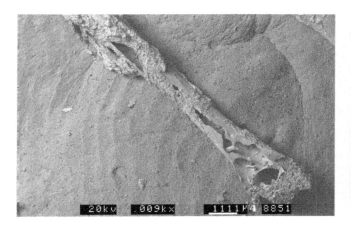

Fig. A.867 SEM microphotograph of a fossil long bone recovered from a carnivore coprolite, probably from a wolf, in Vanguard Cave. This shows the extreme destruction that can be caused by digestion, as A.871. (Fernández-Jalvo and Andrews 2000)

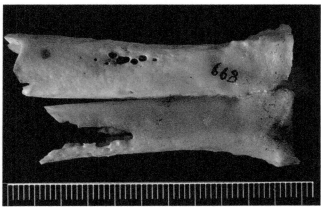

Fig. A.870 Digestion of a modern large bone by a vulture (*Aegypius occipitalis*) perforated by the strong gastric juice action, as in A.816. A.J. Sutcliffe collection

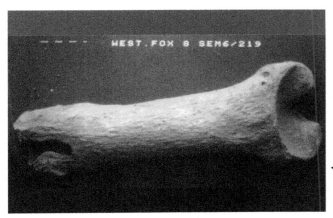

◀ **Fig. A.868** SEM microphotograph of a modern rodent phalanx heavily digested (A.865) by red fox, *Vulpes vulpes*. Much surface tissue has been lost and the edges of the bone are smoothed and rounded. Field width equals 8.9 mm

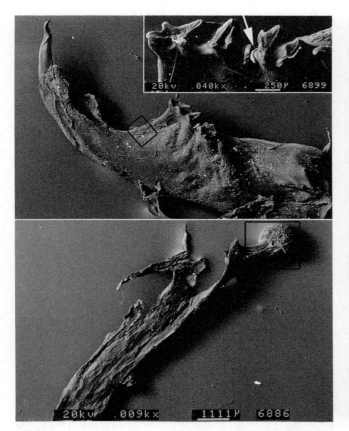

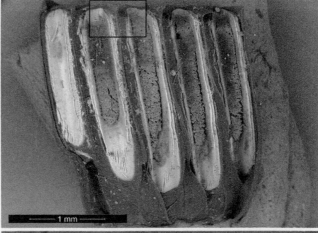

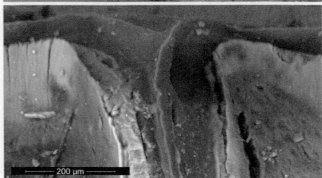

Fig. A.871 Top: SEM microphotograph of a modern rodent mandible recovered from a crocodile scat. Specimen provided by Dan Fisher. The black square on the diastema of the mandible is seen in A.763. Digestion is extremely high. The inset at top right shows a detail of the *in situ* molars. Note the enamel has been almost completely removed by digestion and small islands of enamel are all that remains on the wear facet (arrow). The tip of the incisor has no enamel and appears curled (compare with HCl experiment seen in A.802). Bottom: SEM microphotograph of a rodent femur recovered from a crocodile scat. See A.764 that shows a close up view of the strong corrosion on the femur head produced by crocodile gastric juices. Also note the lack of rounding on the bone edges. Much of the mineral element of the bone has been lost and the unsupported shape of the bone is distorted. (compare with A.765). Specimen provided by Dan Fisher

Fig. A.872 SEM microphotograph of experimental digestion on modern arvicolid molar. The tooth was exposed to hydrochloric acid for 4 h. Note the enamel has been lost along the salient angles (see A.841), but dentine remains almost unaffected and not rounded (see A.800 and A.801). Below is the detail of the square near the occlusal surface

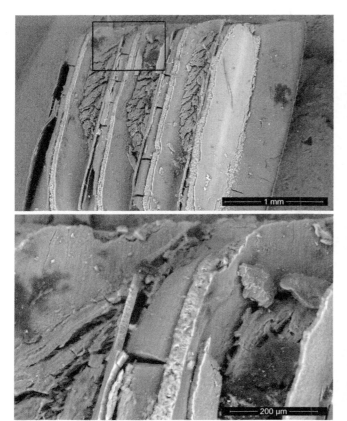

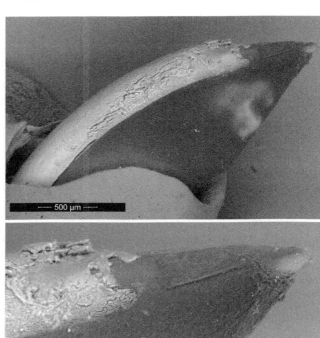

Fig. A.873 SEM microphotograph of experimental digestion on a modern arvicolid molar. The tooth was exposed to hydrochloric acid for 4 h and later immersed in enzymes (protease) for 23 h. Note the enamel has been lost along the salient angles, and the dentine is rounded (see A.841, Text Fig. 8.3). Below is shown a detail of the occlusal end of the salient angle

Fig. A.874 SEM microphotograph of of experimental digestion on a modern rodent upper incisor. The tooth was exposed to HCl for 4 h and later immersed in enzymes (protease) for 23 h. Note the incisor lost the enamel on the very tip of the tooth. Below is shown a detail of the tip of the incisor (see A.818 and A.819)

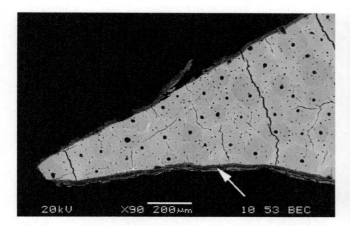

Fig. A.875 SEM microphotograph of a sectioned modern bone fragment regurgitated by hyena. Note the fringe of grayish color all around the bone cross-section (white arrow) shows the depth of penetration of digestion into the compact bone. The bone histology in the interior of the bone is unmodified, with only some fissures across the bone. Chemical analysis of this bone fragment shows differences in chemical composition (see Text Fig. 8.5)

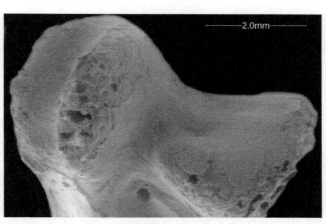

Fig. A.877 SEM microphotograph of the distal end of a femur of *Anas crecca* from the taxidermic collection of the MNCN prepared using enzymes. There is superficial penetration and perforations on the articular bone similar to the effects produced by digestion (see A.864 and A.397). Cracking is also observed on this specimen (see A.766). Similarities between digestion and intense enzymatic activity have been shown by (Denys et al. 1995; Fernández-Jalvo and Marin-Monfort 2008)

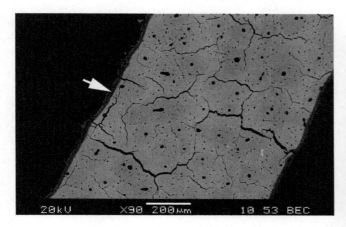

Fig. A.876 SEM microphotograph of a sectioned modern bone fragment that was regurgitated by hyena. Note the fringe of grayish color all around the bone section (white arrow). Chemical analysis shows (see Text Fig. 8.5) differences between the internal cortical bone not damaged by digestion and the external fringe damaged by digestion (see Fernández-Jalvo et al. 2002)

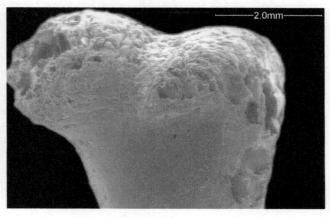

Fig. A.878 SEM microphotograph of the proximal end of a femur of *Anas crecca* from the taxidermic collection of the MNCN prepared using enzymes. Prolonged exposure to enzymes may penetrate the bone cortical layer and perforate the bone surface, similar to the effects produced by digestion (see A.864, A.397 and A.399)

Fig. A.879 Modern red deer vertebra from Riofrío. Roots may affect the bone surface and penetrate into the bone through chemical corrosion produced by the symbiotic combination of roots and bacteria, or of roots and fungi (A.229)

Fig. A.880 SEM microphotograph of a modern vascular plant root partially removed from the bone surface that acts as substrate. The roots of most vascular plant species enter into symbiosis with mycorrhizal fungi. A large range of other organisms, including bacteria, are also closely associated with roots. The area corroded by the root (double headed arrow) is wider than the size of the root (see also A.228 and A.229)

Fig. A.881 SEM microphotograph of root marks on a fossil rodent bone showing the characteristic cracking in the interior of the corroded area made by the root (A.779, A.781, A.782 and A.883). Field width equals 2.8 mm

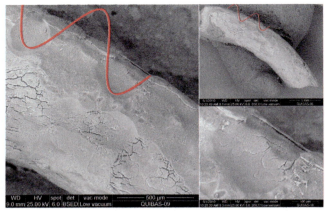

Fig. A.882 SEM microphotographs of root marking on a fossil rodent (see A.426, A.427 and A.428) incisor from Quibas (Murcia, Spain) Early Pleistocene. The site is a karstic system with wide connections to the open air environment, although the sediment is heavily cemented and indications of high relative humidity. This tooth surface is heavily affected by plant roots, the shape of which is not branched, but sinuous, hollowing both the enamel and the dentine (specimen provided by A. Cuadros)

Fig. A.883 SEM microphotograph of a fossil rodent incisor corroded by roots affecting both the enamel and dentine (A.882 and A.884) and showing internal cracking (A.425)

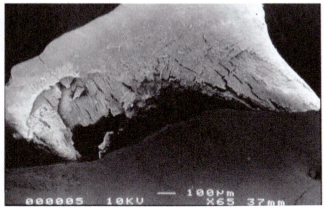

Fig. A.884 SEM microphotograph of the base of a rodent molar with enamel, dentine and tooth root (A.423) affected by plant roots and cracked, from Monte di Tuda (Corsica), see A.230 and A.231. This is a cave shelter, and sediments dug are partly exposed to the open air environment

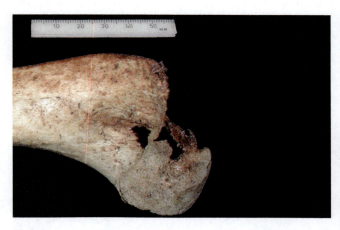

Fig. A.886 Distal end of a juvenile monitored horse femur exposed on the ground for 11 years from Neuadd 2, (pH 4) showing corrosion of the articular surface and epiphysis. The bones of this specimen were on the ground but for the most part were covered by dense ground vegetation consisting of grasses and ericaceous plants such as *Calluna vulgaris*. Relative humidity was beneath the vegetation cover was monitored and found to be constantly high (Andrews and Armour-Chelu 1998). See Text Fig. 8.1 and A.902

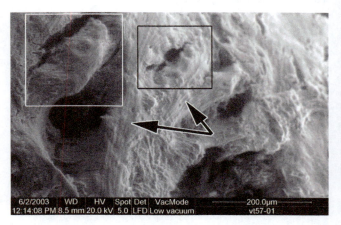

Fig. A.887 SEM microphotograph of the bone surface of the distal end of juvenile monitored horse femur exposed on the ground for 11 years (ND9 from Neuadd 2, pH 4, A.886). The heavily corroded bone surface viewed on the SEM (white left square) is similar in effects to digestion (see A.744) and exposing some histological traits (e.g.: osteons, black square) lamellae (arrows). Fernández-Jalvo et al. 2010

◀ **Fig. A.885** SEM microphotograph of the base of a rodent molar with enamel, dentine and tooth root affected by roots; from Monte di Tuda (Corsica) see also A.230 and A.231. Detail of the corroded enamel and dentine

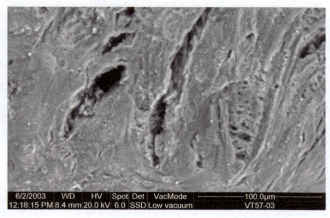

Fig. A.888 SEM microphotograph of the bone surface of the distal end of juvenile monitored horse femur exposed on the ground for 11 years (ND9, Neuadd 2, pH 4, A.887). Another area of the bone surface and corrosion that is similar to digestion (see Chap. 7 organic cracking)

Fig. A.889 Top: modern monitored horse scapula exposed on the ground for 11 years at Neuadd 2 (pH 4) covered by algae. The bones of this skeleton were on the surface of the ground but for the most part were covered by dense ground vegetation consisting of grasses and ericaceous plants such as *Calluna vulgaris*. The bone surface is strongly corroded (Andrews and Armour-Chelu 1998 see Text Fig. 8.1). Bottom: SEM microphotograph detail of surface corrosion similar to digestion cracking (A.745)

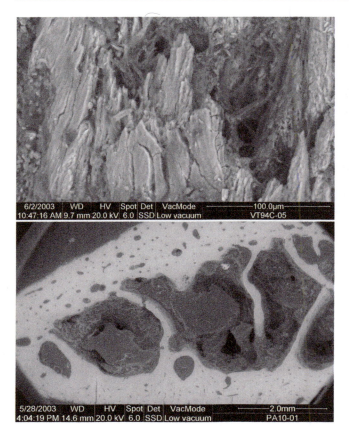

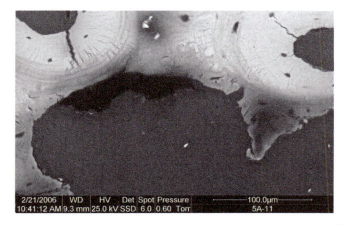

Fig. A.890 SEM microphotograph bone surface of monitored horse scapula from Neuadd 25 exposed for 10 years (ND21). This specimen was in a shallow pond that dried up periodically, exposing the bone on the surface of wet mud. The bone surface is covered by algae (see A.264) that produced heavy exfoliation and corrosion on the surface. The bottom photograph shows high penetration of algae through cancellous tissue and histological features, extending into the interior of the scapula

Fig. A.892 Modern radius of sheep monitored and collected from a boggy area fed by a spring, calm water conditions, Neuadd 24 (pH 5.4). The bone surface appears completely corroded on both sides (see A.447 and A.452)

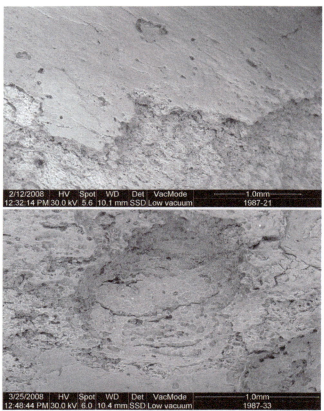

Fig. A.893 SEM microphotograph of the sheep radius monitored and collected from a seasonal spring in Neuadd (A.892). The bone surface is extensively damaged, with some areas more heavily perforated and other areas still preserving the outer cortical tissue. On the bone surface we have observed abundant diatoms as seen at A.447 and A.452, where diatoms appear concentrated inside the corroded areas

Fig. A.891 SEM microphotograph of monitored horse scapula (ND21) from Neuadd 25 after 10 years exposure. Corrosion is evident by the irregular surface and enlarged canaliculi on the corroded edge (see A.265, A.274 and A.436)

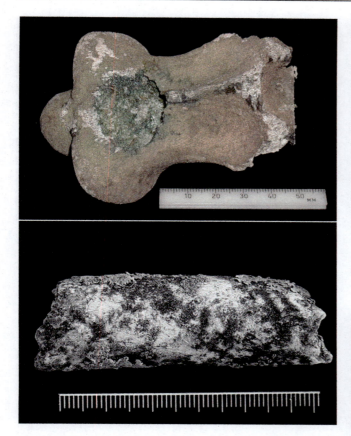

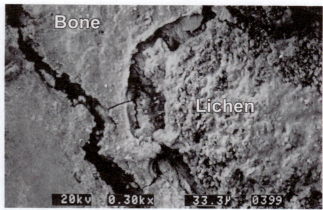

Fig. A.896 SEM microphotograph of lichen on surface and corroded bone underneath and limited by the edge of the lichen. The area where lichen was growing has homogeneous reduction of thickness (see A.440 and A.895). A.J. Sutcliffe collection

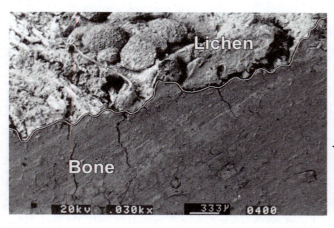

Fig. A.894 Top: axis of young monitored horse from Neuadd 2 exposed on the ground (pH 4) for 14 years, corroded in the area covered by lichen. The bones of this specimen were on the surface of the ground, but for the most part they were covered by dense ground vegetation consisting of grasses and ericaceous plants such as *Calluna vulgaris*. Bottom: long bone shaft fragment heavily covered by lichen from A.J. Sutcliffe's subarctic collection, photographed below A.895, A.896 and in A.437 and A.439

Fig. A.897 Sheep monitored humerus exposed for 17 years from Neuadd 19. The bone was resting on thick leaf litter and beneath a light tree canopy and a seasonally dense ground canopy of nettles (*Urtica*). The bone was completely covered by moss (most now removed), but moss does not have deep penetration capacity. The arrow points to superficial penetration of moss into cancellous tissue of the humerus head (see A.274 and A.898). Soil pH was 4.2

◀ **Fig. A.895** SEM microphotograph of a modern bone section showing superficial corrosion by lichen growth. Lichen still covers the bone surface at the top of the image, and cracks penetrate from this into the bone (A.439). The edge of the bone shows an irregular line adapted to the shape of the lichen's spore bodies. A.J. Sutcliffe collection (A.894 bottom and A.437 top)

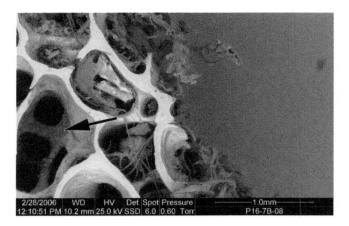

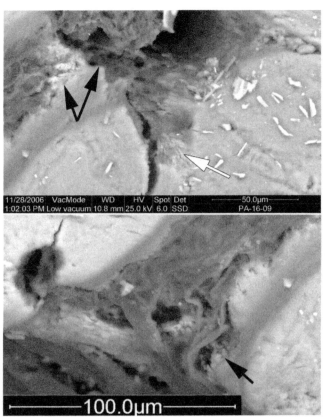

Fig. A.898 SEM microphotograph of the sectioned end of a proximal sheep monitored humerus from Neuadd 19 exposed for 17 years. The photograph shows superficial penetration of moss into cancellous tissue of the humerus head, indicated by a white arrow on A.897. The moss penetrates less than 1 mm into the bone, as shown on the second row of cancellous tissue (black arrow) which is completely empty

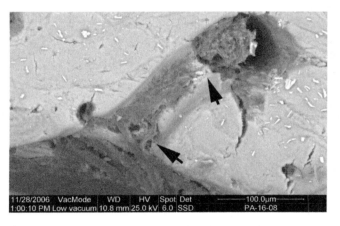

Fig. A.900 SEM microphotograph of a transverse section of the shaft of sheep monitored humerus (ND4 from Neuadd 19) exposed for 17 years on the ground (pH 4.2) resting on thick leaf litter and beneath a light tree canopy and a seasonally dense ground canopy of nettles. The bone was completely covered by moss when it was collected. Both pictures show damage on the bone surface underneath the moss, both by penetration (black arrows) and by enlarged canaliculi (white arrow) (see A.256, A.274, A.783 and A.784)

Fig. A.899 SEM microphotograph of a sheep monitored humerus from Neuadd 19 sectioned at the shaft. The picture shows penetration of moss through histological features. Superficial corrosion (black arrows) can be distinguished as result of moss (see A.256, A.274, A.783 and A.784) but penetration is always superficial

Fig. A.901 SEM microphotograph of a transverse section of the shaft ▶ of sheep monitored humerus (ND4 from Neuadd 19) exposed for 17 years on the ground (pH 4.2) resting on thick leaf litter and beneath a light tree canopy and a seasonally dense ground canopy of nettles. The bone was completely covered by moss when it was collected. The section shows little penetration into the bone cortex, for moss is not as destructive as lichen (A.894, A.895 and A.896)

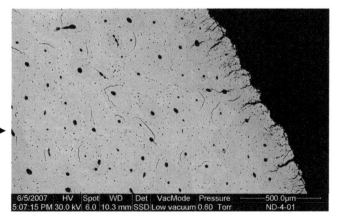

Fig. A.902 Three modern monitored horse limb bones from Neuadd 2. They were exposed under thick ground vegetation in a moorland setting (soil pH 4.0). Corrosion has attacked both proximal and distal ends. This slow corrosion mimics the effects of carnivore "scooping out", an example of which is seen on the zebra femur (left, Collection C.K. Brain) chewed by hyena The bones were exposed for (*1*) 5 years, (*2*) 11 years (*3*) 13 years, all from the same horse skeleton. See Text Fig. 8.1

Fig. A.904 Bone surface corrosion may cover the whole bone surfaces (fossil from Senèze) or corrosion may affect localized areas of bone. This particular corrosion is mainly related to the acidic volcanic sediments (A.805) and the relative immaturity of the individual. Bone surface corrosion, also known as "soil corrosion", is the result of vegetation, humidity and/or "soil acidity" (A.902), but it may also occur in caves (A.353 and A.786)

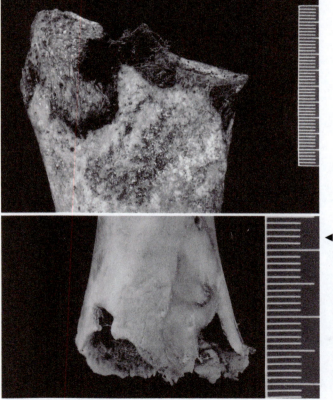

◀ **Fig. A.903** Modern monitored femur (and tibia) exposed by removal of the covering vegetation showing them resting on damp soil. Humidity measurements taken over 12 months showed the relative humidity to be maintained close to 100% regardless of the humidity above the vegetation layer. Soil corrosion is a combination of the acidity of the soil, vegetation and high humidity (Andrews and Armour-Chelu 1998). Observations made at Neuadd experimental field of different monitored carcasses (shown here and in A.805) indicate that putrefaction decay processes combined with corrosion by vegetation in a highly humid substrate cause what is commonly named, "soil corrosion or soil acidity" (see Fernández-Jalvo et al. 2010). Some cases of corrosion may also be produced by microorganisms (A.916), but their activity is not linked to the acidity of the substrate

Fig. A.905 SEM microphotograph of a modern specimen from the MNCN taxidermic collection, prepared using natural maceration. The bone surface had bacteria lying on it (see A.294, A.907 and A.908). After inspection of these specimens, we advised that the use of appropriate equipment (gloves) should be used to handle these types of laboratory-prepared specimens. (Fernández-Jalvo and Marin-Monfort 2008)

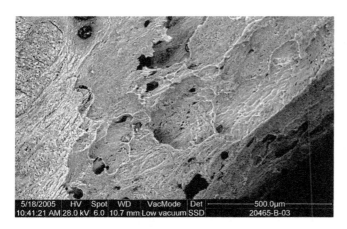

Fig. A.906 SEM microphotograph of the surface of a modern bone prepared using natural maceration methods (see A.294, A.907 and A.908). The irregular surface corresponds to bacterial corrosion that penetrated into the bone (Taxidermic Collection, MNCN)

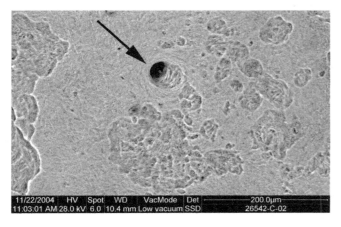

Fig. A.907 SEM microphotograph of damage on the diaphysis on a modern bone prepared using natural maceration showing damage and degradation of the surface by corrosion. The perforation pointed to by an arrow is a nutritient foramen (see A.294, A.907 and A.908). Bacteria may use histological traits to penetrate into the bone. Taxidermic Collection, MNCN

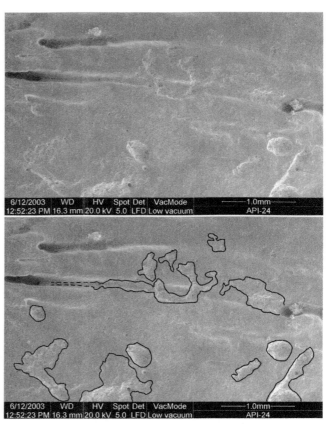

Fig. A.908 SEM microphotograph of damage on the diaphysis of a human bone fragment from Apigliano burial site (Italy). The bone surface shows damage like that seen on taxidermic specimens. Bacterial damage areas (MFD = microscopic focal destruction) are also located around nutrient foramina (A.907). Bacterial damage is shown demarcated below. For further information on the site, see Smith et al. (2002). Specimen provided by C. Smith

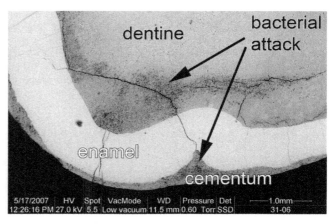

Fig. A.909 SEM micrograph of a tooth transversally cut showing that the dentine and cementum are intensively attacked by bacterial corrosion, but the enamel is not (A.483). Specimen provided by E.M. Geigl

Fig. A.910 Corroded surface bone of a fossil from Concud with distinctive circled whitish areas on the bone surface. SEM analysis has shown that this corrosion corresponds to a new type of taphonomic bioerosion (Pesquero et al. 2010), that peripherally penetrates into the periosteal cortical layer of the compact bone (see A.485 and A.492). Courtesy of D. Pesquero

Fig. A.913 Metapodial of an adult modern monitored horse exposed for 15 years (ND5 from Neuadd 41/1, see also A.292. Rotten surface with a whitish dusty surface. The general appearance to the naked eye is similar to A.910, but the cut section showed characteristic traits of bacterial attack (see A.466) differing from the modifications seen on fossils from Concud (A.491)

Fig. A.911 SEM microphotograph of the bioeroded surface of fossil (fossil from Concud). The bone surface appears corroded, heterogeneously and heavily pitted, with the pits dispersed on the surface, producing increased porosity, but with some intact patches (A.486, A.488 and A.910). Penetration into the bone from the periosteal cortical layer suggests that this was caused by an exogenous agent. Shape and size of this bioerosion tunneling differs from bacterial attack

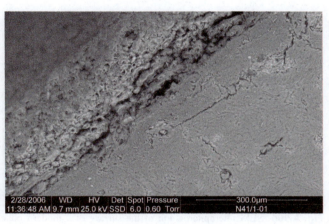

Fig. A.914 SEM microphotograph of an adult modern monitored horse metapodial (ND5, from Neuadd 41/1 exposed for 15 years). This carcass was fully exposed to weathering for 23 years, but some bones were collected after 15 years exposure. There was little or no vegetation cover over the bones. The top layer of the bone (PCL) has a restricted porous layer that seen at higher magnification is identified as bacterial corrosion (A.464 and A.915)

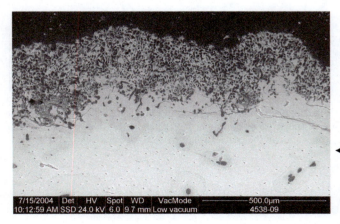

◄ **Fig. A.912** SEM microphotograph of sectioned fossil from Concud showing peripheral bioerosion penetrating from the surface of the bone. Penetration is superficial (less than 300 microns deep) having a random distribution which is independent of the bone histological features (no relationship with Havers canals and osteons A.489 and A.490). (Pesquero et al. 2010)

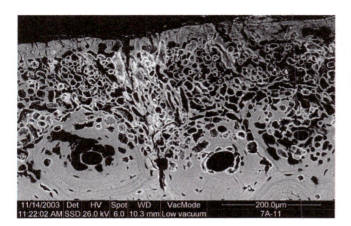

Fig. A.915 SEM microphotograph of a sectioned modern monitored horse metapodial (ND5 from Neuadd 41/1, exposed for 15 years). Bacterial attack is peripheral, but some of the canals of Havers are surrounded by bacterial attack areas (MFD). Bacteria may enter the bone through histological traits (e.g. nutritient foramina, A.907 and A.908). In order to confirm the origin of bacteria, surface examination is needed. However, the small area that can be analyzed does not always provide conclusive results

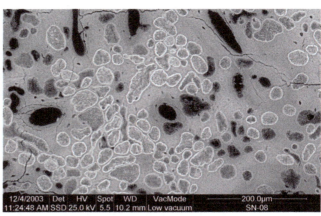

Fig. A.918 SEM microphotograph of a sectioned fossil bear ulna from Sima de los Huesos, Atapuerca (MNCN Collection). This photograph shows intense bacterial attack on the surface of the bone that caused the superficial pitting shown in A.473. This intensity is relatively homogeneous (compare this photograph with A.474 and A.475) from the outer cortical layer to the marrow cavity

Fig. A.916 Mandible of immature fossil bear from Atapuerca (MNCN collection) showing irregular pitting. SEM analyses of a transverse section of another fossil bear bone (A.473) of similar age at death and from the same site has identical corrosion on the surface, and it appears that this corrosion is produced by bacterial attack

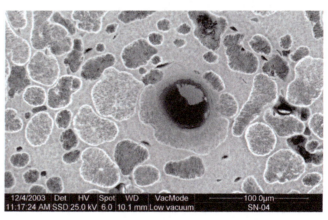

Fig. A.919 SEM microphotograph of a sectioned fossil bear ulna from Sima de los Huesos, Atapuerca (MNCN Collection) The arrangement of bacterial patches accords to histological traits of the bone (around Havers canals and osteons A.475 and A.459 bottom). All this may suggest that body putrefaction occurred before skeletonization. Bacterial attack might be favored by the humid and constant temperature and homogeneous post-depositional cave environment

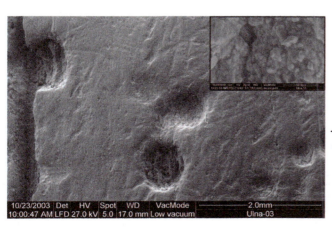

◀ **Fig. A.917** SEM microphotograph of immature fossil bear ulna (A.473) from Atapuerca (MNCN collection). The bone surface may not give clear information of the etiology of the modification, particularly when consolidants or other preservatives have been used. Top right square, at the interior of the round depressions shows an agglomerate of dust mixed with the consolidant that hinders taphonomic interpretations based on the bone surface. In such cases, fossils have to be cut in order to obtain conclusive results (A.918 and A.919)

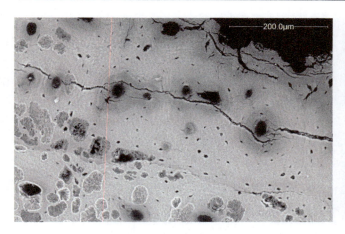

Fig. A.920 SEM microphotograph of a cross-section of a fossil bear bone from Sima de los Huesos, Atapuerca (MNCN Collection) which was attacked by bacteria. The distribution is homogeneous except at the periosteal cortical bone edge (PCL), where the intensity is lower and some localized areas show that bacteria reached the outside (forming pits on the bone surface, A.916 and A.473)

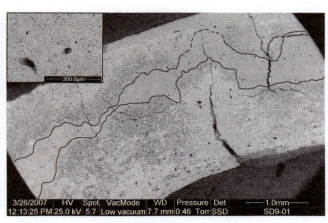

Fig. A.923 SEM microphotograph of a transverse section of a Neolithic fossil sheep bone. The bone is heavily corroded by bacteria on the PCL and ECL, except a fringe of absence or reduced bacterial activity at the MCL (see Text Fig. 2.2). The small box shows a detail of random bacterial distribution around the canals of Havers in this specimen. Specimen provided by E.M. Geigl

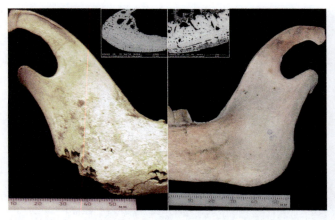

Fig. A.921 Correlation of superficial modifications and histology is not simple. Left: modern monitored sheep mandible (ND27 from Neuadd 30) shows corroded surface at the inferior angle. The section seen on SEM microphotograph (small box), shows intact histology. Right: modern monitored sheep mandible (ND20 from Neuadd 15/4, see A.471) has intact bone surface, but the histology is heavily affected by bacteria on the inferior border and the alveolar sockets. providing relevant environmental information (A.472, Fernández-Jalvo et al. 2010)

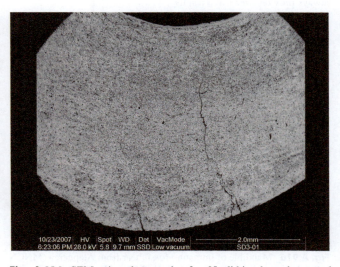

Fig. A.924 SEM microphotograph of a Neolithic sheep bone and extensive bacterial attack throughout the bone section. Heavy bacterial attack and MFD around Havers canals (e.g. A.475 small box, A.459 bottom) may suggest endogenous bacteria. A low incidence of or randomly dispersed bacterial attack (A.467) may be due to scavenging, which reduces putrefaction, or to environmental conditions, which may be unsuitable for bacterial activity. Specimen provided by E.M. Geigl

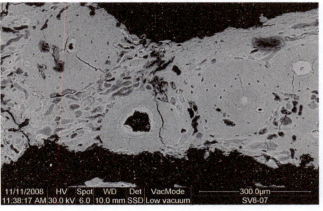

◀ **Fig. A.922** SEM microphotograph of a Neolithic fossil bone showing bacteria entering into the bone through the cortex and surrounding osteons. This is a close-up view of the same specimen photographed in A.926 bottom picture. Exogenous bacteria may also gain access to the interior through nutrient foramina (A.907 and A.908) and enter into the Haversian system. In such cases, exogenous and endogenous bacteria are difficult to distinguish. Specimen provided by E.M. Geigl

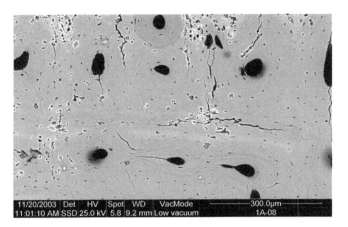

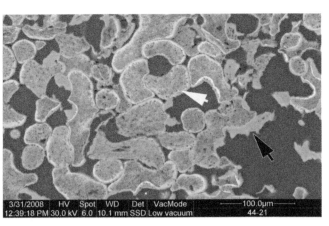

Fig. A.925 SEM microphotograph of a modern monitored sheep tibia (ND16 from Neuadd 1) buried in acidic soil (pH 3.5) for 6 years. The low intensity of bacterial attack and its broad distribution of destructive foci (scattered, dispersed, non-histologically arranged and/or affecting periosteal and endosteal layers, but not medial layers of cortical bone) may suggest a late (exogenous) bacterial attack (A.467) probably from the soil (Fernández-Jalvo et al. 2010)

Fig. A.927 SEM microphotograph of a sectioned fossil bone from Azokh Cave. This fossil bone has a crackled surface aspect, similar to A.788. The SEM photograph of the section shows intense bacterial attack, and the original bone (black arrow) is almost "eaten away". Corrosion that caused that crackled surface (possibly from guano) apparently dissolved the bone, but bone re-deposition by bacteria resisted dissolution (white arrow)

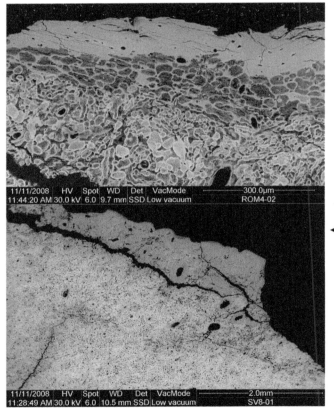

◄ **Fig. A.926** SEM microphotograph of two Neolithic fossil bones. Bacteria have intensively affected these specimens, but the outer edge of the periosteal cortical layer (PCL) is not reached by bacteria (see A.465). This is not infrequent and it is important, because geochemical and biochemical protocols of sample preparation reject this layer to prevent contamination. Systematic rejection of the outer layers may destroy undamaged tissues containing original information. Traditionally, microbial degradation has been considered to degrade DNA macromolecules (Päabo et al. 2004). Furthermore, traces of environmental agents that favor preservation/modification of original molecules of bones during fossilization (e.g. weathering) are removed with the outer edge of the PCL. Periosteal, endosteal and medial layers of bones need more detailed geo- and biochemical studies. Specimen provided by E.M. Geigl

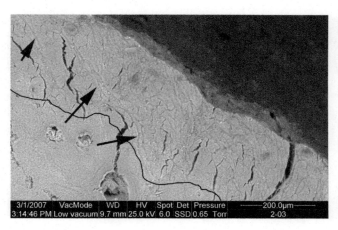

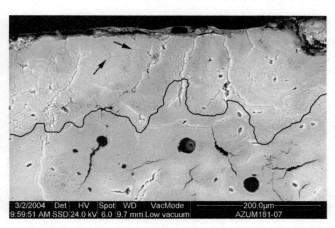

Fig. A.928 SEM microphotograph of a transverse section of a modern bone cultured with fungi (*Chaetomium*) for 5 years at the laboratory. Fungal tunneling (black arrows) is abundant on the periosteal cortical layer (PCL, see Text Fig. 2.2). Fungal tunneling is known as Wedl tunnels (according to Hackett 1981). A crumbly irregular texture can be distinguished at the PCL of the bone (A.928, A.929 and A.930)

Fig. A.931 SEM microphotograph of a section of a fossil bone from Azokh Cave. The periosteal cortical layer appears crumbly (solid line) and Wedl tunnels (black arrows) can be distinguished (A.929 and A.930). Wedl tunneling is more difficult to distinguish in fossils than bacterial attack

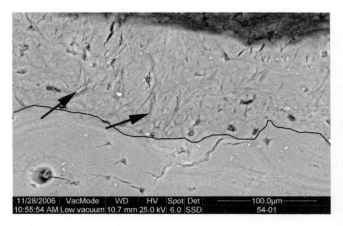

Fig. A.929 SEM microphotograph of a transverse section of a modern monitored sheep vertebra (ND18) exposed on the surface for 6 years from Neuadd 1, in a protected area (pH 3.5). Wedl tunnels are present in the periosteal cortical layer (black arrows) and pitting. An irregular fringe of crumbly irregular texture, very similar to damage obtained from fungi cultures in laboratory on bones (A.928 and A.933), can be distinguished at the PCL of the bone (Fernández-Jalvo et al. 2010)

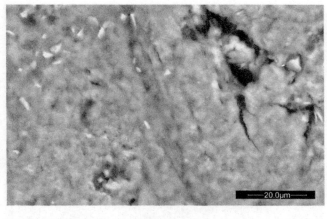

Fig. A.932 SEM microphotograph of a transverse section of a modern bone treated for 5 years with laboratory cultured fungi (*Chaetomium*). Detail of Wedl tunneling (see also A.283). This is an incipient Wedl tunnel, narrower than described in the literature where they are considered to be greater than 10 microns wide. These tunnels are, however, consistent with all specimens cultured in the laboratory and it confirms that Wedl tunnels are definitively made by fungi

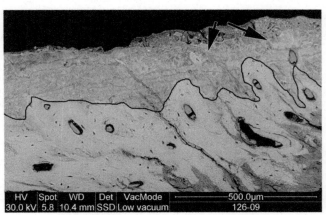

◄ **Fig. A.930** SEM microphotograph of a sectioned fossil bone from Azokh Cave. The periosteal cortical layer appears crumbly in a similar way to bones with laboratory cultured fungi (A.928 and A.933) and modern bones from Neuadd (A.942). Wedl tunnels in this specimen have been re-mineralized (black arrows) see also Text Fig. 8.2

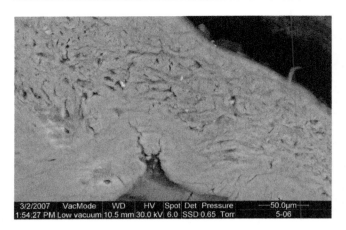

Fig. A.933 SEM microphotograph of a transverse section of a modern bone cultured with fungi (*Mucor*) for 5 years showing crumble texture at the PCL as well as Wedl tunnelling (see A.289 cultured, A.290 from Neuadd). Note a more pitted texture on the outer fringe

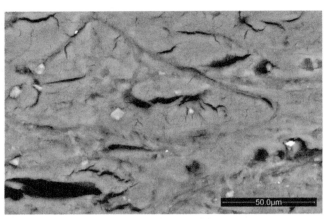

Fig. A.936 SEM microphotograph close up view of previous transverse section A.934 and A.935, of a modern bone cultured with fungi (*Cunninghamella*) for 5 years. Detail of Wedl tunneling. Note also the pitting

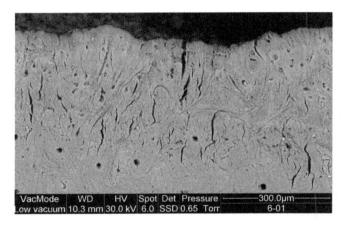

Fig. A.934 SEM microphotograph of a transverse section of a modern bone cultured with fungi (*Cunninghamella*) for 5 years showing crumbled and crackled texture at the PCL as well as Wedl tunneling (A.928) and pitting (A.936 and A.941)

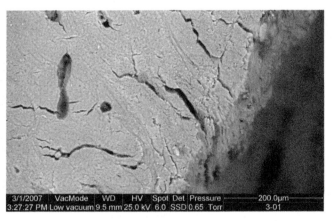

Fig. A.937 SEM microphotograph of a transverse section of a modern bone cultured with fungi (*Penicillium*) for 5 years showing crumbly irregular texture at the PCL as well as Wedl tunnelling (see A.289, A.290 Neuadd)

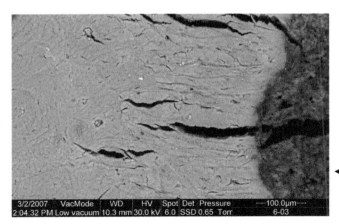

◄ **Fig. A.935** SEM microphotograph close up view of previous transverse section A.934, of a modern bone cultured with fungi (*Cunninghamella*) for 5 years showing crumbled texture at the PCL as well as Wedl tunneling

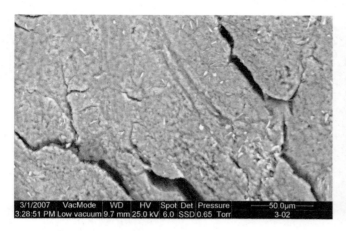

Fig. A.938 SEM microphotograph close up view of previous transverse section A.937, of a modern bone cultured with fungi (*Penicillium*) for 5 years. Detail of Wedl tunneling

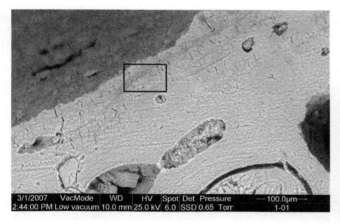

Fig. A.939 SEM microphotograph of a sectioned modern bone cultured with fungi (*Coelomycetes*) for 5 years showing relatively crumbled texture and cracks at the PCL edge of the bone, as well as Wedl tunneling (box outlined is enlarged in A.940)

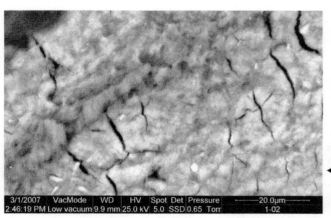

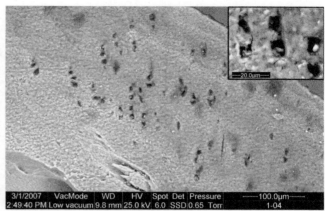

Fig. A.941 SEM microphotograph of a sectioned modern bone cultured with fungi (*Coelomycetes*) for 5 years. No characteristic fungal damage, as seen in previous specimens, was identified in this bone. Wedl tunneling is not recognized but a distinctive pitted surface affects the PCL of the bone (A.929 and A.934). Small box on top right is a detail of these pits that could be anomalous bone resorption or a pathology. Further, laboratory experiments have to be done to confirm these pits are made by fungi

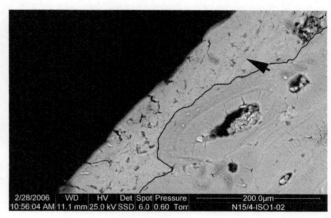

Fig. A.942 SEM microphotograph of a modern monitored sheep vertebra from Neuadd 15-4. In this transverse section, a crumbly irregular and pitted texture (A.928 and A.942), together with Wedl tunneling (black arrow), can be distinguished, affecting the histological outer circumferential lamellae of the bone (solid line)

◀ **Fig. A.940** SEM microphotograph close up view of previous A.939 transverse section of a modern bone cultured with fungi (*Coelomycetes*) for 5 years showing crumbly texture at the PCL edge of the bone. This shows a detail of Wedl tunneling

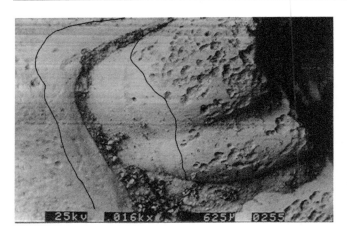

Fig. A.943 SEM microphotograph of dental enamel modification on a fossil tooth from Rusinga Island. The Miocene sediments are derived from an alkaline volcano. Much of the enamel is heavily corroded, but it is absent from an area between the tooth and the jaw along the alveolar margin. This may be due to retention of soft tissue (the gum) at the alveolus which resisted decay (see A.946) and which may have protected the base of the tooth

Fig. A.945 SEM microphotograph in detail of the modification on a fossil tooth from the Songhor Miocene alkaline sediments. This corrosion is similar to damage referred in Concud also with alkaline sediments. Corrosion has a characteristic honeycomb pattern, and the agent may be a microbial mat or biofilm that covered the mandible. Corrosion appears superficial and heavily affects enamel. Bone is less affected as in Concud, and dentine was also observed to be less damaged (A.502)

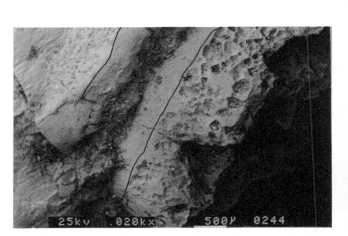

Fig. A.944 SEM microphotograph of similar modification on another fossil tooth from Songhor. The Songhor sediments are highly alkaline. Similar to the Rusinga jaw A.943, corrosion is absent on a fringe between the alveolar margin and the jaw (solid lines). Late retention of the gum on this part (A.946) is frequent and might provide protection. The unknown agent that produced this corrosion had to act before decay of soft tissues or soon after this. See similarities with A.494

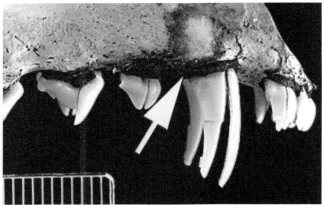

Fig. A.946 Skull of modern fox. The teeth are heavily cracked by weathering. Part of the gum (white arrow) is retained after the skull has lost most soft tissues (A.656). This situation might have happened in Songhor and Rusinga specimens. Remains of the gum probably were still present when corrosion affected these jaws (A.944). The agent that produced the corrosion at these fossil sites remains unknown, as also in Concud (A.493)

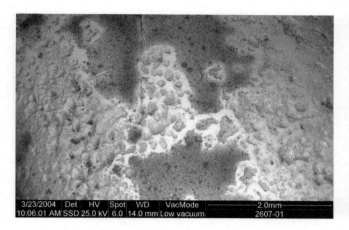

Fig. A.947 SEM microphotograph of intense pitting on the enamel of a fossil tooth from Concud alkaline lakeshore site. Similar corrosion has been observed on several teeth from other sites related to calm water (Neolithic – Charcognier) and (Miocene – Songhor, Rusinga, see A.943 and Los Casiones A.499). Corrosion affects mainly enamel, although dentine and bone show milder corrosion. This corrosion has a characteristic honeycomb shape

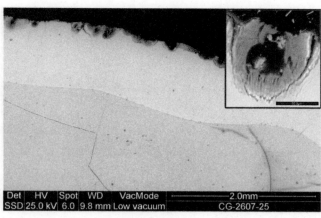

Fig. A.949 SEM microphotograph of a cut section of a Miocene tooth from Concud. Corrosion is superficial, mainly affecting the enamel peripherally. The small box on top right side shows a detail of one of the rounded conical regular perforations that penetrates superficially into the enamel, possibly by the action of biofilm growth. The pattern of this modification is distinctive and penetrates only a few microns into the specimen (A.495)

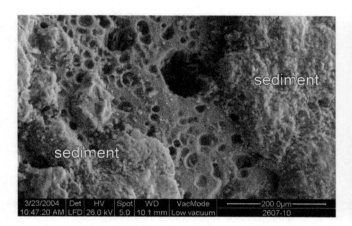

Fig. A.948 SEM microphotograph of a fossil tooth of *Hipparion* from Concud, showing intense pitting on the enamel. Traits of selective action on enamel (rather than dentine or bone), superficial damage (A.495) and repetitive shape pattern, suggest to us this is a bioerosion maybe produced by a microbial mat or biofilm

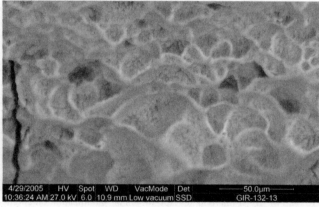

Fig. A.950 SEM microphotograph of a fossil tooth from Charcognier (Neolithic, France, riverbank deposits) showing the modification found on enamel surface. In contrast to bacterial attack, the enamel is affected and the modification has a honeycomb shape (A.496, A.497, A.498, A.499 and A.500). Specimen provided by E.M Geigl

Part IV

Modification by Loss of Bone Tissue or Skeletal Elements

Chapter 9
Breakage and Deformation

Breakage of vertebrate skeletal elements can be assessed both as fracture patterns and fragmentation. Fracture patterns are defined by the morphology of breaks (Villa and Mahieu 1991). Fragmentation is defined in terms of the numbers and types of fragments into which bones have been broken (Villa and Mahieu 1991), and it may be expressed quantitatively as a fragmentation index (Andrews and Bello 2006). Deformation refers to bone that has been distorted out of shape but not broken in the process.

All measures of breakage of skeletal elements in assemblages of fossil bone should distinguish between breakage before burial, during burial, after burial, after preservation in sediment (diagenesis), and during and after excavation and preservation. The agents operating at these five stages are listed below.

Agents and Processes

Agents causing breakage operate at the death of an animal, soon after death, or after deposition of skeletal remains. Predation by carnivores or butchery by humans are examples of initial causes of breakage; scavenging and trampling are examples of breakage soon after death; and compression or sediment action are later examples of breakage, although not necessarily the last, as breakage can also occur during the present day exposure of fossils, their collection and/or screening. Skeletal elements that go into the ground as complete bones may be broken after burial but the fragments remaining in anatomical position until becoming separated by exposure to recent erosion, by screening the sediment or by careless excavation. Dry or wet screening in general is potentially destructive on already broken bones, and it may even cause breakage of previously intact fossil bones. Any analysis of breakage based on fossils from screening should include for comparison some careful excavation of part of the fossil assemblage while still in place in the sediment.

Agents causing deformation and/or changes in the original shape are the same, but they produce different effects at different stages according to the substrate and environment (wet/dry) of deposition/burial:

Trampling by large animals A.1007
Chewing by carnivores A.1013, A.1021 and A.1022
Chewing by herbivores A.1035 and A.1038
Scavenging by raptors A.997, A.998 and A.1032
Butchery by humans A.979, A.984, A.992, A.1005 and A.1006
Sediment pressure and/or movement A.952, A.956 and A.958

Characteristics

The characteristics of broken bone depend very much on the type of bone, on the type of substrate and on the pressure imposed on it by the agent involved. The type of bone includes both the skeletal element involved, its developmental age (mature or immature), and the time since death of the individual. The condition of the bone, whether green or dry, results in different breakage patterns, but this is imperfectly correlated with bone age as much depends on the depositional microenvironment. Characteristic breakage patterns are produced for different skeletal elements, for immature and adult bone, and for fresh as opposed to dry (old) bone. Element shape is also important, as was seen for transport (Voorhies 1969), and small compact bones such as carpals and tarsals, with few projecting processes, tend to be relatively complete (Darwent and Lyman 2002). Patterns of breakage are often not diagnostic, and it is difficult to distinguish between many of the agents producing the breakage. The greatest difference is that between post-depositional breakage resulting from sediment pressure or impact, operating at a time when bones have lost much of their organic

© Springer Science+Business Media Dordrecht 2016
Yolanda Fernández-Jalvo and Peter Andrews, *Atlas of Taphonomic Identifications: 1001+ Images of Fossil and Recent Mammal Bone Modification*, Vertebrate Paleobiology and Paleoanthropology, DOI 10.1007/978-94-017-7432-1_9

matter and have dried out, and other agents which have maximum impact when bones are still green or fresh.

Morphology of Breaks

The classification of break morphology of vertebrate bone is based on Villa and Mahieu (1991) and it is described in terms of fracture outline, fracture angle, fracture edge and the extent of survivorship of the shaft circumference of long bones.

Fracture Outline

The orientation of fractures can be curved or pointed (A.1008) (angle oblique to the long axis of the bone and often ending in a sharp point); transverse (perpendicular to the long axis of the bone) (A.956) or mixed. Curved or spiral breaks usually occur on fresh or green bone, and they can often be attributed to human action or carnivore chewing (Fig. 9.1, see also Fig. 4.7). Other agents cannot be ruled out; for example bones may be broken if an animal falls to its death, producing spiral breaks on its bones; transport in high energy conditions of bones or carcasses soon after death may produce spiral breaks; and finally, heavy trampling over a newly dead carcass may also produce spiral breaks. None of these agents, however, is likely to produce the abundance of curved breaks seen in carnivore or human butchery assemblages. Transverse breaks are typical on mineralized bone and are more likely to be the result of post-depositional and post-burial factors such as sediment movement or compaction. Transverse breaks can also be caused on subfossil bones by falling blocks before burial (Villa and Mahieu 1991), trampling when the bone was shallowly buried, diagenetic modifications during fossilization, sediment compaction and by local micro-faulting.

Fracture Angle

The angle between the break and the bone cortical surface is the fracture angle, and it may be oblique (either acute or obtuse), perpendicular, or both for fractures that have variable angles (i.e., intermediate). Oblique angles occur on green bone, mixed on dry and transverse on buried bone, and their presence parallels the fracture outline.

Fracture Edge

The margin of the fracture may be smooth or jagged. Smooth fractures are usual on green bone, jagged on dry bone, and the indications are the same as that seen above for fracture outline. These three features (fracture outline, angle and edge) can be combined to distinguish between green bone fracture and dry stick fracture, with all that this implies about the age of the bones at time of breakage.

It appears that the morphology of breaks on small mammal bone is similar to that of large mammal bone, for which these features were first described. No systematic work has been done on small mammal bone breakage, but all three features of breakage described above have been observed, resulting from similar processes, although acting at a smaller scale. A sample of broken small mammal bones is shown in Fig. 9.2.

Peeling

A characteristic type of breakage, where the edges are "heavily" jagged was described by White (1992) as peeling. Peeling leaves a roughened surface with parallel grooves or fibrous texture, and it is produced "when fresh bone is fractured and peeled apart similar to bending a small fresh twig from a tree branch between two hands" (White 1992: 140; Fig. 9.1).

Shaft Circumference of Limb Bones

The extent to which the circumference of shafts of limb bones is intact is used to describe the completeness of limb bone shafts. They can have complete circumference for at least a portion of the shaft, more than half complete, and less than half complete (A.1015). Fossils with complete circumferences have not been split down the middle by any agent, whereas the other two categories both show a degree of splitting which is a general indication of carnivore action, but both may have spiral or stepped breaks at either end of the shaft (A.1014). The more extreme the splitting of the shafts, and the greater the number of split bones, the greater the jaw strength of the carnivore (relative to the size of its prey) causing the splitting. Hyenas produce the most extreme degree of splitting of long bones of large mammals, while broken bone assemblages with low frequency of splitting may be due to post-depositional damage. Splitting

Fig. 9.1 Four techniques used by humans to break the bone. Bones showing conchoidal scars are produced by any of the three techniques using stones. Top left: direct percussion against a rock (anvil) using a chopping tool (hammerstone) which has been experimentally seen (Rovira-Formento 2010) to be more effective and less destructive to bone than using a normal stone as seen in Fig. 4.8. Top right: breaking the bone against a stone, and bottom left: breaking the bone throwing a cobble on the bone. Bottom right: there are other techniques to break bones, such as using both hands and not a stone or cobble. This results in peeling as described by White (1992). Drawings by Javier Fernández-Jalvo, copied with permission

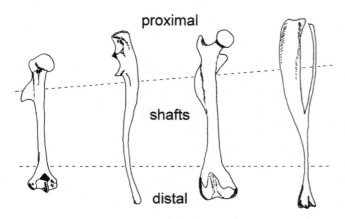

proximal

shafts

distal

Fig. 9.2 Diagrams of the four major limb elements for small mammals showing how the breakages of the elements are defined. Small mammals predated by small carnivores have similar types of breakage to large mammals

of limb bones also occurs as a result of human butchery to obtain access to the marrow, and in a butchered bone assemblage this may be shown by consistent patterns of breakage. Some skeletal elements are more resistant to splitting than others and this needs to be taken into account.

The morphology of breaks on small mammal bones is similar to that on large mammals (Hoffman 1988). Spiral breaks are common, but the splitting of limb bones when it occurs renders most skeletal elements unidentifiable (A.1021). Some predators of small mammals produce no breakage of the bones of their prey, for example some species of owl, typified by barn owls (Andrews 1990). Secondly, there are some owls that produce intermediate and variable breakage. Thirdly are the diurnal raptors and mammalian carnivores that produce up to 100% breakage. The range of values for a sample of predators studied in Andrews (1990) is shown in Table 9.1 and the breakage categories used in the table are taken from Andrews (1990). It is generally possible to identify the breakage patterns for small mammal predators into one of the three categories, but based on breakage alone it is usually not possible to identify the predator to species.

Fragmentation

Fragmentation of bones is also based on Villa and Mahieu (1991) and is described in terms of surviving breadth/length ratios, the proportions of skeletal elements surviving relative to the original intact bone, the proportions of articular and shaft elements (A.1020) and the proportions between isolated teeth and teeth in jaws.

Breadth/Length Ratios of Fragments

The ratios of breadth to length of bone fragments may be used to measure fragmentation, with narrower fragments indicating greater comminution of the bone assemblage. It is nearly always linked with increased splitting of bone, which may be due to carnivore or human action. On the other hand, if the breakage is so extreme as to leave many small fragments, it may be impossible to reconstruct the pathway by which the breakage occurred.

Extent of Shaft Fragmentation

The extent of fragmentation is measured by the proportion of skeletal elements remaining in relation to the original intact bones (Lyman 1994a).

- Limb bone shafts and ribs can be defined as complete or almost complete shaft, one half to three quarters of shaft, one quarter to one half of shaft, and less than one quarter of shaft, excluding the articular ends (Hoffman 1988).
- Mandibles for many mammal species are divided into complete, ascending ramus missing, inferior border of body missing, and in some cases only the diastema remaining (Andrews 1990).
- Maxillae are divided into complete, with zygomatic and nasal processes intact, and maxillary body only remaining.
- Axial bones such as the scapula are seen as intact, with scapular spine remaining, or glenoid process only remaining.
- Vertebrae are either intact or body only remaining.

These data can be summed in a breakage index, whereby completeness is measured on an arbitrary scale of 1 to 10, with 1 signifying less than 10% of bones complete, 2 signifying 10–20% bones complete, and so on up to 10 signifying 90–100% complete (Andrews et al. 2005). Although such an index can be useful in characterizing the degree of comminution of a bone assemblage, analysis of the fragmentation extent by itself does not indicate much about the process by which breakage occurs, and in particular it does not distinguish between pre-depositional and post-depositional breakage. It provides a general measure of fragmentation, and for human burials a measure of disturbance, and more specific conclusions must be based on the morphology of the breaks, as listed above.

An even simpler method of measuring fragmentation is to record the size distribution of bone fragments or specifically

Table 9.1 Breakage patterns of major skeletal elements in small mammals by 11 different predators

		Barn owl	African eagle owl	European eagle owl	Tawny owl	Little owl	Kestrel	Hen harrier	Mongoose	Genet	Coyote	Red fox
Humerus	Complete	99	65	82	53	33	44	22	30	33	7	0
	Proximal	0	5	7	7	33	4	7	29	13	38	8
	Shaft	0	6	0	12	16	27	39	9	10	17	9
	Distal	1	20	11	28	16	25	32	32	44	26	83
Ulna	Complete	97	91	97	69	100	32	60	8	54	25	0
	Proximal	3	7	3	31	0	52	40	92	46	75	67
	Shaft	0	2	0	0	0	8	0	0	0	0	33
	Distal	0	0	0	0	0	8	0	0	0	0	0
Femur	Complete	97	81	83	52	12	20	20	12	12	0	0
	Proximal	1	17	12	22	64	48	40	52	51	42	53
	Shaft	2	2	3	6	12	24	20	13	20	28	21
	Distal	0	1	2	20	12	7	20	23	17	30	26
Tibia	Complete	98	78	86	85	33	31	22	37	57	0	0
	Proximal	1	5	9	7	8	29	22	25	27	90	67
	Shaft	1	14	0	4	50	25	33	38	16	10	33
	Distal	0	0	5	4	8	14	22	0	0	0	0

Definitions of breakage are shown in Fig. 9.2

diaphysis fragments (Villa and Mahieu 1991). The alternative completeness index records the proportion of identifiable specimens (NISP) that are complete (Dodson and Wexlar 1979; Andrews 1990; Lyman 1994b). When NISP equals the minimum number of elements (MNE) fragmentation is zero, and the extent to which NISP is greater than MNE is another measure of fragmentation (Lyman 1994b).

Relative Proportions of Articular and Shaft Fragments

This relationship provides another index of breakage, since articular ends are preferentially destroyed by carnivore ravaging, leaving a preponderance of shaft fragments. Loss of articular ends is not likely to arise as a result of trampling, for although they are weaker than shafts, and may be more easily broken by sediment pressure or trampling, this is offset by the fact that they are more readily identified even as small fragments.

Relative Proportions of Isolated Teeth and Teeth in Jaws

This relationship provides an index indicating the extent to which mandibles and maxillae are broken (Andrews 1990). Large numbers of isolated teeth and low numbers of upper and lower jaws indicates bone loss, the teeth being stronger and more resistant to damage than bone. For example, if a fossil assemblage has 84 isolated molars, and there are 12 molars per individual (the number in many rodent species), it would be expected that 7 mandibles and maxillae would have been present in the original assemblage. If in fact there are just 5 jaws, and between them they have 10 teeth still in place in the jaws, the ratio between the actual number of teeth, 84 plus 10, and the number expected, 5 times 12, gives an index of the excess of teeth: 94 present compared with 60 places in the jaws. This provides a measure of the fragmentation and destruction of the bones of the jaws.

Deformation

Deformation refers to bone that has been distorted out of shape but not broken in the process. It can usefully be distinguished between distortion of articular ends, of limb diaphyses, and rounded bones such as skulls. Post-mortem distortions must be distinguished from bone pathologies (Piper and Valentine 2012).

Modification of Articular Ends

The ends of bones may be altered beyond recognition without actually being broken (A.865) (Brain 1969). Chewing of the ends of bones by predators or scavengers is common, and several species produce taxon-specific types of modification. Human chewing, for example, produces a characteristic splayed effect on the ends of bones (Fig. 9.3) that has also

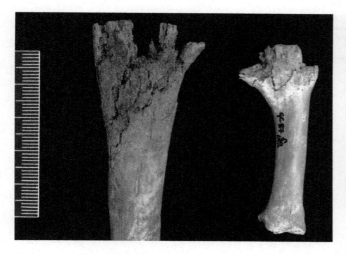

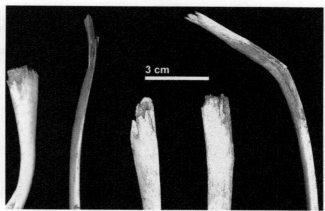

Fig. 9.4 Narrow bones like ribs may be deformed by human chewing, the bones bending but not breaking

Fig. 9.3 Suid rib on the left and a second left human metatarsal on the right GC87-30 from Gough's Cave. The suid rib was experimentally chewed by humans. The extensive deformation of the proximal ends, with depressed flakes of bone and splaying of the ends, is extremely similar in both bones, and may indicate human chewing on the metatarsal from Gough's Cave, reinforcing the evidence of cannibalism found in this site (Andrews and Fernández-Jalvo 2003)

been seen on fossils of Gough's Cave (Andrews and Fernández-Jalvo 2003). The splayed effect is due to the fact that humans have low crowned crushing teeth which crush the ends of bones rather than breaking them. Deformed ends of thin long bones (ribs and vertebral spines) are produced by chewing with flat occlusal surfaces with little cusp relief. Bending the ends of thin bones occurs when pushing the bone up or down using the hands and holding the ends between the upper and lower cheek teeth, and this is the easiest way to break the bone and suck the marrow (Fig. 9.4). Narrow bones like ribs are deformed and bent but not broken (Fig. 9.4), and this has also been seen on bones chewed by chimpanzees after they have killed and eaten monkeys (Pobiner et al. 2007).

Medium to large sized herbivores also chew bone, and they produce an even more characteristic pattern (A.1035 and A.1038). Initially the ends of bones are marked with extensive scoring (linear marks, see Chap. 3), but with more extensive chewing the middle of the articular ends is penetrated while leaving the sides intact, thus producing forked ends. A similar effect may be produced by heavy rodent gnawing (A.1033), which is easily distinguished from large herbivore chewing by the linear marks made by the rodent incisors (see Chap. 3). Carnivores, whether as predators or scavengers, also chew the ends of bones, removing all or parts of the articular ends, and they produce damage similar

to that observed for corrosion in wet or vegetated substrates (see Chap. 8). The damage may be so great as to completely destroy the articular end and epiphysis ("scooping out"), leaving a gnawed end with linear marks and perforations from the animal's teeth.

Modification of Diaphyses

Long bones may be deformed by carnivore action and by sediment compression (A.958), and evidence from sediments may be necessary to distinguish this process. Ribs eaten or chewed by humans may be slightly deformed as a result of chewing (A.1001) (Fig. 9.4), but more commonly long bone deformation comes about through being compressed by trampling or by sediment movement (A.957 and A.959). Digestion and corrosion may also alter diaphyses (A.867 and A.871).

Distortion of Rounded Bones

Rounded bones like skulls may be distorted by sediment pressure (A.963). More compact skeletal elements, such as carpals and tarsals, which are also somewhat rounded, are more resistant to breakage and distortion (Lyman 1994a). Digestion may produce distortion and deformation of the shape of the element digested, and at extreme degrees of digestion, the original shape may be completely changed. Digestion may also produce deformations by chemical corrosion and collapse of the dentine (see Chap. 8).

Atlas Figures

A.951–A.1050

Fig. A.954 Metapodial fossil bones at Concud showing diagenetic breakage in situ (transverse breakage). The two parts of the fossil are displaced while still buried; micro-faults in the sediment have produced movements of the two ends of the fossil (black arrow). Compare with A.953. Courtesy of D. Pesquero

Fig. A.951 Hominin fossil clavicle from TD6-Gran Dolina (Atapuerca). Lack of displacement of fragments suggests no post-burial transport or no reworking (A.952, A.955 and A.956). The fracture angle (Villa and Mahieu 1991) is the angle between the fracture surface and the bone cortical surface. Obtuse or acute angles are commonly associated with green bone fractures, while 90° angles are associated with dry or fossil permineralized bone fractures. Photo M. Bautista

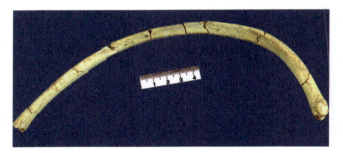

Fig. A.955 Fossil rib from Senèze, a volcanic crater lakeshore (maar) site. The rib was broken diagenetically (post-burial) with transverse breaks but all fragments in situ indicating the absence of post-burial transport or reworking processes (A.951, A.952 and A.953)

Fig. A.952 Fossil rib from TD6-Gran Dolina (Atapuerca). Sediment pressure produced diagenetic transverse breakage. This type of breakage is produced when bones are fossilized and buried already (A.951 and A.955). Photo M. Bautista

◄ **Fig. A.953** Fossil vertebra and long bone from Batallones, Spain (Late Miocene, 10 Ma). Sediment pressure produces diagenetic breakage, transverse to the length of the fossil bone (A.951, A.952, A.955 and A.956). This type of breakage is produced when bones have dried out and buried, and the lack of displacement of fragments is evidence of lack of reworking processes or post-burial transport. Transverse breakage also occurs after fossilization

Fig. A.956 Fossil juvenile metatarsal of *Metacervoceros* from Senèze. The broken edge and surface of the break is flat and smooth, broken at right angles to the length of the bone. All these traits are characteristic of diagenetic breakage due to sediment compression when the fossil was buried (A.951, A.952 and A.953). The surface pitting on this fossil is mainly related to the immaturity of the individual and the acidic environment of the lakeshore site (volcanic sediments)

Fig. A.958 Fossil bone at Langebaanweg (A.1078). The deposits include riverine and lagoonal fine sediments, and sediment compaction leads to breakage and deformation of bones during diagenesis. The broken fragments are in relative position but not moved by any transport. A similar distortion may occur soon after deposition if the bone is resting in soft mud and one end is depressed by trampling (A.957 and A.959). In both cases, the bone was probably broken after shallow burial when embedded in soft plastic clays. Courtesy of C. Denys

◄ **Fig. A.957** Sediment deformation of mandibular and postcranial skeleton (fossil ribs and vertebrae) of a Sivathere (an extinct giraffe) from the Late Miocene site of Langebaanweg (A.1078). The deposits include riverine and lagoonal fine sediments, and sediment compaction may lead to breakage and deformation, with bending the bone in one or more places to conform with the shape of the sediments. This is likely to occur relatively soon after burial. Deformation is the response of bone under the long-term action of an applied force (the weight of sediment or a stone) producing an irreversible plastic deformation of the bone (A.960). A shorter application of force (e.g., a falling block striking the bone) may produce the more immediate response of breakage. Top photo A. Louchart/J. Carrier

Fig. A.959 Top: sediment deformation and breakage of fossil long bones and vertebra at Langebaanweg (A.1078). Apart from deformation due to sediment compression (A.957), some bones at the site have also local depressions (black arrow) that were probably produced by trampling by other animals. Bottom: trampling deformation of a sivathere metapodial. The local depressions suggest that the main deformations took place when bones were shallowly buried in the aquatic environment of the site, and water or wet sediment were important factors in producing these plastic deformations. Trampling in this instance may be considered to be a biotic agent, because it is not the bone pressed against the sediment that modifies the bone (producing trampling marks), but the actual hoof pressing on the bone that breaks and deforms the trampled bone

Fig. A.961 When applying a sudden stroke of 2.5 kiloNewtons on a wet bone or a bone immersed in water, the bone deforms as shown. Similarities can be seen with A.959. These types of experiments are not natural and have to be highly controlled to get conclusive results to obtain a pattern. But the fact of having mechanized equipment allows us to control some of the parameters and be able to repeat the experiment and the results

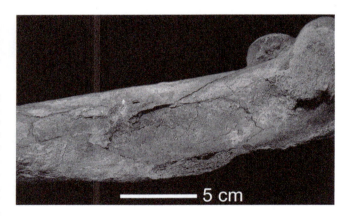

Fig. A.962 Deformation by compression of a fossil left femur of *Allohippus* recovered from Senèze. The fossil was resting across a stone when it was recovered. Most bones affected by deformation are thin bones like ribs or vertebrae, while long bones have been observed to be deformed only by large stones (A.975). The very fine plastic brown clay sediment, typical of lake deposits, that covers most fossils at this site, favored the preservation of broken fragments in place

◀ **Fig. A.960** Experiments have been performed with a materials testing device to simulate breakage and deformation (A.957). Experimental protocol includes testing dry and wet bones, fine to coarse substrates, or immersed in water, and this has allowed us to simulate deformations like the rib seen here under constant and slow compression

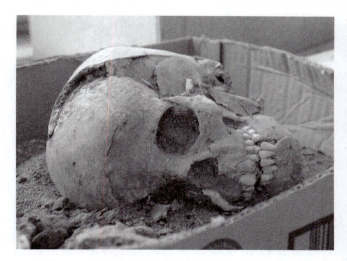

Fig. A.963 Deformation of a human skull from Çatalhöyük. The front half of the skull has been displaced downwards relative to the back part. This skull is from an articulated skeleton (see A.1070 and A.1071) buried beneath the floor of an occupied building, and the deformation is due to sediment pressure

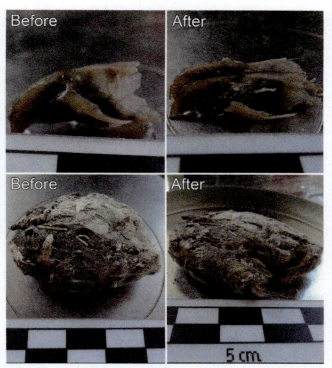

Fig. A.966 Results of compressive experiments on rodent skulls and owl pellets. Top, skulls were immersed in water for several weeks before the experiment (as in A.960). Plasticity of bones when wet is extremely high compared with skulls compressed dry, and the skull is slightly flattened but still articulated. Bottom, pellets react in almost the same way whether wet or dry, having a high plastic resistance to deformation. The pellet is strongly flattened after compression

Fig. A.964 Fossil from Langebaanweg site. The bone is almost totally destroyed but with some teeth remaining close to anatomical position (see A.741 and A.957). Courtesy of C. Denys

Fig. A.967 Experimental falling blocks at Riofrío. The block shown in the picture fell from a height of 1.5 m. The deer bones have been broken across their diaphyses (A.968, A.969, A.970, A.971 and A.972)

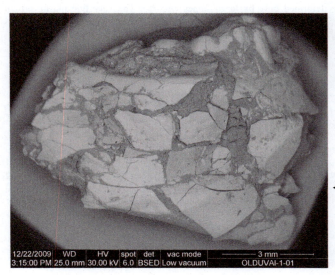

◄ **Fig. A.965** Rodent mandible from Olduvai FLK-NW site (level 20). The site is a lakeshore with alkaline sediments derived from alkaline volcanoes. One of the taphonomic features at this site are fossils deformed by compression (at least for small mammals). This mandible was heavily compressed and deformed with fragments of bones attached together, indicating compression occurred after burial, or when still protected by the pellet (see A.966 and A.975 bottom)

Fig. A.968 Experimental falling blocks at Riofrío. The block shown in the picture fell from a height of 1.5 m. The deer scapula was broken after several tries (A.969, A.970, A.971 and A.972; and A.967)

Fig. A.971 Experimental falling blocks at Riofrío. The humerus still contained marrow after several months exposed on the ground. The spiral break is characteristic of breakage of fresh bone. (A.967, A.968, A.969, A.970 and A.972)

Fig. A.969 Experimental falling blocks at Riofrío. A block falling from a height of 1.5 m broke the metapodial across the mid-shaft, and several flakes split off after several attempts (A.967, A.968, A.970, A.971 and A.972)

Fig. A.972 Top: experimental falling blocks at Riofrío. A block fell from a height of 1.5 m, detaching a lateral flake of bone from the edge of the ulna, displaced by the falling block. This can be quite common as direction is not completely under the control of the subject when letting the blocks fall. Bottom: experimental falling blocks at Riofrío. The bone surface shows a characteristic set of grooves. This is the result of friction of the stone against the bone during its fall. It is also noticeable that the metapodial was this time positioned on a stone as, without it, the bone was being pushed into the soil at every attempt to break it. Also notice a small lateral flake of bone detached from the edge of the bone. (A.967, A.968, A.969, A.970 and A.971)

Fig. A.970 Experimental falling blocks at Riofrío. Breakage of the humerus was harder and needed more strength. The humerus was finally broken after several attempts. (A.967, A.968, A.969, A.971 and A.972)

Fig. A.973 Fossil right metacarpal of *Allohippus* from Senèze. The bone was struck by a falling block while exposed on the surface. This impact probably caused the scar (or flake) on the bone surface (top figure, A.972, A.990 and A.992), and weakness after impact in this part of the bone induced bone fracture when sediment compacted and compressed the fossil. The outline and angle of fracture, smooth, curved and oblique, suggests weakness was produced when the bone was fresh and rich in collagen, but it only broke when already buried (bottom figure) as this was recorded as a single fossil in the excavation record. Possibly the transverse fissure next to the actual curved impact break formed later (A.956)

Fig. A.975 Fossil bones from Senèze, where falling blocks fell down a steep slope and trampling by animals coming to drink to the lakeshore damaged bones. These two cases are the result of big blocks that impacted and deformed the bone surface (A.962). The sediment is a fine and plastic lake clay that encloses and supports the bones, virtually encasing them in natural casts. In these conditions, all small broken fragments are preserved together

Fig. A.974 Fossil right radio-ulna of *Eucladoceros* from Senèze. Similar to A.973 top, this bone was also struck and broken by a falling block producing a superficial scar. This time, however, the impact did not strike a weak area on the bone surface and did not break the bone, probably because of higher robustness of the radio-ulna compared to metapodials

Fig. A.976 Fossil bone from Senèze, where falling blocks may also produce impact marks (white arrow) on the edges of breakages (as in A.315, A.317 and A.973). This fossil comes from a position in the site where falling blocks fell down a steep slope and trampling by animals coming to drink damaged complete skeletons deposited or buried in shallow deposits in this volcanic maar

Fig. A.977 Fossil bone fragment from Senèze with a peeled surface. Peeling was defined as a roughened surface with parallel grooves or fibrous texture "when fresh bone is fractured and peeled apart similar to bending a small fresh twig from a tree branch between two hands" (White 1992: p. 140). Peeling in Sèneze is not the result of human action, but of trampling or falling blocks while the bone, free of soft tissues, was partially resting on a stone or a hard surface (A.1006)

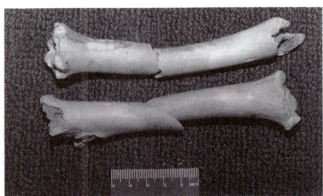

Fig. A.980 Modern femora of sheep broken by Koi people to extract the marrow (A.969, A.971 and A.983). Collection of Koi people at the Transvaal Museum, South Africa (collected by Brain 1967, 1969)

Fig. A.978 Fossil bone fragment of human long bone from Tianyuan Cave Site (Zhoukoudien) broken by gravitational impact of roof fall blocks on exposed bones (A.330 and A.969), affecting both sides (medial and lateral)

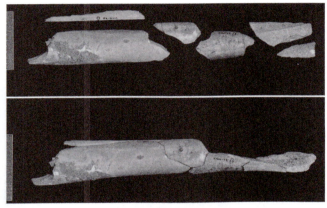

Fig. A.981 Fossil tibia of *Cervus elaphus* from Vanguard Cave broken in 6 pieces to extract the marrow by Neanderthals. These pieces were dispersed in the site and refitted by Cáceres and Fernández-Jalvo (2012). Oblique breakage with curved smooth edges (see A.979 and A.980). Courtesy of C. Stringer

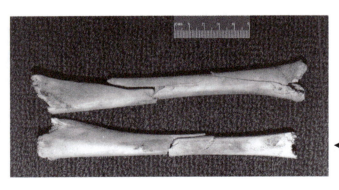

◄ **A.979** Modern tibia of sheep broken by Koi people to extract the marrow (A.969, A.971, A.981 and A.983). Collection of Koi people at the Transvaal Museum, South Africa (collected by Brain 1967, 1969)

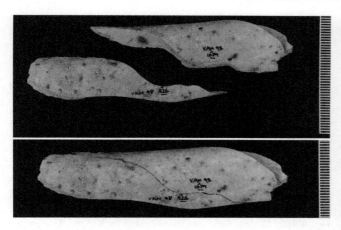

Fig. A.982 Fossil long bone fragments from Vanguard Cave broken by Neanderthals (middle Paleolithic) refitted by Cáceres and Fernández-Jalvo (2012). Oblique breakage, curved smooth edges (see A.979 and A.980). Courtesy of C. Stringer

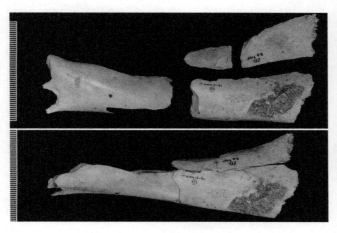

Fig. A.983 Fossil humerus of *Capra ibex* from Vanguard Cave broken in 4 pieces to extract the marrow by Neanderthals. These pieces were dispersed in the site and recovered in two succeeding seasons, and refitted by Cáceres and Fernández-Jalvo (2012). Oblique breakage with smooth edges (see A.979 and A.980). Courtesy of C. Stringer

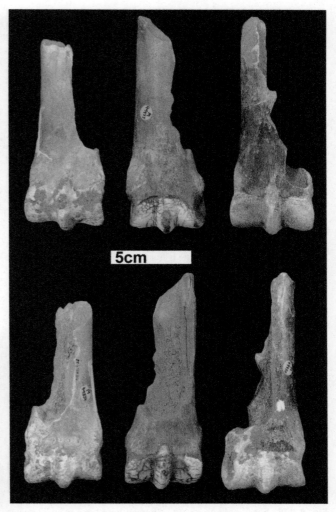

Fig. A.984 Distal fossil metapodials of *Equus ferus*, found in Gough's cave. From left to right ventral view of M49834, dorsal views of M49977 and M49950. Opposite views below. The breakage pattern is repetitive and the shafts are split up to the articular surface with oblique fractures, curved, smooth, and shaft circumference 2. On M49834 there are cut marks on the terminal end of the central ridge of the trochlea and a conchoidal scar on the dorsal side of the oblique fracture (not seen here); M49977 has percussion marks on the central ridge of the trochlea, visible here on the dorsal aspect as discoloration of the articular surface, conchoidal scars along the oblique break, and cut marks on the shaft; M49950 also has cut marks on the shaft along the ridge bordering the post-articular sulcus and conchoidal scars along the oblique break, and extensive percussion marks on the distal (terminal) part of the articular surface (A.985)

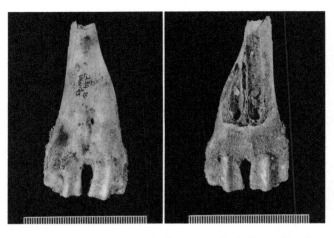

Fig. A.985 Distal end of a fossil metatarsal of *Capra ibex* from Vanguard Cave showing a similar pattern of breakage to that observed on the metapodials of *Equus ferus* from Gough's Cave. The broken edge has oblique fractures, curved, smooth, and shaft circumference 2 (Villa and Mahieu 1991). This breakage pattern appears to be concerned with breakage of the shaft for extraction of marrow and possibly also with disarticulation of the foot (A.984)

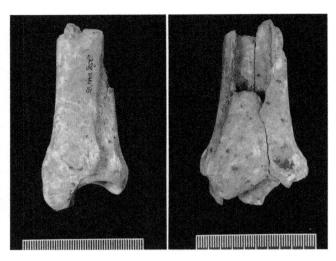

Fig. A.988 Distal fossil tibia *Cervus elaphus* from Vanguard Cave, with characteristic pattern of oblique break, curved and smooth edge and shaft circumference 2, that is consistent with human breakage for marrow extraction (A.987 and A.985)

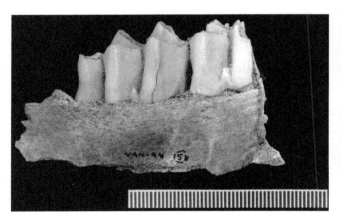

Fig. A.986 Cervid (*Cervus elaphus*) fossil mandible with the inferior border, ascending ramus and symphysis broken, bearing M2–M3. This fossil was found at Vanguard Cave, a site with evidence of Neanderthal occupation. Breakage shown here has a pattern that is frequent in human butchery (Lyman 1994). Compare with natural breakage A.504, A.505 and A.515

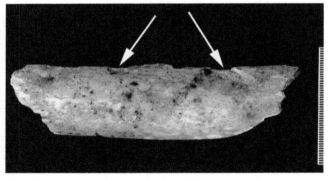

Fig. A.989 Fossil fragment of rib from Vanguard Cave bearing cut marks on the surface and broken. Note the surface is covered by brown spots that indicate direct exposure to fire (embers A.526). The surface is also affected by flakes of bone (fragments splintered off by the impact white arrows, see A.121)

◀ **Fig. A.987** Distal end of fossil tibia of Equid (*Equus ferus*) from Gough's Cave with oblique break, curved and smooth edge that is consistent with human breakage (A.988). The site has evidence of human activity, both on animal remains and on the human remains, with evidence of cannibalism. Animal and human bones were broken and processed in the same manner. Andrews and Fernández-Jalvo (2003)

Fig. A.990 Fossil long bone fragment from Abric Romani site. Direct impact from stones on the bone similar to impacts formed when knapping stones. The black arrow shows the area of direct impact that produced a conchoidal scar at the marrow cavity. The white arrow points to a small adhered bone flake formed during another impact, exposing the marrow as bone flakes became detached (A.991, A.992 and A.993). Courtesy of I. Cáceres

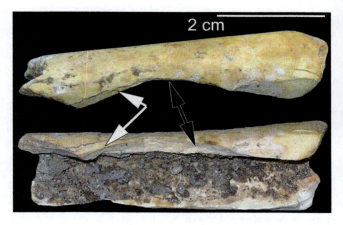

Fig. A.991 Fossil long bone from Cueva de Ambrosio broken by percussion and producing a curved broken edge or conchoidal scar on the bone edge (black double head arrow) and a flake still adhered to the bone (white arrow). This is also the result of direct impact (A.992 and A.993). Specimen provided by S. Ripoll

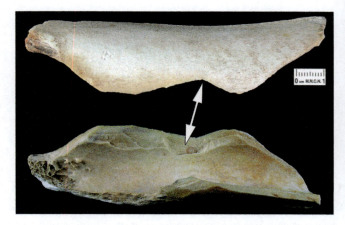

Fig. A.992 Fossil long bone fragment from Cueva Ambrosio that was broken by direct percussion that produced conchoidal scars in the interior of the bone (bottom picture, A.991 and A.993). Specimen provided by S. Ripoll

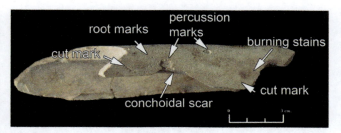

Fig. A.993 Fossil long bone fragment from Abric Romani showing several percussion marks and scars, both marks linked to human activity. There are also cut marks, burning, and conchoidal scars. There is later evidence of root marks produced by vascular plants. The pitting (percussion marks) associated with conchoidal scars indicate breakage is due to direct impact (A.991 and A.992). Courtesy of I. Cáceres

Fig. A.994 Experimental direct percussion with a cobble to extract the marrow. The methods and the results of bone breakage is shown in Text Fig. 9.1. Courtesy of I. Cáceres

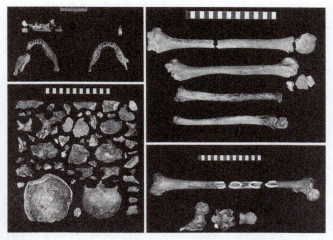

Fig. A.995 Experimental breakage of human modern bones falling a distance of 13 m. All bones were intact before falling. The two skulls were completely smashed, except for the vault and the face. Mandibles were also broken, and three out of six femora broke with spiral breaks on the shaft and two others at the neck (A.996). Humeri had similar patterns with jagged broken edge of the shaft and proximal end. Andrews and Fernández-Jalvo (1997)

Fig. A.996 Remains of modern sheep carcass scavenged by lammergeier or Bearded Vulture (*Gypaetus barbatus*). Sheep carcasses abandoned by local people at Barranco de las ovejas (Huesca, Spain) were carried into the air by the lammergeiers and dropped against the stones of the ravine (A.997 and A.998). Some bones broke in pieces and bone splinters could then be swallowed by these specialized bone-eater vultures

Fig. A.997 Remains of the sheep carcass scavenged by lammergeier or Bearded Vulture (*Gypaetus barbatus*). Skulls also break in the fall (notice the straight breakage line of the inner side of the maxilla), allowing the vultures to get to the brain, tongue and throat (A.996 and A.998)

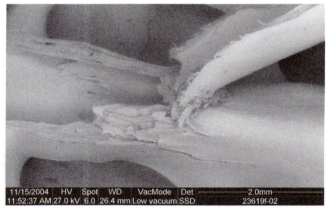

Fig. A.999 Small box bottom right. Broken fossil from Langebaanweg (see A.1078) outside the excavation protected area. Fossil bones are exposed on the ground by seasonal rains and wind. Once fossils are uncovered on the ground and exposed to weathering agents, fossil breakage occurs and bones formerly in articulation (see A.957) may become separated, probably belonging to the same fossil specimen. This is the final stage 5 of weathering as defined by Behrensmeyer (1978)

Fig. A.1000 Breakage by preparation techniques of skeletons in museums using enzymes. Bones may soften and break when handled (see A.693). Taxidermic collection (Fernández-Jalvo and Marin-Monfort 2008)

◄ **Fig. A.998** Remains of the sheep carcass scavenged by lammergeier or Bearded Vulture (*Gypaetus barbatus*). The skeleton is almost completely in anatomical connection after a high fall (arrow) indicated by the broken distal metapodial (white arrow A.996)

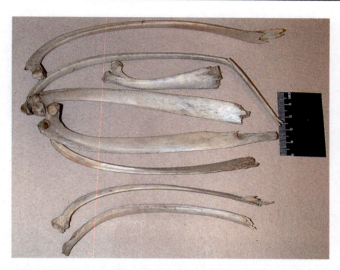

Fig. A.1001 Breakage and deformation of modern ribs by chewing by Koi people (collection housed at the Transvaal Museum, South Africa, collected and described by Brain 1969). The bent shape at the end of thin bones is characteristic of human and chimpanzee chewing and is frequent in this ethnological collection. The shape results from bending bones by holding part of the bone between the upper and lower cheek teeth and pulling or pushing the bones with the hands (A.1002, A.1003 and A.1004, Text Fig. 9.4)

Fig. A.1003 Breakage and deformation of the ends of modern ribs during human chewing by Koi people (collection housed at the Transvaal Museum, South Africa, collected and described in Brain 1969). This shape results from bending bones by holding part of the bone between the upper and lower cheek teeth and pulling or pushing the bones with the hands (A.1001, A.1002, A.1004, Text Fig. 9.4)

Fig. A.1002 Breakage and deformation of modern ribs during human chewing by Koi people (collection housed at the Transvaal Museum, South Africa, collected and described by Brain 1969). Jagged edges and peeling are frequent. This shape results from bending bones by holding part of the bone between the upper and lower cheek teeth and pulling or pushing the bones with the hands (A.1001, A.1003 and A.1004)

Fig. A.1004 Bone from Azokh I, historic level. Thin bones like ribs may be deformed by human chewing, where the ends are bent, but not broken (A.1003). From Azokh Cave, Unit I, historic level

Fig. A.1005 Modern pig rib on the left and a second left human fossil metatarsal on the right: GC87-30 from Gough's Cave. The pig rib was experimentally chewed by humans (Text Fig. 9.3). The extensive deformation of the proximal ends, with depressed flakes of bone and splaying of the ends, is characteristic of human and chimpanzee chewing. The similarity to the fossil from Gough's Cave may indicate human chewing and reinforces the evidence of cannibalism at this site (Andrews and Fernández-Jalvo 2003; White and Toth 2004; Fernández-Jalvo and Andrews 2011)

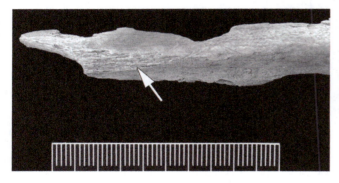

Fig. A.1006 Fossil bone from Gough's cave, a left human ulna (GC87-209) with extensive peeling at the proximal end. Peeling is an area of roughened exfoliated surface (white arrow) produced when a fresh bone is snapped in two and part of the surface bone is peeled away. "Peeling shape" may also be caused by trampling or falling blocks (A.977)

Fig. A.1008 Experimental breakage by trampling at Riofrío. The proximal end of a large mammal femur broken by stepping on it during experimental trampling by four people running across the surface (A.1009 and A.1012)

Fig. A.1009 Recent breakage by deer in Riofrío. Stampedes are frequent at Riofrío, and herds running across the plains trample and break bones resting on the surface of the ground, as seen here and experimentally (A.1008 and A.1012)

◄ **Fig. A.1007** Trampling experiment in natural conditions at Riofrío. Bones resting on the ground were trampled by four people running across the surface. Skulls, flat bones are the most affected, as shown here (A.967, A.968, A.969, A.970 and A.971)

Fig. A.1010 Experimental breakage by deer trampling at Riofrío. A variety of long bones, scapulae, metapodials and mandibles were marked and dispersed to be trampled by deer-herds grouped in an enclosure for veterinary control. The picture shows a modern femur of red deer left on a side of the enclosure that channeled the herds of deer which trampled the bones (A.1009)

Fig. A.1012 Experiments performed with a materials testing device to simulate trampling on dry, weathered and fresh bones, as well as wet bones immersed in water for weeks. Experiments use a relatively narrow tool, applying 2.5 kilonewtons in a sudden movement. Bone breakages obtained experimentally have close similarities with naturally trampled bones by deer from Riofrío (see A.1009 and A.1011 research in progress)

Fig. A.1011 Broken bones from Riofrío broken by herds of deer trampling bones resting on the ground. As seen experimentally, nasal bones and skulls (top) may easily be broken. Long bones (bottom) show a characteristic shape when broken by intense trampling. These experiments and observations are still in progress (A.1007 and A.1009)

Fig. A.1013 Modern human femoral fragments from Kajiado chewed and broken by spotted hyaenas, *Crocuta crocuta* (A.1014, A.1015 and A.1016). Differences between chewed fragments broken by carnivores and other taphonomic agents (e.g. trampling) are established by accompanying marks, the site context and the general taphonomic traits of the bone association, for example, puncture tooth marks in chewed bones, or linear marks on trampled bones

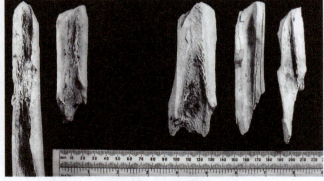

Fig. A.1014 Examples of spiral breaks on modern human bones from Kajiado resulting from spotted hyena (*Crocuta crocuta*) scavenging (A.1013, A.1015 and A.1016). The breaks are acute at one end and obtuse at the other, both have smooth edges Kajiado (Tanzania)

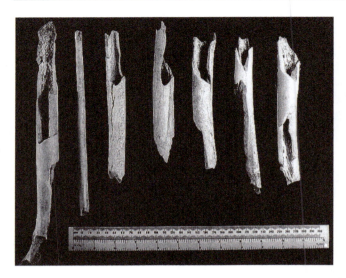

Fig. A.1015 Modern human bones from the den of a spotted hyena (*Crocuta crocuta*) at Kajiado showing spiral breaks and oblique breaks, but all with part of the circumference of the shaft entire (A.1013, A.1014 and A.1016)

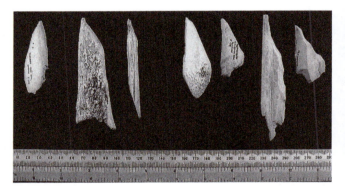

Fig. A.1016 Fragmentary modern human bones from the den of a spotted hyena (*Crocuta crocuta*) at Kajiado showing spiral breaks, mostly acute angled (A.1013, A.1014 and A.1015)

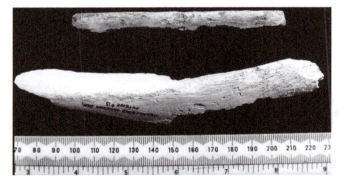

Fig. A.1017 Modern bone flakes chewed by spotted hyena (*Crocuta crocuta*) at Ngorongoro lakeside den showing rounding of the ends of the bone (A.599). A.J. Sutcliffe collection

Fig. A.1020 Modern fragments of bone chewed by wolf (*Canis* ▶ *lupus*). The size of these fragments is smaller than those resulting from chewing by hyenas. Courtesy G. Haynes. Compare with A.1021, A.1022 and A.1023

Fig. A.1018 Surface corrosion due to soil acidity of bones collected from a spotted hyena den (*Crocuta crocuta*) in Ngorongoro Crater, Tanzania. Surface corrosion by urine and organic activity weaken and soften the bones that may then become fragile and easily broken or almost destroyed (A.789)

Fig. A.1019 Fossil bones from Concud. Top: conchoidal breakage by carnivore chewing (A.395 and A.396). Chewing by carnivores, especially hyenids and canids, may produce conchoidal scars on the edge of the broken fragment, similar to the modifications humans produce by using stones against the bone surface (direct percussion). Scars resulting from carnivore chewing tend to be smaller than those struck by humans A.332. Bottom: flaked bones due to tooth impact may mimic conchoidal scars on bones struck by humans to reach the marrow. In this case a non-human process was involved, probably a carnivore, as this is a Miocene site in Spain (Concud 7 My BP) before humans were present. Courtesy of D. Pesquero

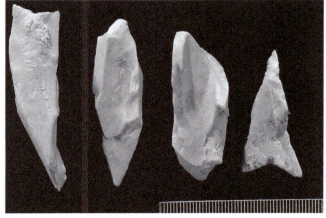

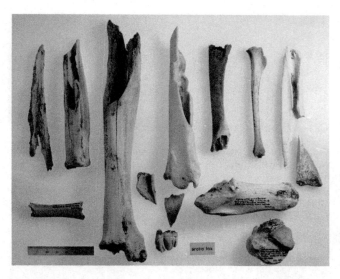

Fig. A.1021 Fragments of modern bones chewed by modern arctic fox (*Alopex lagopus*) from a collection of bones outside their den. The small size of arctic foxes hinders their capacity to break bones and they rarely produce splinters from broken bone. They more commonly eat small mammals and insects (Andrews 1990). Compare with A.1020, A.1022 and A.1023

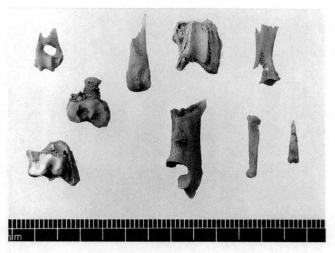

Fig. A.1023 Modern broken bones from a coyote scat (*Canis latrans*). Prey species are mainly rabbits. (Andrews 1990). Compare with A.1020, A.1021 and A.1022

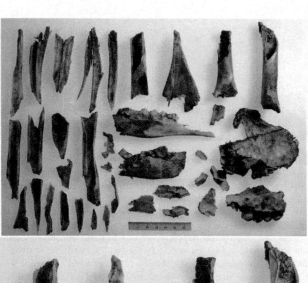

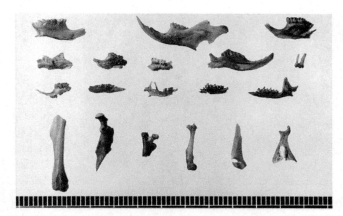

Fig. A.1024 Modern small mammals predated by small carnivores such as genets (*Genetta genetta*) have similar types of breakage on their bones to that present on large mammal bones. (Andrews 1990). Compare with A.1023

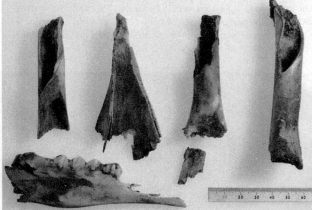

◄ **Fig. A.1022** The remains of two lion kills of a juvenile wildebeest. The modern broken bones have oblique breaks with acute smooth edges and on the right some gnawed ends of bones. The shaft circumferences are mostly still intact, but with extensive breakage of the ends of the bones. Lions are not adapted to break large or adult bones because their teeth are weaker than canids and hyenids. However lions can break bones of very young animals, such as these juvenile wildebeest, and splinters are relatively frequent, although bigger than those made by canids and hyenids. Both specimens courtesy of P. Jones. Compare with A.1020, A.1021 and A.1023

Fig. A.1025 SEM microphotograph of a modern monitored rodent femur head chewed by soricids (see Andrews 1990, for details of the find, and A.174 and A.179 for general and close up views). Field width equals 5.4 mm

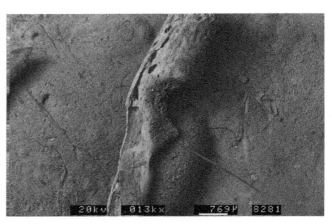

Fig. A.1028 SEM microphotograph of a small mammal proximal end of ulna showing cracking, breakage and deformation by digestion (see A.867 and A.615)

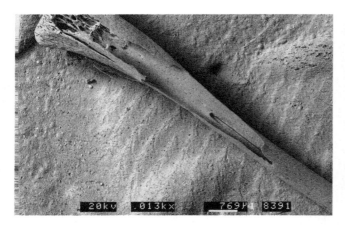

Fig. A.1026 SEM microphotograph of a modern bone broken and digested by weasel *Mustela nivalis*. Broken bone edges are rounded by acidic gastric juices indicating that the bone was broken before digestion (A.614 and A.616). Extreme digestion (see A.897)

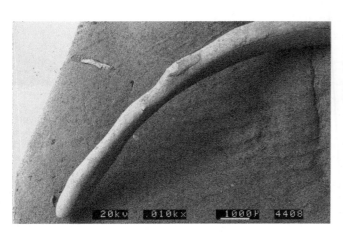

Fig. A.1027 SEM microphotograph of deformation of a small mammal incisor by extreme digestion, so great that the shape is almost unrecognizable. See Text Fig. 8.4, A.836, A.838 and A.839

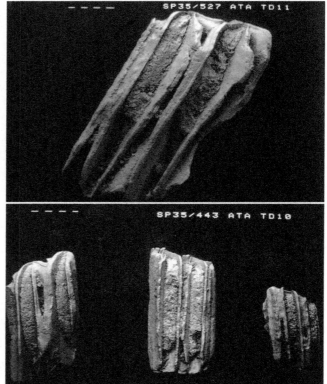

Fig. A.1029 SEM microphotographs of several fossil vole molars from different units of Atapuerca Gran Dolina site. All show extreme digestion, with enamel disappearance and dentine collapse, as observed in incisors (A.838 and A.839), making the shape of the molars almost unrecognizable (A.844). Top: field width equals 3.6 mm. Bottom: field width equals 5.2 mm

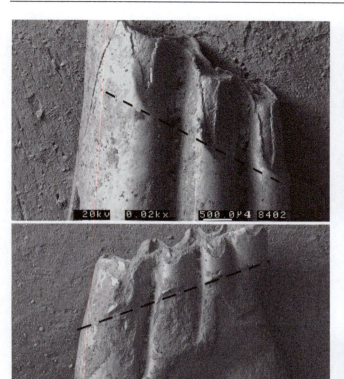

Fig. A.1032 Remains of fallow deer recently eaten by vultures at Riofrío. Vultures may cause considerable breakage of bones, especially cranial elements and flat bones (A.403, A.404, A.405 and A.406). Notice individual punctures on the nasal bone inside the square and breakage on the anterior part of the skull shown by an arrow

Fig. A.1030 SEM microphotograph of vole molars eaten by modern mongoose *Herpestes* sp. Breakage took place during ingestion and before digestion, as indicated by rounded edges of breakage (black lines) more evident on the top figure, but also seen on the specimen at the bottom (Andrews 1990). See other modifications by mongoose species in A.394, A.849 and A.1031

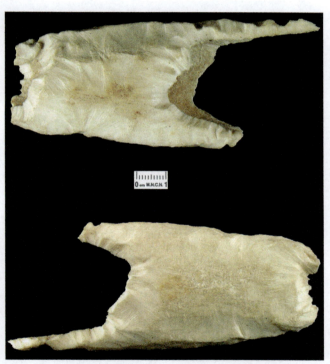

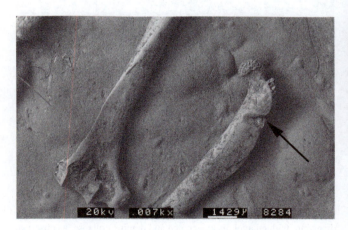

Fig. A.1031 SEM microphotograph of small mammal long bones broken before digestion by modern mongoose *Herpestes* sp. This is indicated by rounding on the broken edge of the humerus shaft (lateral left margin), and a rounded tooth mark near the proximal end of the femur (black arrow). See close up view in A.173

Fig. A.1033 Long bone fragments gnawed by rodents. There was probably oblique breakage at the ends of the bones made by carnivores and later gnawed and reshaped by rodents. In this case the forked shape produced is similar to that of large herbivore chewing (compare with A.1034 and A.1040), but the broken edges have abundant narrow grooves transversal to breakage that reveal rodent gnawing (A.187 and A.199)

Fig. A.1034 Weathered bone chewed by deer from Riofrío. Notice that both ends are chewed, and cracking and flaking by weathering is interrupted by chew marks. Chewing therefore post-dated the weathering (A.165 and A.166). Chewing by herbivores is not for feeding purposes, but to gain access to minerals deficient in their diet (osteophagia)

Fig. A.1038 Modern long bone of sheep chewed by sheep from Neuadd, producing the characteristic forked shape of the end of the bone (A.166, A.169, A.1035 and A.1040)

Fig. A.1035 Deformation of the distal ends of modern deer metapodials chewed by deer from Riofrío leaving the characteristic forked end (A.1038, A.1039 and A.1040). Not only deer, but camels, giraffes, cows (A.166), and sheep have this behavior

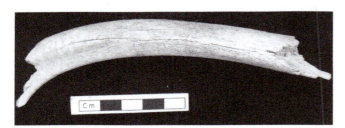

Fig. A.1036 The ends of ribs are chewed by deer at Riofrío in the same way as for long bones. Notice on the left end of the bone that grooves made by the herbivore are superimposed on cracks previously produced by weathering. See close-up view in A.165

Fig. A.1039 Modern bone fragment collected from Riofrío chewed by deer. At the park of Riofrío, the rangers feed the animals, adding extra minerals, so that they probably do not need the extra minerals from the bone (Cáceres et al. 2011). It may be that this behavior in Riofrío is more intense than in natural conditions (A.166, A.169, A.1035 and A.1040). Courtesy of I. Cáceres

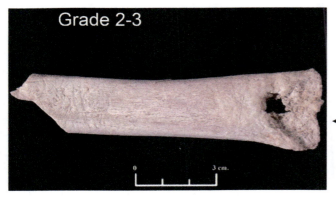

◄**Fig. A.1037** Modern bone from Riofrío chewed by deer. Ends of modern bones are chewed be deer using the molars and premolars, making the hole seen here. Continued chewing of the bone produces a characteristic forked shape (A.1035, A.1039 and A.166). This is a relatively frequent behavior among herbivores to obtain minerals that are deficient in their diet (Cáceres et al. 2011). Courtesy of I. Cáceres

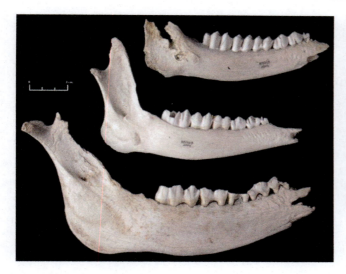

Fig. A.1040 Modern mandibles. Chewing not only affects long bones, but also mandibles, antlers or any relatively long piece of bone. Herbivores take the end of any skeletal element that they can hold in the mouth and chew bone (A.166, A.169, A.1035 and A.1039). Courtesy of I. Cáceres

Fig. A.1041 Pathological deformation by an abscess on a deer mandible from Riofrío. The abscess may have been produced by the advanced wear of the molars due to the intense osteophagia observed in the Riofrío park bone (A.166, A.1035 and A.1040)

Fig. A.1042 Pathological deformation by arthrosis in an ulna articulation of deer from Riofrío (A.1043 and A.1044). Pathology has to be distinguished from taphonomic modifications

Fig. A.1043 Pathological deformation (unformed osseous growth) on the articular end of a humerus from the same individual (A.1042) of deer from Riofrío

Fig. A.1044 Pathological deformation on the articular end of an ulna and radius from Riofrío (compare with A.1042). Arthrosis is present with unformed bone growing on the ulna-radius connection and loss of osseous tissue on elbow at the proximal end of the ulna

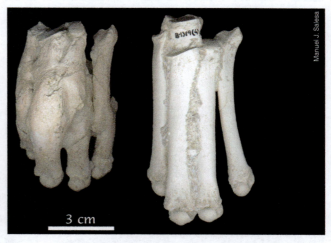

Fig. A.1045 Fossil bones. On the left, a trauma produced by an accident on four metapodials of *Machairodus*, when the animal was alive. The break was badly fused and produced marked deformation of the foot. For comparison on the right are intact bones from Batallones, Spain (A.953). Courtesy of M. Salesa

Fig. A.1046 Fossil bones. Pathology on a mounted skeleton of horse from Senèze on exhibit in the museum of the Université Claude-Bernard Lyon 1. The proximal end of the right metacarpal has heavy osteoarthritis (see also A.1047, A.1048, A.1049 and A.1050)

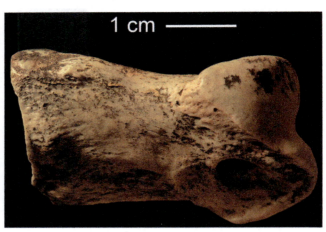

Fig. A.1049 Fossil bones. Pathology at Senèze. Second phalanx of *Eucladoceros ctenoides* having moderate bone overgrowth on the dorsal side of the distal end in connection with the third phalanx. (See also A.1046, A.1047, A.1048 and A.1050)

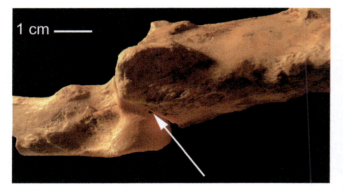

Fig. A.1047 Fossil bones. Pathology at Senèze. Advanced osteoarthritis on the joint between phalanx I and phalanx II of *Metacervoceros rhenanus*. Both phalanges have strong outgrowths on the margins of the joints and the distal end of phalanx I (on the left) which also has a perforation that can be distinguished in this picture (white arrow). See also A.1046, A.1048, A.1049 and A.1050)

Fig. A.1050 Fossil bones. Pathology at Senèze. Second phalanx of *Allohippus* has anomalous bone growth on the palmar side. Pathology result of anomalous bone growth, injuries or diseases has to be distinguished from taphonomic modifications. (See also A.332, A.1043, A.1045 and A.1046)

◀ **Fig. A.1048** Fossil bones. Pathology at Senèze. Mild osteoarthritis on the distal end of a fourth metacarpal of *Dicerorhinus etruscus*. (See also A.1046, A.1047, A.1049 and A.1050)

Chapter 10
Disarticulation and Completeness

Degree of disarticulation and skeleton completeness can be measured in fossil assemblages to determine aspects of its taphonomic history (Todd and Frison 1992). They are also useful tools when dealing with human burials for the same reason (Andrews et al. 2005). They will only be mentioned briefly since they are not strictly speaking taphonomic modifications that can be illustrated here, but there are some general points that can usefully be raised. Two types of agents are instrumental in disarticulation of animal skeletons, natural processes and human action. Taphonomic work has focused on both, for they have processes in common, and several agents have been identified in the disarticulation process, namely predation, scavenging, trampling, transport, soil activity, and sediment movement. The action of humans as predators or scavengers parallels that of their animal equivalents to a certain extent, and much has been written on the degree of selection of skeletal elements by humans. This is discussed in detail by Lyman (1994a) and will not be entered into here, but there is another aspect to human action that will be briefly described, and that is the act of burial that came into being over the past tens of thousands of years.

Agents and Processes

Predation, A.1051
Scavenging, A.1054 and A.1056
Transport before burial, A.1058, A.1066 and A.1069
Trampling before and after burial, A.1063 and A.1064
Burial and the effects of soil activity, A.1075
Sediment and water movement, A.1066
Bacterial decay, A.1075
Human burial, A.1070 and A.1071

Characteristics

Disarticulation is the separation and dispersal of skeletal elements from their anatomical position in the skeleton. Predation by carnivores or first access of carnivores to a complete carcass frequently results in some degree of disarticulation and/or loss of skeletal elements. This is often followed by scavenging, and either or both may lead to further disarticulation of the skeleton and transport before burial. Separation by itself is not enough to justify this term, for the bones of skeletons in undisturbed conditions may lose their articular connection but still remain in anatomical position, as happens regularly in human burials. Movement of elements away from anatomical position is consequent on disarticulation, and both the degree of separation of elements, and the distance moved by contiguous elements, provide information about the degree of disturbance during preservation. When movement of elements has been so great that they can no longer be associated with a skeleton, this also becomes a measure of completeness of the skeleton.

Patterns of Disarticulation

The work of Toots (1965) and Hill (1979), Hill and Behrensmeyer (1984) has documented the pattern of disarticulation for a number of large mammalian herbivores. This work shows that there is a sequence of disarticulation of skeletons similar in outline although differing in details, depending on the species, size of the animal and environmental factors. First to separate from an exposed mammal carcass is the scapula, quickly followed by separation of the mandible from the skull (A.1064). Next follows disarticulation of the shoulder joint and of the

© Springer Science+Business Media Dordrecht 2016
Yolanda Fernández-Jalvo and Peter Andrews, *Atlas of Taphonomic Identifications: 1001+ Images of Fossil and Recent Mammal Bone Modification*, Vertebrate Paleobiology and Paleoanthropology, DOI 10.1007/978-94-017-7432-1_10

skull from the vertebral column. The hind limbs separate next, and once detached from the skeleton, both forelimbs and hindlimbs disarticulate in similar fashion, from distal (metacarpals and phalanges) to proximal. The vertebral column remains intact for considerably longer than any other part of the skeleton, disarticulating in the order cervical, thoracic, lumbar and sacrum.

The pattern of disarticulation may differ in some respects when it is the product of butchery. At the Garnsey bison kill site, disarticulation was extensive, although the bone assemblage had been subject to fluvial disturbance (Speth 1983). The most commonly articulated elements were found to be feet and lower portions of forelimbs (from the radius down), and this differs from carnivore-ravaged bone assemblages (Speth 1983). It also differs from the mammoth butchery site at Colby (Frison and Todd 1986), where the phalanges were subject to the greatest destruction. There were also occasional short portions of vertebral column at both the Garnsey and Colby sites, with portions of thoracics, lumbars and cervicals being preserved, and this is consistent with naturally occurring disarticulation and also with human disturbance (see below).

Comparable work has not been done on small mammals, not for want of trying, but it has proved difficult to keep track of small mammal bones, particularly when they are exposed on the surface of the ground. They tend to blow away or disappear under mysterious circumstances. One study on small mammal disarticulation and dispersal was carried out under controlled laboratory conditions, and the results of this are shown in Fig. 10.1 (Armour-Chelu and Andrews 1994). Three rodent carcasses were placed on the surface of an intact block of forest soil (pH 7.9) protected from external influences. The bodies decayed rapidly and started to disarticulate within 51 days, with the bones of the extremities being first to go. They were dispersed down the soil profile (Fig. 10.1) by the actions of a single earthworm (*Lumbricus terrestris*). The metapodials and phalanges were quickly followed by limb bones, both fore and hind, and they were also dispersed by the earthworm. The skull and vertebrae remained in articulation on the carcasses for 8 months. This pattern is similar to that of large mammals, and after 11 months, disarticulation was complete and the vertebrae were dispersed by earthworm burrowing.

Korth (1979) like other authors (Wolf 1973; Shotwell 1955; Voorhies 1969) related the distributions of transported bone assemblages experimentally based on the resistance of anatomical elements to movement by flowing water (apart from relative orientations and rounding traits of bones). Two aspects of bones are important here, the relative density of the bone (Lyman 1994a) and its shape. Some elements have protuberances that may help to anchor the bones to the bottoms of streams, and others have large flat surfaces that have a sail effect, such as scapulae, which catch the water flow. Voorhies (1969) grouped anatomical elements of sheep and coyote that were moved with increasing water energy. These sequential groups were adapted by other authors experimenting with other taxa, and in general terms it could be said that ribs and probably phalanges are the easiest to move while teeth, mandibles and probably skulls are moved last, moving at a high rate of energy calculated by Voorhies at 152 cm/s.

Small mammals were included in some experiments (Dodson 1973), and the water velocity needed to move their bones was calculated between 6 and 35 cm/s. Finally, Behrensmeyer (1975) tried to establish a model to predict the rate of energy needed to move certain anatomical elements based on a formula to establish the hydrodynamic capacity of each individual bone, and similar attempts applied to small mammals were made by Korth (1979) and Denys (1983). Moving water may not only select and transport bones, but it may also orient bones according to the rate of energy and the bone type.

Gravitational dispersal may also organize and select bones, this time due to the facility to move down slope (A.1069). We have observed that skulls and rounded bones may be dispersed furthest down slopes, and long bones tend to lie transversally to the inclination of the slope. Trampling by large sized animals may enhance down slope dispersal, and trampling of animal carcasses before and after burial has an impact on bone dispersal that has long been underestimated. On open sites, bones may be distributed tens of meters from the site of death of an individual (A.1065), but where animal movement is concentrated along game trails or paths, both the destruction and movement of bone can be very great. Finally, when bones are moved with the sediment, the resulting distribution forms bone breccias.

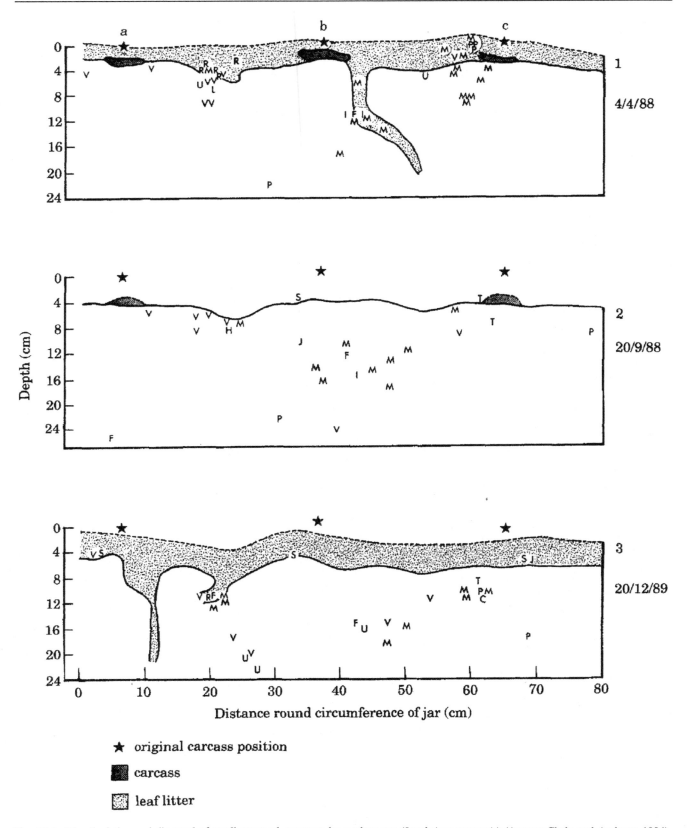

Fig. 10.1 Disarticulation and dispersal of small mammal carcasses by earthworms (*Lumbricus terrestris*) (Armour-Chelu and Andrews 1994). Three vertical soil profiles showing three stages of carcass decomposition and dispersal of the bones. The experiment started on 8th January, 1988: stage 1, April 1988 after 84 days; Stage 2, September 1988, after a further 136 days; and Stage 3, December 1989 after a further 81 days. The vertical intrusions of soil humus down the profile are along earthworm burrows. Additional leaf litter was added in early 1989 to provide food for the earthworms. Positions of bones are indicated by letters: F femur, H humerus, I innominate, L partial limb, M metapodial, P phalnges, R ribs, S skull, T tibia, U ulna and V vertebrae

Human Burials

A special case may be made for human burials. Where a series of skeletons is available for study, it has been found useful to use an ordinal index as suggested in Chap. 9 for breakage. One index applies to disarticulation, and a second to degree of completeness, and they have been applied to an assemblage of human burials at Çatalhöyük (Andrews et al. 2005). The disarticulation index is on a scale of 1–10, with 1 representing a fully disarticulated skeleton and 10 and a fully articulated skeleton, with stages in between as follows:

10 fully articulated A.1070
9 disarticulation of hyoid bone and/or the patella
8 disarticulation of scapula and/or mandible and skull A.1071
7 disarticulation of forelimbs
6 disarticulation of hindlimbs
5 disarticulation of phalanges and metapodials
4 disarticulation of cervical vertebrae
3 disarticulation of thoracic vertebrae
2 disarticulation of lumbar vertebrae
1 isolated bones A.1072

The completeness index is similarly on a scale of 1–10, with 10 representing a complete skeleton and 1 a single bone, with stages in between as follows:

10 complete
9 missing some phalanges, patella or hyoid
8 missing skull in addition
7 missing two limbs, but skull may be present
6 missing skull and two limbs
5 groups of 21–30 bones
4 groups of 11–20 bones
3 groups of 3–10 bones
2 single complete limb bone or cranium
1 fragment of skull or limb bone

The state of preservation of associated skeletons and/or bones can also be measured using three preservation indexes: the Anatomic Preservation Index (API), the Bone Representation Index (BRI) and the Qualitative Bone Index (QBI) based on the work of Bello (Bello and Andrews 2006, Andrews and Bello 2006). The Anatomical Preservation Index (API) is a preservation score assessing the quantity of osseous material present. It is an elaboration of a previous index proposed by Dutour (1989) and expresses the ratio between the score of preservation (i.e., how much is preserved) for each single bone and the total anatomical number of bones in the skeleton. The scores of preservation have been arranged in the following six classes:

Class 1: 0% of bone preserved;
Class 2: 1–24% of bone preserved;
Class 3: 25–49% of bone preserved;
Class 4: 50–74% of bone preserved;
Class 5: 75–99% of bone preserved;
Class 6: 100% of bone preserved.

The Bone Representation Index (BRI; cf. Dodson and Wexlar 1979) measures the frequency of each bone in the sample. It is the ratio between the actual number of bones removed during excavation and the total number of elements of the skeleton that should have been present (vertebral column, costal cage, hands and feet can be counted as single elements [Bello and Andrews 2006]): BRI = 100 × Σ Nb observed/Nb theoretical.

The state of preservation of cortical surfaces can be evaluated by the Qualitative Bone Index (QBI), which is the ratio between the sound cortical surface and the damaged cortical surface of each single bone (Bello et al. 2003). The scores of preservation of cortical surfaces have been arranged in the following six classes:

Class 1: 0% of sound cortical surface;
Class 2: 1–24% of sound cortical surface;
Class 3: 25–49% of sound cortical surface;
Class 4: 50–74% of sound cortical surface;
Class 5: 75–99% of sound cortical surface;
Class 6: cortical surface completely sound.

In the case of human burial assemblages the major taphonomic agents are humans themselves, although other factors may come into play as well. Burials at the Neolithic town of Çatalhöyük in Turkey show a range of preservation from complete and fully articulated skeletons to individuals represented by single bones. We have recognized five kinds of grave with human burials from Çatalhöyük. These are graves with single primary burials, not disturbed in any way; graves with articulated skeletons but with the skull removed, in some cases with cut marks on the atlas vertebra showing how the skull was removed; graves with possible double burials, where the bodies have either been buried together or

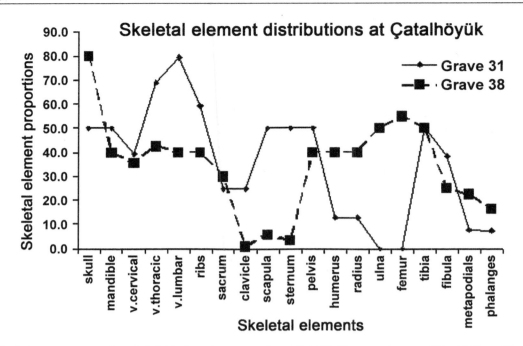

Fig. 10.2 Skeletal element representation of skeletal element representation at Çatalhöyük, comparing a multi-burial Grave 38 which contained 12 individuals, with secondary burial in Grave 31. The skeletal element analysis shows most bones well represented in the Grave 38 assemblage, with cranial and postcranial elements all present. The secondary burials in Grave 31 contained four contemporaneous burials in a shallow grave just below floor level, and the skeletal element analysis shows that skulls and partially articulated vertebrae were the only common elements. Most limb elements were absent

so close in time that the earlier burial would still have been intact when the later burial was interred; graves with multiple disturbed skeletons, where the grave has been opened more than once for later interments; and graves with secondary burials, where there has been either movement of some body parts from a temporary location to their present resting place or a time lapse in the burial, or both.

Disturbance of human burials at Çatalhöyük is seen in the multiple burials. Grave 38 contained 12 individuals, nine fully disarticulated (taphonomic index 9–12), two in partial articulation (taphonomic index 22–28), and one complete and fully articulated (taphonomic index 38, all index scores out of 40) (A.1072) (Andrews et al. 2005). The sequence of events here was the burial of the nine individuals over an unknown period of time, each burial disturbing the one before. Close to the end of the use of this grave, two more bodies were buried, disturbing still further the earlier burials.

Soon after this, while these two bodies were only partly decayed, the grave was opened for the last time. The two partly articulated individuals were removed, and the last body put in place, and then the two partly decayed bodies were replaced, one at the feet of the final individual and one alongside the final body, which was still complete and fully articulated when excavated. The composite skeletal element proportion of these 12 individuals is shown in Fig. 10.2, where it is seen that most limb elements are 40–60% represented, but some, such as patellae, sacrum and sternum are poorly represented. Distal limb elements of the hands and feet are also under-represented, and it is assumed that the missing bones were lost or destroyed during the various periods of disturbance.

In another multiple burial at Çatalhöyük there is evidence of delay between time of death and final burial without movement from one burial place to another. The final

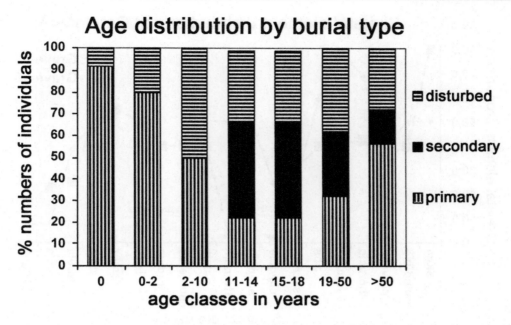

Fig. 10.3 Age distribution of the 62 burials beneath the floor of a single house at Çatalhöyük. Individuals of all ages were found in primary burials, where there was only one body in the grave. Individuals of all ages were also found in disturbed graves, that is graves that had been opened more than once to add additional bodies. Only late adolescents and adults were found in the secondary burials

interment was a complete skeleton lacking only the skull (burial 1466) which was probably removed some time after burial (Andrews et al. 2005). A second example is from grave 31 at Çatalhöyük, which has four individuals represented almost entirely by axial skeletons and skulls. The four sets of vertebrae and ribs were found to be partly articulated and aligned parallel to each other in an extremely shallow grave cut just below floor level. The analysis of skeletal elements shows that vertebrae and ribs are the only skeletal elements well represented (Fig. 10.2). The loss of nearly all the limb bones from the four individuals and the presence of late-disarticulating axial skeletons indicates that these four bodies were exposed or buried elsewhere and most of the early-disarticulating elements were lost. At a stage when the axial skeletons were still partially articulated, they were moved to their present burial place, in the process losing most of the limb bones which either had separated from the axial skeleton or were removed by scavengers. In addition, the individuals concerned were all young adult or late adolescent (Fig. 10.3), in contrast to the mixed age burials elsewhere. Adults and adolescents are the individuals most likely to be away from the town for protracted periods of time, herding, collecting firewood, tending crops, hunting etc., and it is likely that if they died far from the town, their final burial was delayed until they could be brought back. This is evidence of secondary burial.

Atlas Figures

A.1051–A.1080

Fig. A.1051 Impala hunted and eaten by a leopard in Samburu Park (Northern Kenya). Notice the leopard has penetrated the carcass of its prey from the region of the pelvis, affecting the femora and ribs (see A.1054 and A.1056) as described by Blumenschine (1986), who documented this sequence of access. Courtesy of M. Antón

Fig. A.1052 Top: inside this baobab tree in Tanzania there is a barn ▶ owl nest occupied for many years by the barn owls. When the tree decays and disappears, the bone assemblage will appear forming a rounded circle. Bottom: detail of small mammal bones from barn owl decayed pellets inside the baobab of previous picture

Fig. A.1053 SEM microphotograph of five fossil small mammal vertebrae in anatomical connection (from Atapuerca, TD8). In caves, bones may be cemented by the high carbonate content of the water. This facilitates skeletal elements remaining together in anatomical position (see A.1055 and A.1056). Field width equals 7.2 mm

Fig. A.1055 Disarticulation of a monitored fox skeleton at Neuadd. Both limbs on the uppermost side of the body have been taken by scavengers (compare with A.1056), but otherwise the skeleton is intact (Andrews and Armour-Chelu 1998)

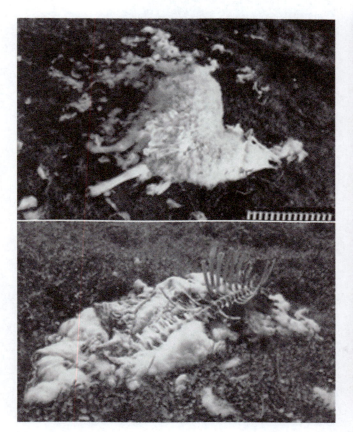

Fig. A.1054 Disarticulation of monitored sheep carcass at Neuadd. Above is the intact carcass 1 day after death. The carcass has been scavenged, with entry in the abdominal region as in A.1051, and the partial skeleton of an unborn lamb is present in the carcass. Below is the same carcass 4 months later with only the axial skeleton remaining. No trace of the unborn lamb was seen. The skull had moved some way down the slope from the carcass. (Andrews and Armour-Chelu 1998)

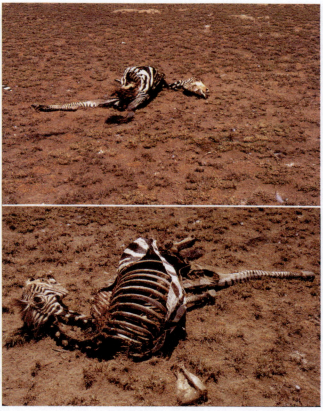

Fig. A.1056 Skeleton of zebra at Olduvai Gorge. Around the skeleton there are abundant excrements and feathers of vultures which scavenged the carcass. The absence of forelimbs suggests a second set of scavenger, probably hyenas (compare with A.1055)

Fig. A.1057 Hyena den with accumulation of bones. The species is *Crocuta crocuta*. Hyenas bring anatomical elements to their den, where cubs and other clan members are waiting. Skulls and limbs are the most frequent elements taken by hyenas and brought to their dens (see A. 1013, A.1016, A.1017 and A.1018). Notice the ground is covered by coprolites (bottom left corner) and possibly fungi (white spots just above). Photo J. Sutcliffe

Fig. A.1059 Riofrío skeletal dispersal monitoring. An adult male deer was found dead in the park. A square (4 × 4 m) was marked around the carcass. A few months afterwards, the bones were all dispersed, probably by foxes, and some of the dispersed bones are indicated here by inserts (Cáceres et al. 2009). Compare with A.1060

Fig. A.1058 Outside of a monitored brown hyena (*Hyaena brunnea*) den in Tswalu Kalahari Reserve. Dispersal of bones brought by hyenas to the den located just behind the photographer. Red squares point to scats left by hyenas. Limb bones (with or without the scapulae) and skulls (with or without mandibles) the easiest elements to be removed from the carcasses and the easiest elements to be transported in the mouth for some distance to the den (see A.1057). The feeding place from where most remains come from is about 2 km distant

Fig. A.1060 Bone dispersal of a fallow deer at Riofrío. Dispersal was mainly by vultures across a wide area in this open land. The inserts show different parts of the animal. The head and scapula were located on the right of the picture, the skin (inside out and twisted), on the left, and the limbs and vertebrae were in the center. Isolated bones were dispersed across this area. The rangers observed the presence of vultures scavenging the carcass, and remains of scats and feathers confirmed their activity. The presence of several individuals of vultures and the openness of the area favor a wider dispersal of the skeleton than usual (A.1059)

Fig. A.1061 Monitored deer skeleton with soft parts covered by dermestids (black spots). Dermestids are small Coleoptera beetles and their larvae feed on decayed substances. They are scavengers of carcasses, and their action produces disarticulation as they eat remains of ligaments and fur, sometimes scratching into and on the surfaces of bone (A.206, A.211 and A.410). Photo Z. San Pedro

Fig. A.1063 Trampling dispersal of bones by deer at Riofrío. Compare to the site of Draycott (see A.1064 and the text). The same type and degree of dispersal has also been observed when stone artefacts were laid out and dispersed by the deer in another experiment

Fig. A.1062 Wildebeest in Ngorongoro. Herds follow game trails or lines of least resistance, and trampling of bones in these situations may be intense. The capacity for herd animals to disperse bone is great, especially in areas where they congregate (A.1063 and A.1064)

Fig. A.1064 Draycott (UK), natural dispersal by animal trampling, in ▶ this case, cows. Dispersal was monitored for 8 years, from shortly after the time of death. Some bones dispersed to the left of this picture into a small cave entrance, and the others were broken by trampling and dispersed down slope to the right and either buried or destroyed. Linear marks made by trampling have been studied (see Text Fig. 3.3). Some of the buried bones has root marks

Fig. A.1065 Horse skeleton dispersed by herds of herbivores in Botswana (see A.1063 and A.1064)

Fig. A.1067 Bird pellets soaked in water or in environments with high humidity (or recently regurgitated) are quickly destroyed by water movement. In contrast, pellets that have become dried out may remain complete for longer periods, and they are resistant to wetting even after days in water streams (Fernández-Jalvo and Andrews 2003)

Fig. A.1068 Gravitational dispersal. An adult male fallow deer died near the top of this slope at Riofrío and its skeleton was dispersed down this strong slope. In this case the bones have almost reached the river (see A.1069)

Fig. A.1066 Long bone in a relatively fast flowing seasonal stream. It was not moved by the water but was oriented along the direction of flow of the water. This is a femur of deer with the distal end upstream and the proximal end (downstream) chewed and eaten. (Inset, lateral view). Photo Z. San Pedro

Fig. A.1069 Bone dispersal at Barranco de las Ovejas. There is shape ▶ and size selection of bones by gravitational action (A.1068). At the bottom of a steep slope, sheep skulls and rounded stones accumulate while long bones and vertebral columns remain near the top of the slope. In addition there were complete skeletons which had been dropped on the stones by Lammergeiers or Bearded Vulture (*Gypaetus barbatus*)

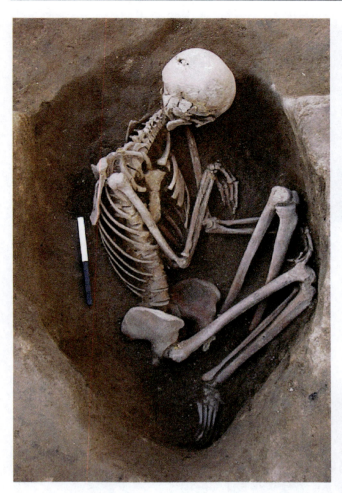

Fig. A.1070 Human burial at Çatalhöyük, with the skeleton intact and all elements preserved and in articulation (A.1071 and A.1072). This was a primary burial under the floor of a house and was not disturbed by any subsequent burials. This skeleton scores the maximum of 10 on the disarticulation scale and 10 on the completeness scale (see text). (Andrews et al. 2005)

Fig. A.1072 Multiple burial 38 at Çatalhöyük with 12 individuals placed in the same grave at different periods. The last body to be buried in the grave was a fully articulated skeleton (disarticulation index 10), already removed before this picture was taken. Two partially articulated skeletons (index 5–6) were disturbed by the last burial, and bones from 9 other individuals (index 1–2, see Chap. 10) had been buried before and been disarticulated by these last three burials. The intact skeleton of a child is in a separate grave cut at bottom left (Andrews et al. 2005)

Fig. A.1073 Wildebeest carcass at Tswalu Kalahari Reserve (South Africa) at the feeding place (see A.1058). In dry hot environments bones may remain in anatomical connection for long periods retained by the skin. The skin also protects the bones from weathering

◄ **Fig. A.1071** Human burial at Çatalhöyük with all elements present and in articulation except for the skull (A.1072 and A.1070). This was removed after burial and partial decay of the body, with the atlas preserving several cut marks made when the skull was detached from the body. This skeleton scores 8 on the disarticulation scale and 8 on the completeness scale (see Chap. 10 and Andrews et al. 2005)

Fig. A.1074 Process of burial of a mandible of deer in Riofrío (see A.1063)

Fig. A.1076 Digger ants: small mammal bones brought together with gravel by the ants near the fossil site of Langebaanweg. The action of digger animals (e.g. ants, termites, badgers, rabbits) is as a potential cause of reworking

Fig. A.1077 Rain and erosion expose fossils at Langebaanweg (A.999). This is especially frequent in Africa and places where tropical rains are particularly strong. Exposed parts of fossil bones may also become weathered in a manner difficult to distinguish from pre-burial weathering, but the parts of the fossil remaining buried are not weathered

Fig. A.1075 Some levels at Azokh Cave have fossils with strong diagenetic modification caused by bat guano. Corrosion has seriously damaged the bone surface and histology, and it has been especially harmful to the preservation of DNA and collagen. Corrosion occurred when fossils were already buried and acid fluids (rich in phosphates, sulfates and carbonates dissolved from the bat guano) percolated through the sediment. Fossils present at the central area of the site, where bat guano was thickest, are severely altered (bottom left inset) where bat guano was deposited. The sediment in the central area acquired a crumbly grey texture and limestone boulders became soft and had a distinct bright white-yellowish color. In contrast, near the cave wall, the sediment remained unaltered and fossils are structurally unaltered (top right inset). The bottom figure shows the fossils from the insets above in situ. The black lines show the limit of diagenetic corrosion by bat guano

Fig. A.1078 Langebaanweg excavation area (A.999). The excavators have left fossils exposed as part of an in situ palaleontological museum. The site has yielded abundant fossils of animals now extinct dated at 5 Ma (Matthews 2004). Fossils provide evidence of many taphonomic surface modifications (A.740) and deformations related to wet environments (A.957 and A.959). Courtesy of C. Denys

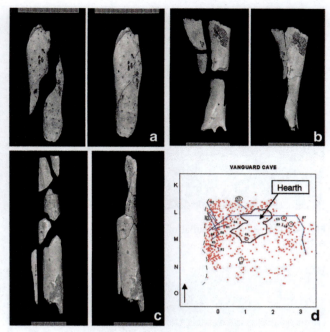

Fig. A.1080 Fossils around a hearth in Vanguard Cave could be refitted (A.981, A.982 and A.983), providing temporal and spatial information about the human occupation of the site (Cáceres and Fernández-Jalvo 2012)

◀ **Fig. A.1079** Excavation of fossil assemblages exposes fossils to the atmosphere, resulting in their drying out and weathering. Fossils exposed in the course of excavation need to be recovered as quickly as possible, for they start to degrade as soon as they are uncovered. Fossil from Azokh Cave heavily corroded (A.788)

Part V
Conclusions

Chapter 11
Why Taphonomy?

Correct identification is crucial to the interpretation of how and why animal bone assemblages accumulate in archaeological and paleontological sites, but just as important is the interpretation placed on the identifications. Although there are gaps in our knowledge, we have been able to identify and link many taphonomic modifications with the processes and agents that produced them. The question we will address in the last chapter is how to make use of this information, and we will do this by looking at several different approaches to taphonomic data.

At its most fundamental level, a question not infrequently asked is: why bother with taphonomy? The implication here is that there are more interesting questions in evolutionary biology than simply asking about the ways in which plants and animals become preserved in the prehistoric record. A common conclusion is that even if an assemblage is biased during the time of its accumulation, this in itself does not affect the species present. Even if this were true, as soon as associations of animals are considered, and their phylogenetic and environmental status, the ways in which they came to be associated are of extreme importance. Many prehistoric sites include species that are not contemporaneous but came together because there were reworking processes acting in the past, bringing together animals from different environments or even from different geological periods. In other cases, the species found together in a fossil site may in fact have been living at the same time but in different habitats, and they have been brought together by a transporting agent, such as a carnivore or a river. In the closely allied subject of forensic anthropology, it is axiomatic that the easy answer is not always the correct one. Forensic scientists examine every possible line of evidence when examining a crime scene, for the consequences of missing a vital piece of information are of enormous significance to the process of law. In our view, similar caution has to be observed in paleontology, where nothing can be assumed until specific investigations are carried out, in this case taphonomic studies.

In order to understand evolutionary biology and past ecosystems at greater resolution, we need to know as much as possible about the environment in which both individual species and whole communities lived and were fossilized. Two fundamental but apparently obvious principles that are sometimes forgotten, are firstly that fossils do not have metabolic functions and so are not biological or even paleobiological entities, although they bear paleobiological information (Fernández-López 1982); and secondly, that while they may have lost much of their biological information during fossilization, fossil associations provide information on past environments that otherwise may have been lost (Behrensmeyer 1984; Fernández-López 1991). Fossils provide a compendium of information created during fossilization that needs to be decoded by taphonomic methods.

We would argue, therefore, that taphonomy is basic to our understanding of individual species, communities and past environments and hence to our grasp of evolutionary processes. If there is a failing in taphonomic studies, it is that taphonomists have failed to get this message across, and that far from being a negative subject, taphonomy adds greatly to the information content in fossil assemblages. The identification and description of fossil animals and plants provides a snap-shot of a particular point in the past, but the taphonomic approach adds to this all the biological and physical events and processes that have acted from time of death, during fossilization and up to the present, a period usually much longer than the lifetimes of the biota from which the fossil assemblages were derived.

The prehistoric record can never be complete, either in terms of species representation or skeletal element representation. With respect to species representation, taphonomy can inform investigators about species that were in fact present (by the evidence they leave as trace fossils) even though they are not skeletally represented at the site. Considering skeletal element representation, when an animal is in a place suitable for preservation, only parts of it are preserved, rarely its complete skeleton, and even more rarely its soft tissues. It is obvious that absence of elements from a fossilized skeleton does not mean it lacked those bones in life, and it is the task

© Springer Science+Business Media Dordrecht 2016
Yolanda Fernández-Jalvo and Peter Andrews, *Atlas of Taphonomic Identifications: 1001+ Images of Fossil and Recent Mammal Bone Modification*, Vertebrate Paleobiology and Paleoanthropology, DOI 10.1007/978-94-017-7432-1_11

of taphonomy to understand in what way, and to what degree, a prehistoric bone assemblage is incomplete and why some parts of the skeleton are present and some are not.

Determination of the extent to which a prehistoric bone assemblage has been altered depends on its mode of accumulation, the time taken for it to accumulate, and fossilization and/or diagenetic processes. The first might entail transport from one place to another, or the intervention of predators, and the second entails time averaging, the length of time it takes for an assemblage to accumulate and be buried and preserved. The third depends on the sedimentary environment present at the place of preservation. All of these processes leave traces on the bones, and the correct identification of taphonomic processes depends on accurate description of the modifications resulting from these processes. The comparative approach based on modern analogues and laboratory experiments helps to identify the taphonomic process and agent, and it is this aspect that we seek to address. This is an important and necessary first step. There is much more that can be done, however, and in undertaking this compilation we have shown where some of the strengths and weaknesses of taphonomic research are to be found. For example, little is known about the link between plant roots and root marks on bones, but clearly if root marks could be related to the plants that made them they would be a valuable resource in identifying vegetation types where no actual remains of the vegetation have been preserved. Another example is the difficulty in identifying tooth marks with the predators that made them, and at present we cannot even reliably identify predator size, let alone species, when faced with fossil tooth marks. How indeed can we distinguish between predation and scavenging in the fossil record? It is unusual to find direct evidence of predation on large mammals, such as puncture marks on the face that suggest asphyxia. Multiple modifications may further confuse the issue, such as the evidence left by bacterial attack on digested bones. There are, however, good indications from evidence of primary and secondary access to carcasses that may distinguish between predation and scavenging, for example by body part representation affected by tooth marks or cut marks (Blumenschine 1986; Lyman 1994a; Domínguez-Rodrigo and Piqueras 2003). These and other issues will no doubt be the subject for future research for many years, but as our comparative database increases and becomes more sophisticated it will be possible to produce a more complete and accurate compendium of taphonomic images.

While it is the case that assemblages of prehistoric species must be treated as no more than a sample of the species pools present in the past, absence of a species from an assemblage does not necessarily mean that that species was absent from the source assemblage. For example, absence of carnivores from a bone assemblage accumulated by carnivores is not unusual, for carnivores rarely die next to their prey. Predators do, however, significantly alter the species

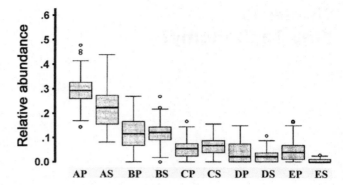

Fig. 11.1 Boxplots showing the ranges of relative abundances of mammal species for 61 modern faunas (Soligo 2001). The boxplots show the relative abundances of 10 ecomorphological categories of mammals for the 61 faunas, each with median (horizontal line through box), 25th and 75th percentiles, range excluding outliers (whiskers) and outliers (circles). The ecological categories are as follows: A < 1 kg, B 1–10 kg, C 10–45 kg, D 45–180 kg, E > 180 kg. *P* primary consumers, *S* secondary consumers

and skeletal elements present in an assemblage because of prey selection. This fact highlights another object of taphonomic research: to discover in what way and to what degree an assemblage of species in the prehistoric record is biased. In this case, taphonomic modifications on prehistoric bone provide some information, but it is the community structure of the faunas, or lack of it, that provides evidence on species loss or species mixing from more than one habitat.

In a study of body mass, Damuth (1982) used the relative abundance of species in relation to body mass to assess the level of preservation of the community structure in assemblages of fossil mammals. Energetic and metabolic regularities present in mammalian communities constrain the variability possible in mammalian communities, and departure from the expected set of regularities can be a measure of taphonomic bias. For example, proportions of small and large mammals in a broad range of habitats is approximately 80:20; the upper size limits of mammals with specialized feeding habits such as rodents and soricids is limited by their adaptations. The extent of the departure from this degree of variability is a measure of degree of taphonomic bias, for it is unlikely that past communities diverged from this norm, at least for the Neogene. In a refinement of Damuth (1982), Soligo (2001, Soligo and Andrews 2005) assessed the composition of fossil assemblages with respect to the size and dietary adaptations of the species they contain. Ten ecovariables were assessed based on primary and secondary consumers divided across five size classes, and the combined distributions of the 10 variables were plotted for 61 modern faunas from a wide variety of habitats (Fig. 11.1). Most faunal distributions were found to be within a limited range of variation with few outliers. Having set this standard applying to a large sample of present day animal communities, fossil faunas could be compared with these distributions to see if they fit in. If they do, it indicates that the fossil fauna is a functionally coherent assemblage, but if they do not fit within the total range of modern faunas, it indicates a functional imbalance, a departure from the energetic and metabolic constraints common to living faunas and brought about most probably by taphonomic bias.

It is possible that early mammalian faunas were not bound by the same ecological constraints as they are today. There are minor variations in metabolic adaptations in living mammals, and greater variations could be expected in the ancestral stages of mammalian evolution, but with the appearance of most living families of mammal in the Neogene such variations from present day patterns would be minimal. In many cases departure from the norm can be attributed to specific taphonomic biases through the ecological combinations of body mass and dietary guild. For example, many of the PlioPleistocene faunas from Africa lacked small primary consumers (Soligo and Andrews 2005), a situation without parallel in any modern African habitat from desert to tropical rain forest, and whatever the interpretation of the paleoenvironments of these faunas, it is clear that they are taphonomically biased against these ecotypes. The Olduvai bed I faunas were the only Pleistocene faunas from Africa not to have a serious ecological

imbalance, which adds weight to the interpretations of Fernández-Jalvo et al. (1998).

This highlights another area of taphonomic research in which little work has been done. This is to quantify the extent to which taphonomic bias may influence the structure of prehistoric animal or plant assemblages. All such assemblages may be considered incomplete to some degree, since some species are more likely to be preserved than others. Some efforts have been made to model the potential effect of the resultant taxonomic bias by removing species from modern comparative faunas using rarefaction or by adding species in order to simulate the effects of faunal mixing or time averaging (e.g., Dreyer 1984; Foote 1992; Robb 2002; Andrews 2006). These studies targeted specific elements of the assemblages, such as large or small specimens, to estimate the extent to which their loss affects the ecological signal of the assemblage.

One recent example has introduced species bias to test the effect of species loss and species mixing on modern faunas of known habitat (Andrews 2006). Ecological data from 19 faunas from a range of tropical African habitats were ordinated, and habitat-based ordination scores were calculated by averaging the habitat scores for all species in the habitats. The ordination method used was weighted averages

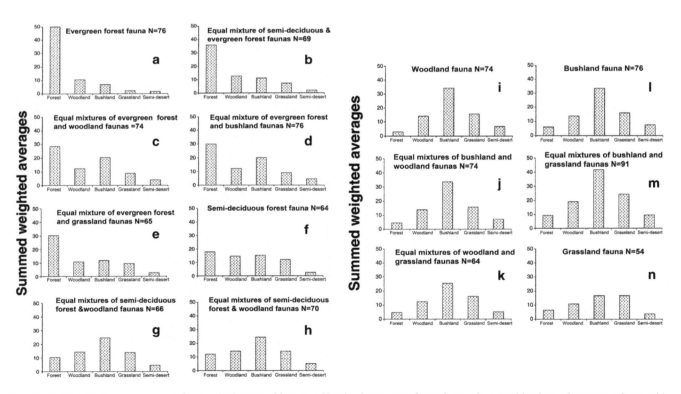

Fig. 11.2 Weighted averages scores for equal mixtures of faunas. **a** Unmixed evergreen forest faunas; **b–e** combinations of evergreen forest with semi-deciduous forest, woodland, bushland and grassland faunas respectively; **f** unmixed semi-deciduous forest fauna; **g, h** combinations of semi-deciduous forest with woodland and bushland faunas respectively; **i** unmixed woodland fauna; **j, k** combinations of woodland with bushland and grassland faunas; **l** unmixed bushland fauna; **m** mixture of bushland and grassland faunas; **n** grassland fauna

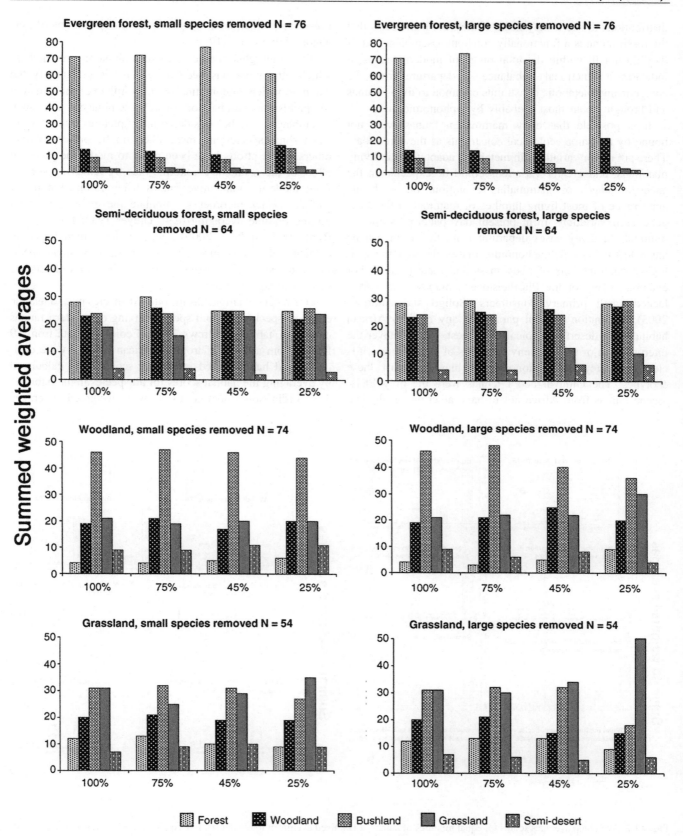

Fig. 11.3 Weighted averages scores for modern faunas with species progressively removed, from left to right on each bar chart: 100% (no reduction), 75, 45 and 25% of species remaining. The column of figures down the left side shows the effects of removal of small-bodied species, while those on the right show the effects of removal of large-bodied species

(Gauch 1989), which provided the base line patterns for the 19 African habitats. The ordination scores of a series of faunas were combined, either on the basis of equal numbers from pairs of different habitats, or on the basis of one quarter to three quarters of species from different habitats, and the significance of the resulting differences in ecological composition could be statistically tested (Fig. 11.2). Some of the mixed assemblages gave predictably mixed signals, but in many cases the ordination scores were still clear and the sources of the mixtures could be estimated. For example, when a tropical forest mammal fauna is mixed with equal numbers of species from a woodland fauna, it looks, as expected, like a mixture of the two but with the forest element still dominant (Andrews 2006). On the other hand, if it is mixed unequally, at 25% forest species and 75% woodland species, the forest element is lost. If the proportions are reversed, the forest signal is retained. This example was for a fauna from a high diversity evergreen forest habitat, but when a less diverse forest fauna from a semi-deciduous forest is mixed with woodland or bushland fauna, it becomes indistinguishable from woodland or bushland, and this remains the case whether the mixing is equal between habitat types or is unequal (Andrews 2006).

In a second set of analyses, the cumulative scores for each habitat were progressively reduced to 25% of their original values by eliminating either small (<1 kg) or large (>90 kg) mammals (Fig. 11.3). For example, eliminating small mammals from a modern tropical forest fauna decreases the forest elements in the fauna to such a degree that it becomes harder and harder to detect its forest affinities. Conversely, the elimination of large mammals from the same fauna had little effect, and even when it is reduced to 25% of its former size, it still has recognizable forest affinities (Andrews 2006). Woodland faunal assemblages on the other hand show little difference with loss of both large and small species, but grassland faunas actually increase their grassland signal with loss of either large or small species (Andrews 2006). If there is taphonomic evidence for under- or over-representation of large or small species, for example from evidence of predation or transport, these bias differences between habitat types need to be taken into account (see below).

These data formed the basis of comparison with two of the faunas from Pliocene deposits at Laetoli, both with evident biases in favor of large mammals and loss of some small mammal species. One fauna was found to have a marked resemblance to a mixture of semi-deciduous forest and woodland species, and the other to a mixture of 75% bushland species and 25% grassland species. Species loss of either large or small species does not seem to have affected the ecological signal, for when small mammals were omitted from a multivariate analysis of one of the faunas, it produced no change (Kovarovic et al. 2002), which is consistent with the lack of change in the species loss models in woodland faunas.

Whether an accumulation of a prehistoric animal assemblage is the function of hominin or non-hominin activity is a question of much importance and one that taphonomic studies can address. Carnivore-ungulate ratios and mean flake lengths from carnivore-ravaged assemblages have been used in combination in an effort to identify predator species responsible for large mammal assemblages (Thackeray 2007). The contrast was made between the combination of low carnivore-ungulate ratios and extensive fragmentation of bone suggesting hominin activity (Brain 1981) with the opposite condition indicating carnivore activity (Klein 1975; Thackeray 1979).

Another attempt to assess the extent to which predators affected the composition of mammal assemblages has been directed specifically at small mammals (Andrews 1990). This study investigated links between predator and prey for small mammals. Most predators have specific hunting profiles, even the most opportunistic: by size (large or small), by activity pattern (nocturnal or diurnal), by habit (terrestrial or aerial), and to some extent by habitat (wooded or non-wooded); and their prey assemblages can be compared with actual small mammal assemblages present on the ground to measure selection bias specific to each predator (Andrews 1990). The summed biases for 18 predators are shown in Fig. 2.39 of Andrews (1990), and since many predators also produce characteristic patterns of taphonomic modification on the bones of their prey, the search for these patterns in fossil faunas may enable the identification of the type of predator, even the species in some cases, responsible for a small mammal fauna. Assuming the hunting bias for the living predator was the same in the past as it is in the present, the bias can be incorporated into the reconstruction of the fossil fauna. As with the approach of Damuth (1982), this method requires intensive scrutiny of the available fossil material.

Concluding Remarks

A definitive taphonomic work that showed the importance of closer inspection of the evidence in the fossil record has been provided by Bob Brain (1981). From the previously held view that australopiths were the hunters of the animals found mixed with their fossils, Brain showed that the small puncture marks on both animal and hominin bone, which were not seen by the taphonomically untrained eye, were produced by predators. As a result he was able to show that australopiths were not the hunters that brought bones to their cave but rather were the hunted. They were part of the prey of large predators, an insignificant part of their food chain, and this great taphonomic work turned many conclusions about early hominin behavior on their head.

We have tried with this book to provide tools for helping investigators identify modifications recorded on bones, whether from archaeological and/or paleontological sites. We have provided indications of how to distinguish modifications, based on our own experience, by means of a large number of illustrations reproduced at high resolution on the online files. These show the similarities and differences of taphonomic modifications linked to known processes and agents. We recognize that this does not solve all the taphonomic issues present at any site, but we hope that the tools provided here will be useful for a first taphonomic approach and for neophyte taphonomists. We hope also that it will be illuminating to the curiosity of the non-taphonomist wishing to identify "strange" scratches, pits or corroded areas on fossils. As with any book we view this as a starting point for the subject and hope that it may attract researchers to consider taphonomy as an essential part of the work they do and to recognize the important perspective and level of understanding it can bring to their studies.

References

Andrews, P. (1990). *Owls, caves and fossils*. London: Natural History Museum, and Chicago: University of Chicago Press.

Andrews, P. (1995). Time resolution of the Miocene fauna from Pasalar. *Journal of Human Evolution, 28*, 343–358.

Andrews, P. (1999). Taphonomy of the Shuwayhat proboscidean: Emirate of Abu Dhabi, western region. In P. Whybrow & A. Hill (Eds.), *Fossil vertebrates of Arabia* (pp. 338–353). New Haven: Yale University Press.

Andrews, P. (2006). Taphonomic effects of faunal impoverishment and faunal mixing. *Palaeogeography, Palaeoclimatology, Palaeoecology, 241*, 572–589.

Andrews, P., & Armour-Chelu, M. (1998). Taphonomic observations on a surface bone assemblage in a temperate environment. *Bulletin of the Geological Society of France, 169*, 433–442.

Andrews, P., & Bello, S. (2006). Pattern in human burial practices. In C. Knusel & R. Gowland R. (Eds.), *The social archaeology of funerary remains* (pp. 14–29). Oxford: Oxbow Books.

Andrews, P., & Cameron, D. (2010). Rudabanya: Taphonomic analysis of a fossil hominid site from Hungary. *Palaeogeography, Palaeoclimatology, Palaeoecology, 297*, 311–329.

Andrews, P., & Cook, J. (1985). Natural modifications to bones in a temperate setting. *Man, 20*, 675–691.

Andrews, P., Cook, J., Currant, A., & Stringer, C. (1999). *Westbury cave*. Bristol: Western Academic Specialist Publishers.

Andrews, P., & Ersoy, A. (1990). Taphonomy of the Miocene bone accumulations at Pasalar, Turkey. *Journal of Human Evolution, 19*, 379–396.

Andrews, P., & Evans, E. M. N. (1983). Small mammal bone accumulations produced by mammalian carnivores. *Paleobiology, 9*, 289–307.

Andrews, P., & Fernández-Jalvo, Y. (1997). Surface modifications of the Sima de los Huesos fossil humans. *Journal of Human Evolution, 33*, 191–217.

Andrews, P., & Fernández-Jalvo, Y. (2003). Cannibalism in Britain: Taphonomy of the Cresswellian (Pleistocene) faunal and human remains from Gough's cave (Somerset, England). *Bulletin of the Natural History Museum, London, 58*, 59–81.

Andrews, P., & Ghaleb, B. (1999). Taphonomy of the Westbury cave bone assemblages. In P. Andrews, J. Cook, A. Currant & C. Stringer (Eds.), *Westbury cave* (pp. 87–126). Bristol: Western Academic Specialist Publishers.

Andrews, P., & Jenkins, E. (2000). The taphonomy of the small mammal faunas. In L. Barham (Ed.), *The Middle Stone Age of Zambia, South Central Africa* (pp. 57–62). Bristol: Western Academic and Specialist Press.

Andrews, P., Molleson, T., & Boz, B. (2005). The human burials at Çatalhöyük. In I. Hodder (Ed.), *Inhabiting Catalhoyuk: Reports from the 1995–1999 seasons* (pp. 263–489). Cambridge: Cambridge University Press.

Andrews, P., & Turner, A. (1992). Life and death of the Westbury bears. *Acta Zoologica Fennica, 28*, 139–149.

Andrews, P., & Van Couvering, J. A. H. (1975). Environments in the East African Miocene. In F. S. Szalay (Ed.), *Contributions to Primatology*, Vol. 5, (pp. 62–103). Basel: Karger.

Andrews, P., & Whybrow, P. (2005). Taphonomic observations on a camel skeleton in a desert environment in Abu Dhabi. *Paleontologia Electronica 8*(1): 23A, 17p, 1.5MB.

Armour-Chelu, M., & Andrews, P. (1994). Some effects of bioturbation by earthworms (Oligochaeta) on archaeological sites. *Journal of Archaeological Science, 21*, 433–443.

Armour-Chelu, M., & Andrews, P. (1996). Surface modifications of bone. In M. Bell (Ed.), *The experimental earthworks project, 1960–1992* (pp. 178–185). Portsmouth: English Heritage.

Arsuaga, J. L., Martínez, I., Gracia, A., Carretero, J. M., Lorenzo, C., García, N., et al. (1997). Sima de los Huesos (Sierra de Atapuerca, Spain). The site. *Journal of Human Evolution, 33*, 109–127.

Barham, L. (2000). *The Middle Stone Age of Zambia, South Central Africa*. Bristol: Western Academic and Specialist Press.

Barham, L., Pinto Llona, A., & Andrews, P. (2000). The Mumbwa caves behavioural record. In L. Barham (Ed.), *The Middle Stone Age of Zambia, South Central Africa* (pp. 81–148). Bristol: Western Academic and Specialist Press.

Baquedano, E., Domínguez-Rodrigo, M., & Musiba, C. (2012). An experimental study of large mammal bone modification by crocodiles and its bearing on the interpretation of crocodile predation at FLK Zinj and FLK NN3. *Journal of Archaeological Science, 39*, 1728–1737.

Behrensmeyer, A. K. (1975). Taphonomy and paleoecology of the Plio-Pleistocene vertebrate assemblages east of Lake Rudolf, Kenya. *Bulletin of the Museum of Comparative Biology, Harvard, 146*, 473–578.

Behrensmeyer, A. K. (1978). Taphonomic and ecological information from bone weathering. *Paleobiology, 4*, 150–162.

Behrensmeyer, A. K. (1981). Vertebrate paleoecology in a recent East African ecosystem. In J. Gray, A. J. Boucot & W. B. N. Berry (Eds.), *Communities of the past* (pp. 591–615). Stroudsburg: Huchinson-Ross.

Behrensmeyer, A. K. (1983). Patterns of natural bone distribution on recent land surfaces: Implications for archaeological site formation. In J. Clutton-Brock & C. Grigson (Eds.), *Animals and archaeology: Hunters and their prey, British archaeological reports, International Series* (Vol. 163, pp. 93–106). Oxford.

Behrensmeyer, A. K. (1984). Taphonomy and the fossil record. *American Scientist, 72*, 558–566.

Behrensmeyer, A. K. (1991). Terrestrial vertebrate accumulations. In P. A. Allison & D. E. G. Briggs (Eds.), *Taphonomy: Releasing the data locked in the fossil record*, Vol. 9 of Topics in Geology. New York: Plenum Press.

Behrensmeyer, A. K. (1993). The bones of Amboseli. *National Geographic Research and Exploration, 9,* 402–421.

Behrensmeyer, A. K. (2007). Changes through time in carcass survival in the Amboseli ecosystem, southern Kenya. In T. R. Pickering, N. Toth, & K. Schick (Eds.), *Breathing life into fossils: Taphonomic studies in honor of C.K. Brain* (pp. 63–86). Gosport: Stone Age Institute Press.

Behrensmeyer, A. K., Gordon, K. D., & Yanagi, G. T. (1986). Trampling as a cause of bone surface damage and pseudo-cutmarks. *Nature, 319,* 768–771.

Behrensmeyer, A. K., & Hill, A. (Eds.). (1980). *Fossils in the making.* Chicago: University of Chicago Press.

Bell, L. S. (1990). Paleopathology and diagenesis: An SEM evaluation of structural changes using backscattered electron imaging. *Journal of Archaeological Science, 17,* 85–108.

Bell, L. S. (Ed.). (2012). *Forensic microscopy for skeletal tissues.* Dordrecht: Springer.

Bell, L. S. (2012). Identifying post-mortem microstructural change to skeletal and dental tissues using backscattered electron imaging. In L. S. Bell (Ed.), *Forensic microscopy for skeletal tissues* (pp. 173–190). Dordrecht: Springer, Humana Press.

Bell, L. S., Boyde, A., & Jones, S. J. (1991). Diagenetic alteration to teeth in situ illustrated by backscattered electron imaging. *Scanning, 13,* 173–183.

Bell, L. S., & Elkerton, A. (2008). Unique marine taphonomy in human skeletal material recovered from the medieval warship Mary Rose. *International Journal of Osteoarchaeology, 18,* 523–535.

Bell, L. S., Lee Thorp, J. A., & Elkerton, A. (2009). The sinking of the Mary Rose warship: A medieval mystery solved? *Journal of Archaeology Science, 36,* 166–173.

Bell, L. S., Skinner, M. F., & Jones, S. J. (1996). The speed of post mortem change to human skeleton and its taphonomic significance. *Forensic Science International, 82,* 129–140.

Bello, S., & Andrews, P. (2006). The intrinsic pattern of preservation of human skeletons and its influence on the interpretation of funerary behaviours. In C. Knusel & R. Gowland (Eds.), *The social archaeology of funerary remains* (pp. 1–13). Oxford: Oxbow Books.

Bello, S., Thomann, A., Rabino Massa, E., & Dutour, O. (2003). Quantification de l'Etat de conservation des colections ostéoarchéologiques et des champs d'application en anthropologie. *Anthropologie, 5,* 21–37.

Bennàsar, M. L. (2010). *Tafonomía de micromamíferos del Pleistoceno inferior de la Sierra de Atapuerca (Burgos): la Sima del Elefante y la Gran Dolina.* Tarragona: University Rovira i Virgili.

Bermúdez de Castro, J. M., Bromage, T. G., & Fernández-Jalvo, Y. (1988). Buccal striations on fossil human anterior teeth. An evidence for handedness in the middle and early upper Pleistocene. *Journal of Human Evolution, 17,* 403–412.

Binford, L. R. (1981). *Bones. Ancient men and modern myths.* London: Academic Press.

Bischoff, J. L., Aramburu, A., Arsuaga, J. L., Carbonell, E., & Bermúdez de Castro, J. M. (2003). The Sima de los Huesos hominids date to beyond the U/Th equilibrium (>350 kyr) and perhaps to 400–500 kyr: New radiometric dates. *Journal of Archaeological Science, 30,* 275–280.

Blumenschine, R. (1986). Carcass consumption sequences and archaeological distinction of scavenging and hunting. *Journal of Human Evolution, 15,* 639–659.

Blumenschine, R., Masao, F. T., Stansistreet, I. G., & Swisher, C. C. (Eds.) (2012). Five decades after Zinjanthropus and Homo habilis: Landscape palaeoanthropology of Plio-Pleistocene Olduvai Gorge, Tanzania. *Journal of Human Evolution, 63,* No. 2, special issue.

Blumenschine, R. J., Peters, C. R., Capaldo, S. D., Andrews, P., Njau, J., & Pobiner, B. L. (2007). Vertebrate taphonomic perspectives on Oldowan hominin land use in the Plio-Pleistocene Olduvai Basin, Tanzania. In T. R. Pickering, N. Toth, & K. Schick (Eds.), *Breathing life into fossils: Taphonomic studies in honor of C.K. Brain* (pp. 161–180). Gosport: Stone Age Institute Press.

Blumenschine, R. J., & Selvaggio, M. M. (1988). Percussion marks on bone surfaces as a new diagnostic of hominid behaviour. *Nature, 333,* 763–765.

Blumenschine, R. J., & Selvaggio, M. M. (1991). On the marks of marrow bone processing by hammerstones and hyenas: Their anatomical patterning and archaeological implications. In J. D. Clark (Ed.), *Cultural beginnings* (pp. 17–32). Bonn: Dr. R. Habelt GMBH.

Boaz, N. T., & Behrensmeyer, A. K. (1976). Hominid taphonomy: Transport of human skeletal parts in an artificial fluviatile environment. *American Journal of Physical Anthropology, 45,* 53–60.

Bodén, P. (1988). Epipsammic diatoms as borers: An observation on calcareous sand grains. *Sedimentary Geology, 59,* 143–147.

Brain, C. K. (1967). Bone weathering and the problem of bone pseudo-tools. *South African Journal of Science, 63,* 97–99.

Brain, C. K. (1969). The contribution of Namib desert Hottentots to an understanding of australopithecine bone accumulations. *Scientific Papers of the Namib Desert Research Station, 39,* 13–22.

Brain, C. K. (1981). *The hunters or the hunted.* Chicago: University of Chicago Press.

Brothwell, D. (1976). Further evidence chewing by ungulates: The sheep of North Ronaldsay, Orkney. *Journal of Archaeological Science, 3,* 179–182.

Bromage, T. G. (1987). The scanning electron microscopy/replica technique and recent applications to the study of fossil bone. *Scanning Electron Microscopy, 1987,* 607–613.

Bromage, T. G., Bermúdez de Castro, J. M., & Fernández Jalvo, Y. (1991). The SEM in taphonomy research and its application to studies of cut-marks generally and the determination of handedness specifically. *Anthropologie, 29,* 163–169.

Bromage, T. G., & Boyde, A. (1984). Microscopic criteria for the determination of directionality of cutmarks on bone. *American Journal of Physical Anthropology, 65,* 359–366.

Cáceres, I. (2002). Tafonomía de yacimientos antrópicos en Karst. Complejo Galería (Sierra de Atapuerca, Burgos), Vanguard Cave (Gibraltar) y Abric Romaní (Capellades, Barcelona). Dpto. Historia y Geografía. PhD dissertation, Universitat Rovira i Virgili, Tarragona.

Cáceres, I., Bravo, P., Esteban, M., Expósito, I., & Saladié, P. (2002). Fresh and heated bones breakage. An experimental approach. In M. De Renzi, M. Pardo, M. Belinchón, E. Peñalver, P. Montoya & A. Márquez-Aliaga (Eds.), *Current topics on taphonomy and fossilization* (pp. 471–479) Valencia: Ayuntamiento de Valencia.

Cáceres, I., Esteban, M., & Fernández Jalvo, Y. (2007). Mordeduras de herbívoro en el Bosque de Riofrío (Segovia). In M. L. Ramos, J. E. González & J. Baena (Eds.), *Arqueología Experimental en la Península Ibérica: Investigación, Didáctica y Patrimoni* (pp. 59–67). Santander: EXPERIMENTA.

Cáceres, I., Esteban-Nadal, M., Fernández-Jalvo, Y., & Bennàsar, M. L. L. (2008). Disarticulation and dispersal processes of Cervid Carcasses at the Bosque de Riofrío (Segovia, Spain). *Journal of Taphonomy, 7,* 129–141.

Cáceres, I., Esteban-Nadal, M., Bennàsar, M., & Fernández-Jalvo, Y. (2011). Was it the deer or the fox? *Journal of Archaeological Science, 38,* 2767–2774.

Cáceres, I., & Fernández-Jalvo, Y. (2012). Taphonomy of the fossil bone assemblages from the Middle Area in Vanguard Cave. In R. N. E. Barton, C. Stringer & J. C. Finlayson (Eds.), *Neanderthals in context. a report of the 1995–98 excavations at Gorham's & Vanguard Caves, Gibraltar* (pp. 253–265). Monograph Series, Oxford Committee for Archaeology. Oxford: Oxbow Press.

Capaldo, S. D., & Blumenschine, R. J. (1994). A quantitative diagnosis of notches made by hammerstone percussion and carnivore gnawing on bovid long bones. *American Antiquity, 59,* 724–748.

Carbonell, E., Cebrià, A., Allué, E., Cáceres, I., Castro-Curel, Z., Díaz, R., et al. (1996). Behavioural and organizational complexity in the Middle Paleolithic from the Abric Romaní (Capellades, Anoia). In E. Carbonell & M. Vaquero (Eds.), *The last neandertals – The first anatomically modern humans. Cultural change and human evolution: The crisis at 40 Ka BP* (pp. 385–434). Tarragona: Universitat Rovira i Virgili.

Carbonell, E., & Mosquera, M. (2006). The emergence of a symbolic behaviour: The sepulchral pit of Sima de los Huesos, sierra de Atapuerca, Burgos, Spain. *Comtes Rendu Paleoevolution, 5,* 155–160.

Carbonell, E., Bermúdez de Castro, J. M., Parés, J. M., Pérez González, A., Cuenca Bescós, G., Ollé, A., et al. (2008). The first hominin of Europe. *Nature, 425,* 465–470.

Carbonell, E., & Vaquero, M. (Eds.). (1996). *The last Neandertals – The first anatomically modern humans. Cultural change and human evolution: The crisis at 40 Ka BP.* Tarragona: Universitat Rovira i Virgili.

Chazan, M., Ron, H., Matmon, A., Porat, N., Goldberg, P., Yates, R., et al. (2008). Radiometric dating of the earlier stone age sequence in excavation I at Wonderwerk Cave, South Africa: Preliminary results. *Journal of Human Evolution, 55,* 1–11.

Chen, J.-Y., Schopf, J. W., Bottjer, D. J., Zhang, C.-Y., Kudryavtsev, A. B., Wang, X.-Q., et al. (2007). Raman spectra of a Lower Cambrian ctenophore embryo from SW Shaanxi, China. *Proceedings of the National Academy of Science, 104,* 6289–6292.

d'Errico, F., Giacobini, G., & Puech, P. F. (1984). Varnish replicas: A new method for the study of worked bone surfaces. *OSSA, 9–11,* 29–51.

Damas Mollá, L., Aranburu Artano, A., & García Garmilla, F. (2006). Resistance to the diagenetic alteration of shells of *Chondrodonta* sp in red limestones of the Early Aptian – Albian Ereño (Bizkaia) Resistencia a la alteración diagenética de conchas de *Chondrodonta* sp en las calizas rojas del Aptiense – Albiense inferior de Ereño (Bizkaia). *Geogaceta, 40,* 195–198.

Damuth, J. (1982). Analysis of the preservation of community structure in assemblages of fossil mammals. *Paleobiology, 8,* 434–446.

Darwent, C., & Lyman, R. L. (2002). Detecting the postburial fragmentation of carpals, tarsals, and phalanges. In W. D. Haglund & M. H. Sorg (Eds.), *Advances in forensic taphonomy* (pp. 355–377). Boca Raton: CRC Press.

Dauphin, Y., Andrews, P., Denys, C., Fernández-Jalvo, Y., & Williams, T. (2003). Structural and chemical bone modifications in a modern owl pellet assemblage from Olduvai Gorge (Tanzania). *Journal of Taphonomy, 1,* 209–232.

Dauphin, Y., Denys, C., & Denys, A. (1988). Les mecanismes de formation des gisements de microvertebres: Modification de la composition chimique des os et dents de rongeurs issus de peloness de regurgitation de rapace. *Comptes Rendues de l'Academie Scientifique Paris, 307,* 603–608.

Denys, C. (1983). Les Rongeurs du Pliocéne de Laetoli (Tanzanie): Evolution, Paleoécologie et Paleobigéographie. Approche Qualitative et Quantitative. Thèse 3 cycle Université Pierre et Marie Curie, Paris.

Denys, C., Fernández-Jalvo, Y., & Dauphin, Y. (1995). Experimental taphonomy: Preliminary results of the digestion of micromammal bones in laboratory. *Comptes Rendues de l'Academie Scientifique, série II a (Paris), 321,* 803–809.

Denys, C., Williams, C. T., Dauphin, Y., Andrews, P., & Fernández-Jalvo, Y. (1996). Diagenetical change in Pleistocene small mammal bones from Olduvai Bed I. *Palaeogeography, Palaeoclimatology Palaeoecology, 126,* 121–134.

Denys, C., Andrews, P., Dauphin, Y., Williams, C. T., & Fernández-Jalvo, Y. (1997). Towards a site classification: Comparison of the diversity of taphonomic and diagenetic patterns and processes. *Bulletin of the Geological Society of France, 168,* 751–757.

Denys, C., Schuster, M., Guy, F., Mouchelin, G., Vignaud, P., Viriot, L., et al. (2007). Taphonomy in present day desertic environment: The case of the Djourab (Chad) Plio-Pleistocene deposits. *Journal of Taphonomy, 5,* 177–204.

Diefenbach, C. O. DaC. (1975). Gastric function in *Caiman crocodilus* (Crocodylia: Reptilia), Part I: Rate of gastric digestion and pH and proteolysis. *Comparative Biochemistry and Physiology, 51 A,* 267–274.

Díez, J. C., Fernández-Jalvo, Y., Rosell, J., & Cáceres, I. (1999). The site formation (Aurora Stratum, Gran Dolina, Sierra de Atapuerca, Burgos, Spain). *Journal of Human Evolution, 37,* 623–652.

Dodson, P. (1973). The significance of small bones in paleoecological interpretation. *Contributions in Geology, University of Wyoming, 12,* 15–19.

Dodson, P., & Wexlar, D. (1979). Taphonomic investigations of owl pellets. *Paleobiology, 5,* 279–284.

Domínguez-Rodrigo, M., & Barba, R. (2006). New estimates of tooth mark and percussion mark frequencies at the FLK Zinj site: The carnivore-hominid-carnivore hypothesis falsified. *Journal of Human Evolution, 50,* 170–194.

Domínguez-Rodrigo, M., Barba, R., & Egeland, C. P. (Eds.). (2007). *Deconstructing Olduvai: A taphonomic study of the Bed I sites.* Dordrecht: Springer.

Domínguez-Rodrigo, M., & Piqueras, A. (2003). The use of tooth pits to identify carnivore taxa in tooth-marked archaeofaunas and their relevance to reconstruct hominid carcass processing behaviours. *Journal of Archaeological Science, 30,* 1385–1391.

Domínguez-Rodrigo, M., de Juana, S., Galán, A. B., & Rodríguez, M. (2009). A new protocol to differentiate trampling marks from butchery cut marks. *Journal of Archaeological Science, 36,* 2643–2654.

Dreyer, S. (1984). The Theory and Use of Methods for the Study of Mammalian Palaeoecology. PhD dissertation, University of London.

Dutour, O. (1989). *Hommes Fossilies du Sahara: Peuplement Holocènes du Mali Septentrional.* Paris: Edition du CNRS.

Faith, J. T., Domínguez-Rodrigo, M., & Gordon, A. D. (2009). Long distance carcass transport at Olduvai Gorge? A quantitative examination of Bed I skeletal abundances. *Journal of Human Evolution, 56,* 247–256.

Fernández-Jalvo, Y. (1992). Small mammal Taphonomy at the Karstic complex of Atapuerca (Burgos)/Tafonomía de Microvertebrados del Complejo Cárstico de Atapuerca (Burgos). PhD dissertation, Complutense University of Madrid.

Fernández-Jalvo, Y., & Andrews, P. (1992). Small mammal taphonomy of Gran Dolina, Atapuerca (Burgos), Spain. *Journal of Archaeological Science, 19,* 407–428.

Fernández-Jalvo, Y., & Andrews, P. (2000). The taphonomy of Pleistocene caves, with particular reference to Gibraltar. In C. Stringer, R. N. E. Barton & C. Finlayson (Eds.), *Neanderthals on the edge: 150th anniversary conference of the Forbes' quarry discovery, Gibraltar* (pp. 171–182). Oxford: Oxbow Books.

Fernández-Jalvo, Y., & Andrews, P. (2003). Experimental effects of water abrasion on bone fragments. *Journal of Taphonomy, 1,* 147–163.

Fernández-Jalvo, Y., & Andrews, P. (2010). Taphonomy of the human remains from Tianyuandong. In H. Shang & E. Trinkaus (Eds.), *The early modern human from Tianyuan Cave, China* (pp. 205–210). College Station: Texas A&M University Press.

Fernández-Jalvo, Y., & Andrews, P. (2011). When humans chew bones. *Journal of Human Evolution, 60,* 117–123.

Fernández-Jalvo, Y., & Avery, M. D. (2015). Pleistocene micromammals and their predators at Wonderwerk Cave, South Africa. *African Archaeological Review, 32*, 751–791. doi:10.1007/s10437-015-9206-7.

Fernández-Jalvo, Y., & Cáceres, I. (2010). Tafonomía e Industria Lítica: marcas de corte y materias primas. In J. Rodríguez-Vidal, A. Santiago, & E. Mata (Eds.), *Cuaternario y Arqueología. Homenaje a Francisco Giles Pacheco* (pp. 169–177). Servicio de Publicaciones de Diputación de Cádiz y Servicio de Publicaciones. de Universidad de Cádiz.

Fernández-Jalvo, Y., & Perales, C. (1990). Análisis macroscópico de huesos quemados experimentalmente. In S. Fernández-López (Ed.), *Tafonomia y Fosilizacion* (pp. 105–113). Zaragoza: Fernando Catolico, CSIC.

Fernández-Jalvo, Y., Andrews, P., & Denys, C. (1989). Cut marks on small mammals at Olduvai Gorge Bed-I. *Journal of Human Evolution,36*, 587–589.

Fernández-Jalvo, Y., Andrews, P., & Tong, H. (2015). Taphonomy of the Tianyuandong human skeleton and faunal remains. *Journal of Human Evolution, 83*, 1–14. doi:10.1016/j.jhevol.2015.03.010.

Fernández-Jalvo, Y., Andrews, P., Pesquero, D., Smith, C., Marin-Monfort, Sánchez, B., et al. (2010). Early bone diagenesis in temperate environments, Part I: Surface features and histology. *Palaeogeography, Palaeoclimatology, Palaeoecology, 288*, 62–81.

Fernández-Jalvo, Y., Andrews, P., Sevilla, P., & Requejo, V. (2014). Digestion versus abrasion features in rodent bones. *Lethaia,47*, 323–336.

Fernández-Jalvo, Y., Denys, C., Andrews, P., Williams, T., Dauphin, Y., & Humphrey, L. (1998). Taphonomy and palaeoecology of Olduvai Bed I (Pleistocene, Tanzania). *Journal of Human Evolution,34*, 137–172.

Fernández-Jalvo, Y., Díez, J. C., Cáceres, I., & Rosell, J. (1999). Human cannibalism in the early Pleistocene of Southern Europe (Sierra de Atapuerca, Burgos, Spain). *Journal of Human Evolution,37*, 591–622.

Fernández-Jalvo, Y., King, T., Andrews, P., Moloney, N., Ditchfield, P., Yepiskoposyan, L., et al. (2004). Azokh Cave and Northern Armenia. In *Miscelanea en homenaje a Emiliano Aguire Volume IV, Acala de Henares* (pp. 158–168). Alcalá de Henares.

Fernández-Jalvo, Y., King, T., Yepiskoposyan, L., & Andrews, P. (Eds.) (in press) *Azokh Cave and the Transcaucasian Corridor*. New York: Springer.

Fernández-Jalvo, Y., & Marin-Monfort, M. D. (2008). Biostratinomy in museums: Preparation protocols and methods for vertebrate collections of Natural History Museums observed under the Scanning Electron Microscope. *Geobios, 41*, 157–181.

Fernández-Jalvo, Y., Sánchez-Chillon, B., Andrews, P., Fernández-López, S., & Alcalá Martínez, L. (2002). Morphological taphonomic transformations of fossil bones in continental environments, and repercussions on their chemical composition. *Archaeometry,44*, 353–361.

Fernández-López, S. (1982). La evolución tafonómica (un planteamiento neodarwinista). *Boletín de la Real Sociedad Española de Historia Natural, (Geología), 79*, 243–254.

Fernández-López, S. (1991). Taphonomic concepts for a theoretical biochronology. *Revista Española de Paleontología,6*, 37–49.

Finlayson, C., Giles Pacheco, F., Rodríguez-Vidal, J., Fa, D. A., Gutiérrez López, J. M., Santiago Pérez, A., et al. (2006). Late survival of Neanderthals at the southernmost extreme of Europe. *Nature,443*, 850–853.

Fish, E. W. (1950). Chewing of antlers by deer. *British Dental Journal,2*, 299–300.

Fisher, D. C. (1981). Crocodilian scatology, microvertebrate concentrations, and enamel-less teeth. *Paleobiology,7*, 262–275.

Fisher, D. C. (1981). Taphonomic interpretation of enamel-less teeth in the Shotgun local fauna (Paleocene, Wyoming). *Contributions to the Museum of Paleontology, University of Michigan, 25*, 1–25.

Fisher, D. C. (1981). Mode of preservation of Shotgun local fauna (Paleocene, Wyoming) and its implication for the taphonomy of a microvertebrate concentration. *Contributions to the Museum of Paleontology, University of Michigan, 25*, 1–42.

Foote, M. (1992). Rarefaction analysis of morphological and taxonomic diversity. *Paleobiology,18*, 1–16.

Fortea, J., Rasilla, M., García-Tabernero, A., Gigli, E., Rosas, A., & Lalueza-Fox, C. (2008). Excavation protocols of bone remains for Neanderthal DNA analysis in El Sidrón Cave (Asturias, Spain). *Journal of Human Evolution,55*, 353–357.

Frison, G. C., & Todd, L. C. (1986). *The Colby mammoth site: Taphonomy and archaeology of a Clovis Kill in northern Wyoming*. Albuquerque: University of New Mexico Press.

Furness, R. W. (1988). Predation on ground-nesting seabirds by island populations of red deer Cervus elaphus and sheep Ovis. *Journal of Zoology,216*, 565–573.

Gauch, H. G. (1989). *Multivariate analysis in community ecology*. Cambridge: Cambridge University Press.

Gaudzinski-Windheuser, S., Kindler, L., Rabinovich, R., & Goren-Inbar, N. (2010). Testing heterogeneity in faunal assemblages from archaeological sites. Tumbling and trampling experiments at the early-Middle Pleistocene site at Gesher Bernot Ya'aquov (Israel). *Journal of Archaeological Science, 37*, 3170–3190.

Goldberg, P., & Macphail, R. (2000). Micromorphology of sediments from Gibraltar caves: Some preliminary results from Gorham's cave and Vanguard Cave. In J. C. Finlayson, G. Finlayson, & D. A. Fa (Eds.). *Gibraltar during the quaternary* (pp. 93–108). Gibraltar: Gibraltar Government Heritage Monographs.

Gómez, G. (2000). Análisis Tafonómico y Paleoecológico de micro y mesomamíferos del sitio arqueológico de Arroyo Seco (Partido de Tres Arroyos, Buenos Aires, Argentina. PhD dissertation, Universidad Complutense de Madrid.

Grimm, R. J., & Whitehouse, W. M. (1963). Pellet formation in a great horned owl. *The Auk, 80*, 301–306.

Hackett, C. J. (1981). Microscopical focal destruction (tunnels) in excavated human bones. *Medical Science Law, 21*, 243–265.

Hay, R. (1976). *Geology of the Olduvai Gorge. A study of sedimentation in a Semi-arid basin*. Berkeley: University of California Press.

Haynes, G. (1980). Evidence of carnivore gnawing on Pleistocene and Recent mammalian bones. *Paleobiology, 6*, 341–351.

Haynes, G. (1983). A guide for differentiating mammalian carnivore taxa responsible for gnaw damage to herbivore limb bones. *Paleobiology, 9*, 164–172.

Hedges, R. M., Millars, A. R., & Pike, A. W. G. (1995). Measurements and relationships of diagenetic alteration of bone from three archaeological sites. *Journal of Archaeological Science, 22*, 201–209.

Hill, A. (1979). Disarticulation and scattering of mammal skeletons. *Paleobiology, 5*, 261–274.

Hill, A., & Behrensmeyer, A. K. (1984). Disarticulation patterns of some modern East African mammals. *Paleobiology, 10*, 366–376.

Hoffman, R. (1988). The contribution of raptorial birds to patterning in small mammal assemblages. *Paleobiology, 14*, 81–90.

Jans, M. M. (2005). Histological Characterisation of Diagenetic Alteration of Archaeological Bone. *Geoarchaeological and Bioarchaeological Studies* (Vol. 4). Amsterdam: Institute for Geo and Bioarchaeology, Vrije Universiteit.

Jans, M. M., Kars, H., Nielson-March, C. M., Smith, C. I., Nord, A. G., Arthur, P., et al. (2002). In situ preservation of archaeological bone: A histological study with multidisciplinary approach. *Archaeometry, 44*, 343–352.

Jones, C. G. (2012). Scanning electron microscopy: Preparation and imaging for SEM. In L. S. Bell (Ed.), *Forensic microscopy for skeletal tissues* (pp. 1–20). Dordrecht: Springer, Humana Press.

Kibii, M. J. (2009). Taphonomic aspects of African porcupines (*Hystrix cristata*) in the Kenyan highlands. *Journal of Taphonomy, 7*, 21–27.

King, T., Andrews, P., & Boz, B. (1998). Effect of taphonomic processes on dental microwear. *American Journal of Physical Anthropology, 108*, 359–373.

King, T., Fernández-Jalvo, Y., Moloney, N., Andrews, P., Melkonyan, A., Ditchfield, P., et al. (2003). Exploration and survey of Pleistocene hominid sites in Armenia and Karabagh. *Antiquity, 77*, 1–4.

Klein, R. G. (1975). Middle Stone Age man-animal relationships in southern Africa: Evidence from Die Kelders and Klasies River Mouth. *Science, 190*, 265–267.

Klippel, W. E., & Synstelien, J. A. (2007). Rodents as taphonomic agents: Bone gnawing by brown rats and gray squirrels. *Journal of Forensic Science, 52*, 765–773.

Kordos, L., & Begun, D. R. (2002). Rudabánya: A late Miocene subtropical swamp deposit with evidence of the origin of the African apes and humans. *Evolutionary Anthropology, 11*, 45–57.

Korth, W. W. (1979). Taphonomy of micovertebrate fossil assemblages. *Annals of the Carnegie Museum, 48*, 235–285.

Kovarovic, K., Andrews, P., & Aiello, L. (2002). The palaeoecology of the Upper Ndolanya Beds at Laetoli, Tanzania. *Journal of Human Evolution, 43*, 395–418.

López-González, F., Grandal-d'Anglade, A., & Ramón Vidal-Romaní, J. (2006). Deciphering bone depositional sequences in caves through the study of manganese coating. *Journal of Archaeological Science, 33*, 707–717.

Lowe, V. P. W. (1980). Variation in digestion of prey by tawny owl. *Journal of the Zoological Society, London, 192*, 283–293.

Lyman, R. L. (1984). Broken bones, bone expediency tools, and bone pseudo-tools: Lessons from the blast zone around Mount St. Helens. *Washington. American Antiquity, 49*, 315–333.

Lyman, R. L. (1984). Bone density and differential survivorship of fossil classes. *Journal of Anthropological Archaeology, 3*, 159–299.

Lyman, R. L. (1994). *Vertebrate taphonomy*. Cambridge: Cambridge University Press.

Lyman, R. L. (1994). Relative abundances of skeletal specimens and taphonomic analysis of vertebrate remains. *Palaios, 9*, 288–298.

Lyman, R. L. (2008). *Quantitative paleozoology*. Cambridge: Cambridge University Press.

Lyman, R. R., & Fox, G. L. (1989). A critical evaluation of bone weathering as an indication of bone assemblage formation. *Journal of Archaeological Science, 16*, 293–317.

Marchiafava, V., Bonucci, L., & Ascenzi, A. (1974). Fungal osteocalasia: A model of dead bone resorption. *Calcified Tissue Research, 14*, 195–210.

Marin-Monfort, M. D. (2015). Tafonomía de los macromamñiferos del Pleistoceno Medio al Holoceno del Yacimiento de Azokh 1 en el Caucaso (Nagorno-Karabagh). PhD dissertation, Universidad Autónoma de Madrid

Marin-Monfort, M. D., Cáceres, I., Andrews, P., Pinto Llona, A. C., & Fernández-Jalvo, Y. (in press) Taphonomy and site formation of Azokh 1. In Y. Fernández-Jalvo, T. King, L. Yepiskoposyan, & P. Andrews (Eds.), *Azokh Cave and the Transcaucasian Corridor* (pp. xxx–xxx). New York: Springer.

Martill, D. M. (1990). Bones as stones: The contribution of vertebrate remains to the lithologic record. In S. K. Donovan (Ed.), *The Processes of fossilization* (pp. 270–292). New York: Columbia University Press.

Matthews, T. (2004). The taxonomy and taphonomy of Mio-Pliocene and Late Middle Pleistocene micromammals from the Cape West Coast, South Africa. PhD dissertation, University of Cape Town.

Maxbauer, D. P., Peppe, D. J., Bamford, M., McNulty, K. P., Harcourt-Smith, W. E. H., & Davis, L. E. (2013). A morphotype catalog and paleoenvironmental interpretations of early Miocene fossil leaves from the Hiwegi Formation, Rusinga Island, Lake Victoria. *Kenya. Palaeontologia Electronica, 16*(3), 28A.

Michel, L. A., Peppe, D. J., Lutz, J. A., Driese, S. G., Dunsworth, H. M., Harcourt-Smith, W. E. H., et al. (2014). Remnants of an ancient forest provide ecological context for early Miocene fossil apes. *Nature Communications, 5*, 3236. doi:10:1038.

Millard, A. R. (2001). Deterioration of bone. In D. Brothwell & M. Pollard (Eds.), *Handbook of archaeological sciences* (pp. 637–647). Ontario: Wiley.

Molleson, T. (1990). The accumulation of trace metals in bone during fossilization. In N. D. Priest & F. L. Van der Vyver (Eds.), *Trace metals and fluoride in bones and teeth* (pp. 341–365). Boca Ratón: C.R.C. Press.

Molleson, T., & Andrews, P. (1996). Trace element analyses of bones and teeth from Çatalhöyük. In I. Hodder (Ed.), *On the Surface: Çatalhöyük 1993–1995* (pp. 265–270). British Institute of Archaeology at Ankara.

Molleson, T., Andrews, P., & Boz, B. (2005). Reconstruction of the Neolithic people at Çatalhöyük. In I. Hodder (Ed.), *Inhabiting Catalhoyuk: Reports from the 1995–1999 seasons* (pp. 279–531). Cambridge: Cambridge University Press.

Molleson, T., & Cox, M. (1993). *The Spitalfield Project, Volume 2. The Anthropology*. CBA Research Report 86.

Montalvo, C. I., Pessino, M. E. M., & González, V. H. (2007). Taphonomic analysis of remains of mammals eaten by pumas (*Puma concolor*, Carnivora, Felidae) in central Argentina. *Journal of Archaeological Science, 34*, 2151–2160.

Montalvo, C. I., Pessino, M. E. M., & Bagatto, F. C. (2008). Taphonomy of the bones of rodents consumed by Andean hog-nosed skunks (*Conepatus chinga*, Carnivora, Mephitidae) in central Argentina. *Journal of Archaeological Science, 35*, 1481–1488.

Montalvo, C. I., Melchor, R. N., Visconti, G., & Cerdeño, E. (2008). Vertebrate taphonomy in loess-palaeosol deposits: A case study from the late Miocene of central Argentina Taphonomie de vertébrés des dépôts de loess-paléosols: un exemple dans le Miocène supérieur d'Argentine Centrale. *Geobios, 41*, 133–143.

Nielsen-Marsh, C. M., & Hedges, R. E. M. (1999). Bone Porosity and the use of mercury intrusion porosimetry in bone diagenesis studies. *Archaeometry, 41*, 165–174.

Njau, J. K., & Blumenschine, R. J. (2006). A diagnosis of crocodile feeding traces on larger mammal bone, with fossil examples from the Plio-Pleistocene Olduvai Basin, Tanzania. *Journal of Human Evolution, 50*, 142–162.

Noll, M. (1995). Evidence for handedness from cutmark orientation on long bones. *Abstracts of the Palaeoanthropology Society Meetings, 1995*, 28–29.

Nomade, S., Pastre, J. F., Guillou, H., Faure, M., Guérin, C., Delson, E., et al. (2014). ^{40}Ar/^{39}Ar constraints on some French landmark Late Pliocene to Early Pleistocene large mammalian paleofaunas: Paleoenvironmental and paleoecological implications. *Quaternary Geochronology, 21*, 2–15.

Olsen, S. L., & Shipman, P. (1988). Surface modification on bone: Trampling vs. butchery. *Journal of Archaeological Science, 15*, 535–553.

Pääbo, S., Poinar, H., Serre, D., Jaenicke-Després, V., Hebler, J., Rohland, N., et al. (2004). Genetic analyses from ancient DNA. *Annual Review of Genetics, 38*, 645–679.

Pesquero Fernández, D. (2006). Tafonomía del Yacimiento de Vertebrados Miocenos de Cerro de la Garita (Concud, Teruel). PhD dissertation, University Complutense of Madrid.

Pesquero, M. D., Ascaso, C., Alcalá, L., & Fernández-Jalvo, Y. (2010). A new taphonomic bioerosion in a Miocene lakeshore environment. *Palaeogeography, Palaeoclimatology, Palaeoecology, 295*, 192–198.

Pesquero, M. D., Alcalá, L., & Fernández-Jalvo, Y. (2013). Taphonomy of the reference Miocene vertebrate mammal site of Cerro de la Garita (Concud, Teruel, Spain). *Lethaia, 46*, 378–398.

Pesquero, M. D., & Fernández-Jalvo, Y. (2014). Bioapatite to calcite, an unusual transformation seen in fossil bones affected by aquatic bioerosion. *Letahia, 47*, 533–546.

Pickering, T. R., & Egeland, C. P. (2006). Experimental patterns of hammerstone percussion damage on bones and zooarchaeological inferences of carcass processing intensity by humans. *Journal of Archaeological Science, 33*, 459–469.

Pickering, T. R., Toth, N., & Schick, K. (Eds.) (2007). *Breathing life into fossils: Taphonomic studies in honor of C.K. Brain*. Gosport: Stone Age Institute Press.

Pickford, M., & Andrews, P. (1981). The Tinderet Miocene sequence in Kenya. *Journal of Human Evolution, 10*, 11–33.

Pinto Llona, A., & Andrews, P. (1996). Amphibian taphonomy from cave deposits in England. *Comm Reunion Tafonomia y fosilizacion, 1996*, 327–330.

Pinto Llona, A., & Andrews, P. (1999). Amphibian taphonomy and its application to the fossil record of Dolina (Middle Pleistocene, Atapuerca, Spain). *Palaeogeography, Palaeoclimatology, 149*, 411–429.

Pinto Llona, A., & Andrews, P. (2004). Scavenging behaviour patterns in cave bears *Ursus spelaeus*. *Revue de Paleobiologie, 23*, 845–853.

Pinto Llona, A., & Andrews, P. (2004). Taphonomy and palaeoecology of *Ursus spelaeus* from northern Spain. *Symposium International Ours Cavernes. Musée. Lyons, 2*, 163–170.

Pinto Llona, A., Andrews, P., & Etxebarria, F. (2005). *Taphonomy and palaeoecology of bears*, Fundacion Oso de Asturias (pp. 1–562). (Spanish text with English translation (pp. 563–670.))

Piper, K., & Valentine, G. (2012). Bone pathology. In L. S. Bell (Ed.), *Forensic microscopy for skeletal tissues* (pp. 51–88). Dordrecht: Springer, Humana Press.

Pobiner, B. L., DeSilva, J., Sanders, W. J., & Mitani, J. C. (2007). Taphonomic analysis of skeletal remains from chimpanzee hunts at Ngogo, Kibale National Park, Uganda. *Journal of Human Evolution, 52*, 614–636.

Pruvost, M., Schwarz, R., Bessa Correia, V., Champlot, S., Braguier, S., Morel, N., et al. (2007). Freshly excavated fossil bones are best for amplification of ancient DNA. *Proceedings of the National Academy of Sciences of the United States of America, 104*, 739–744.

Rabinovich, R., & Horwitz, L. K. (1994). *An experimental approach to the study of porcupine damage to bones*. Treignes, Belgique: Taphonomie/Bone Modification. Editions du CEDARC.

Raczynski, J., & Ruprecht, A. C. (1974). The effects of digestion on the osteological composition of owl pellets. *Acta Ornithologica, 14*, 1–12.

Raven, P. H., Evert, R. F., & Eichhorn, S. E. (1986). *Biology of plants*. New York: Worth Publishers.

Ripoll, S. (1988). La Cueva de Ambrosio (Almeria, España) y su posicion cronoestratigrafica en el Mediterraneo Occidental. *British Archaeological Research International Series, 462*, 1–596.

Robb, C. (2002). Missing mammals: The effects of simulalted fossil preservation biases on the paleoenvironmental reconstruction of hominid sites. *American Journal of Physical Anthropology, 34*, 132 suppl.

Rovira Formento, M. (2010). *Aproximación experimental a la explotación de huesos largos de grandes animales para la recuperación de la médula ósea y su aplicación arqueológica al registro faunístico del Nivel 3 colluvio de Isernia La Pineta (Molise, Italia)*. Erasmus Mundus Master in Quaternary and Prehistory. University Rovira i Virgili.

Sánchez, V., Denys, C., & Fernández-Jalvo, Y. (1997). Origine et formation des accumulations de microvertebrés de la couche 1a du site du Monte di Tuda (Corse, Holocéne). *Contribution de l'étude taphonomique des micromammifères. Geodiversitas, 19*, 129–157.

Schick, K. D., & Toth, N. (1993). *Making silent stones speak. Human evolution and the dawn of technology*. New York: Simon and Schuster.

Schmidt, C. W., & Uhlig, R. (2012). Light microscopy of microfractures in burned bone. In L. S. Bell (Ed.), *Forensic microscopy for skeletal tissues* (pp. 227–234). Gosport: Springer, Humana Press.

Schmidt, C. W., Moore, C. R., & Leifheit, R. (2012). A preliminary assessment of using a white light confocal imaging profiler for cut mark analysis. In L. S. Bell (Ed.), *Forensic microscopy for skeletal tissues* (pp. 235–248). Gosport: Springer, Humana Press.

Shipman, P. (1981). *Life history of a fossil*. Cambridge: University of Harvard Press.

Shipman, P., Foster, G., & Schoeninger, M. (1984). Burnt bones and teeth: An experimental study of color, morphology, crystal structure and shrinkage. *Journal of Archaeological Science, 11*, 307–325.

Shipman, P., & Rose, J. (1983). Early hominid hunting, butchering, and carcass-processing behaviors: Approaches to the fossil record. *Journal of Anthropological Archaeology, 2*, 57–98.

Shipman, P., & Rose, J. (1984). Cutmark mimics on modern and fossil bones. *Current Anthropology, 25*, 116–117.

Shotwell, J. A. (1955). An approach to the paleoecology of mammals. *Ecology, 36*, 327–337.

Smith, C. I., Nielsen-Marsh, C. M., Jans, M. M. E., Arthur, P., Nord, A. G., & Collins, M. J. (2002). The strange case of Apigliano: Early 'fossilisation' of medieval bone in southern Italy. *Archaeometry, 44*, 405–415.

Soligo, C. (2001). Adaptations and Ecology of the Earliest Primates. PhD dissertation, University of Zurich.

Soligo, C., & Andrews, P. (2005). Taphonomic bias, taxonomic bias and historical non-equivalence of faunal structure in early hominin localities. *Journal of Human Evolution, 49*, 206–229.

Spennemann, D. H. R. (1990). On recognizing and interpreting butchery marks in tropical faunal assemblages, some comments asking for caution. In S. Solomon, I. Davidson, & D. Watson (Eds.) *Problem solving in taphonomy*, Tempus Vol 2.

Speth, J. D. (1983). *Bison kills and bone counts*. University of Chicago Press.

Stiner, M. (1995). Differential burning, recrystallization, and fragmentation of archaeological bone. *Journal of Archaeological Science, 22*, 223–237.

Stringer, C. B. (2000). The Gough's Cave human fossils: An introduction. *Bulletin of the Natural History Museum, London, 56*, 135–139.

Stringer, C. B. (2012). I The Gibraltar Neanderthals and excavations in Gorham's and Vanguard Caves 1995–1998. Introduction and history of the excavation Project. In N. Barton, C. Stringer, & C. Finlayson (Eds.), *Neanderthals in context: A report of the 1995–98 excavations at Gorham's and Vanguard Caves*, Gibraltar (pp. 1–12). Monograph Series, Oxford Committee for Archaeology. Oxford: Oxbow Press.

Stringer, C. B., Finlayson, J. C., Barton, R. N. E. Fernández-Jalvo, Y., Cáceres, I., Sabin, R., et al. (2008). Neanderthal exploitation of marine mammals. *Proceedings of the National Academy of Sciences of the United States of America, 105*, 14319–14324.

Sutcliffe, A. J. (1970). Spotted hyaena: Crusher, gnawer, digester and collector of bones. *Nature, 227*, 1110–1113.

Sutcliffe, A. J. (1973). Similarity of bones and antlers gnawed by deer to human artifacts. *Nature, 246*, 428–430.

Sutcliffe, A. J. (1977). Further notes on bones and antlers chewed by deer and other ungulates. *Deer, 4*, 73–82.

Tappen, M. (1994). Bone weathering in the tropical rain forest. *Journal of Archaeological Science, 21*, 667–673.

Thackeray, J. F. (1979). An analysis of faunal remains from archaeological sites from southern South West Africa (Namibia). *South African Archaeological Bulletin, 34*, 18–33.

Thackeray, J. F. (2007). Hominids and carnivores at Kromdraai and other Quaternary sites in Southern Africa. In T. R. Pickering, K.

Schick, & N. Toth (Eds.), *Breathing life into fossils: Taphonomic studies in honor of C.K. Brain* (pp. 67–73). Gosport: Stone Age Institute Press.

Thompson, C. E. L., Ball, S., Thompson, T. J. U., & Gowland, R. (2011). The abrasion of modern and archaeological bones by mobile sediments: The importance of transport modes. *Journal of Archaeological Science, 38*, 784–793.

Todd, L. C., & Frison, G. C. (1992). Reassembly of bison skeletons from the Horner site: A study in anatomical refitting. In J. L. Hofman & J. G. Enloe (Eds.), *Piecing together the past: Applications of refitting studies in archaeology* (pp. 63–82). British Archaeological Reports, International Series 578.

Tong, H. W., Zhang, S., Chen, F., & Li, Q. (2008). Rongements sélectifs des os par les porcs-épics et autres rongeurs: cas de la grotte Tianyuan, un site avec des restes humains fossiles récemment découvert près de Zhoukoudian (Choukoutien). *L'Anthropologie, 111*, 353–369.

Toots, H. (1965). Sequence of disarticulation in mammalian skeletons. *Contributions in Geology, University of Wyoming, 4*, 37–39.

Trueman, C., Behrensmeyer, A. K., Tuross, N., & Weiner, S. (2004). Mineralogical and compositional changes in bones exposed on soil surfaces in Amboseli National Park, Kenya: Diagenetic mechanisms and the role of sediment pore fluids. *Journal of Archaeological Science, 31*, 721–739.

Turner, A. (1983). The quantification of relative abundances in fossil and subfossil bone assemblages. *Annals of the Transvaal Museum, 33*, 311–321.

Vaquero, M. (1999). Intrasite spatial organization of lithic production in the Middle Palaeolithic: The evidence of the Abric Roman í (Capellades, Spain). *Antiquity, 73*, 493–504.

Villa, P., & Mahieu, E. (1991). Breakage patterns of human long bones. *Journal of Human Evolution, 21*, 27–48.

Villa, P., Sánchez Goñi, M. F., Cuenca Bescós, G., Grün, R., Ajas, A., García Pimienta, J. C., et al. (2010). The archaeology and paleoenvironment of an Upper Pleistocene hyena den: An integrated approach. *Journal of Archaeological Science, 37*, 919–935.

Voorhies, M. R. (1969). Taphonomy and population dynamics of an Early Pliocene vertebrate fauna Knox County, Nebraska. *Contributions in Geology, University of Wyoming, 1*, 1–69.

Weigelt, J. (1927). *Rezente wirbeltierleichen und ihre paläobioloische bedeutung.* Leipzig: Max Weg Verlag.

Weigelt, J. (1989). *Recent vertebrate carcasses and their paleobiological implications.* Chicago: University of Chicago Press.

Western, D. (1973). Cyclical changes in the habitat and climate of an East African ecosystem. *Nature, 241*, 104–106.

White, T. D. (1992). *Prehistoric cannibalism at mancos.* Princeton: Princeton University Press.

White, T. D., & Toth, N. (2004). Carnivora and carnivory: Assessing hominid toothmarks in zooarchaeology. In T. R. Pickering, K. Schick, & N. Toth (Eds.), *Breathing life into fossils* (pp. 281–296). Gosport: Stone Age Institute Press.

Williams, J. (2001). Small mammal deposits in archaeology: A taphonomic investigation of *Tyto alba* (barn owl) nesting and roosting sites. PhD dissertation, University of Sheffield.

Wolff, R. G. (1973). Hydrodynamic sorting and ecology of a Pleistocene mammalian assemblage from California (U.S.A). *Palaeogeography. Palaeoclimatology, Palaeoecology, 13*, 91–101.

Yalden, D. W., & Yalden, P. E. (1985). An experimental investigation of examining kestrel diet by pellet analysis. *Bird Study, 32*, 50–55.

Zheng, W., & Schweitzer, M. H. (1012). Chemical analyses of fossil bone. In L. S. Bell (Ed.), *Forensic microscopy for skeletal tissues* (pp. 153–172). Gosport: Springer, Humana Press.

Figure Index

3-1 Linear-Inorganic – A.1–A.130

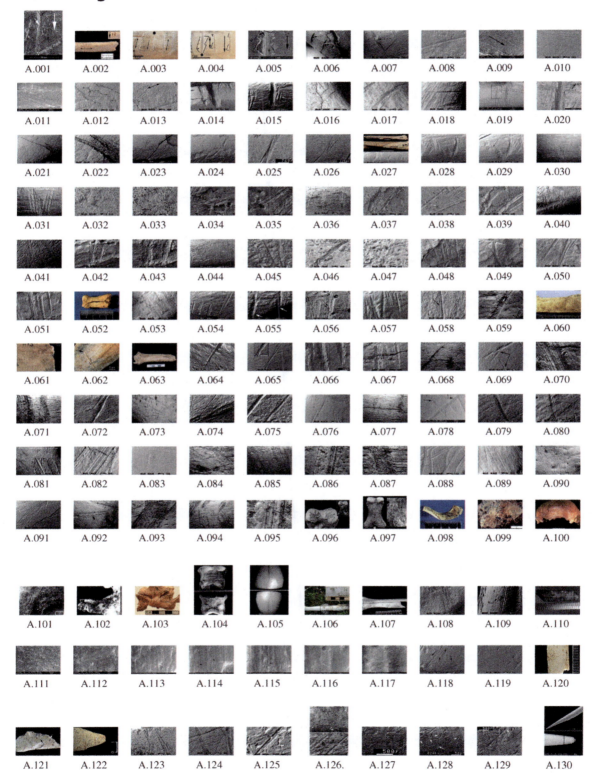

A.001 A.002 A.003 A.004 A.005 A.006 A.007 A.008 A.009 A.010

A.011 A.012 A.013 A.014 A.015 A.016 A.017 A.018 A.019 A.020

A.021 A.022 A.023 A.024 A.025 A.026 A.027 A.028 A.029 A.030

A.031 A.032 A.033 A.034 A.035 A.036 A.037 A.038 A.039 A.040

A.041 A.042 A.043 A.044 A.045 A.046 A.047 A.048 A.049 A.050

A.051 A.052 A.053 A.054 A.055 A.056 A.057 A.058 A.059 A.060

A.061 A.062 A.063 A.064 A.065 A.066 A.067 A.068 A.069 A.070

A.071 A.072 A.073 A.074 A.075 A.076 A.077 A.078 A.079 A.080

A.081 A.082 A.083 A.084 A.085 A.086 A.087 A.088 A.089 A.090

A.091 A.092 A.093 A.094 A.095 A.096 A.097 A.098 A.099 A.100

A.101 A.102 A.103 A.104 A.105 A.106 A.107 A.108 A.109 A.110

A.111 A.112 A.113 A.114 A.115 A.116 A.117 A.118 A.119 A.120

A.121 A.122 A.123 A.124 A.125 A.126. A.127 A.128 A.129 A.130

© Springer Science+Business Media Dordrecht 2016
Yolanda Fernández-Jalvo and Peter Andrews, *Atlas of Taphonomic Identifications: 1001+ Images of Fossil and Recent Mammal Bone Modification*, Vertebrate Paleobiology and Paleoanthropology, DOI 10.1007/978-94-017-7432-1

3-2 Llinear-Org-Animal – A.131–A.226

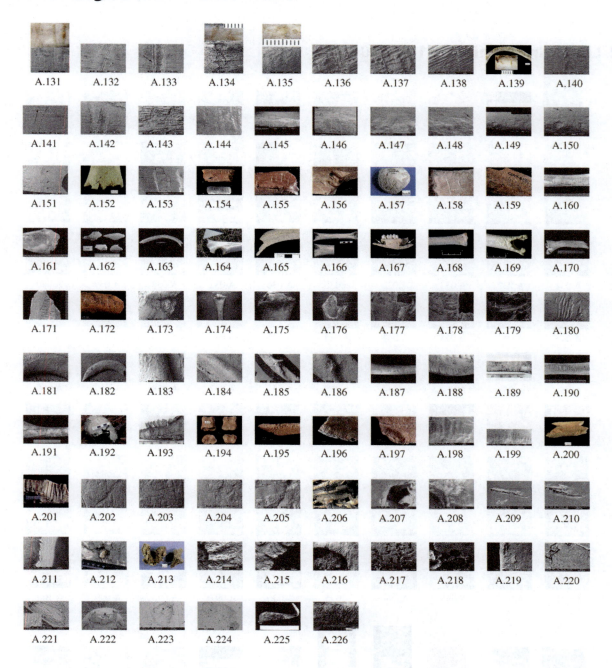

A.131   A.132   A.133   A.134   A.135   A.136   A.137   A.138   A.139   A.140

A.141   A.142   A.143   A.144   A.145   A.146   A.147   A.148   A.149   A.150

A.151   A.152   A.153   A.154   A.155   A.156   A.157   A.158   A.159   A.160

A.161   A.162   A.163   A.164   A.165   A.166   A.167   A.168   A.169   A.170

A.171   A.172   A.173   A.174   A.175   A.176   A.177   A.178   A.179   A.180

A.181   A.182   A.183   A.184   A.185   A.186   A.187   A.188   A.189   A.190

A.191   A.192   A.193   A.194   A.195   A.196   A.197   A.198   A.199   A.200

A.201   A.202   A.203   A.204   A.205   A.206   A.207   A.208   A.209   A.210

A.211   A.212   A.213   A.214   A.215   A.216   A.217   A.218   A.219   A.220

A.221   A.222   A.223   A.224   A.225   A.226

3-3 Linear-Org-Plants – A.227–A.276

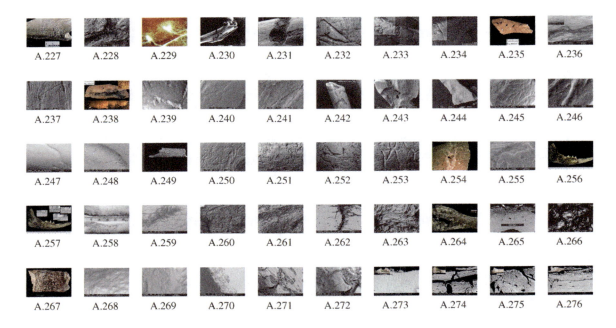

A.227 A.228 A.229 A.230 A.231 A.232 A.233 A.234 A.235 A.236

A.237 A.238 A.239 A.240 A.241 A.242 A.243 A.244 A.245 A.246

A.247 A.248 A.249 A.250 A.251 A.252 A.253 A.254 A.255 A.256

A.257 A.258 A.259 A.260 A.261 A.262 A.263 A.264 A.265 A.266

A.267 A.268 A.269 A.270 A.271 A.272 A.273 A.274 A.275 A.276

3-4 Linear-Org-Microorg – A.277–A.306

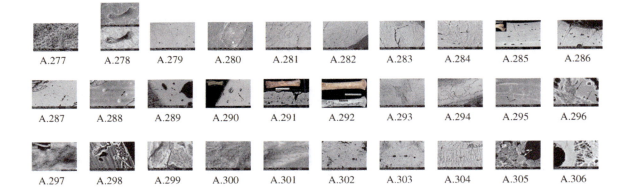

A.277 A.278 A.279 A.280 A.281 A.282 A.283 A.284 A.285 A.286

A.287 A.288 A.289 A.290 A.291 A.292 A.293 A.294 A.295 A.296

A.297 A.298 A.299 A.300 A.301 A.302 A.303 A.304 A.305 A.306

4-1 Pits Perforat-Inorganic – A.307–A.354

A.307   A.308   A.309   A.310   A.311   A.312   A.313   A.314   A.315   A.316

A.317   A.318   A.319   A.320   A.321   A.322   A.323   A.324   A.325   A.326

A.327   A.328   A.329   A.330   A.331   A.332   A.333   A.334   A.335   A.336

A.337   A.338   A.339   A.340   A.341   A.342   A.343   A.344   A.345   A.346

A.347   A.348   A.349   A.350   A.351   A.352   A.353   A.354

4-2 Pits Perforat-Organic-Animal – A.355–A.422

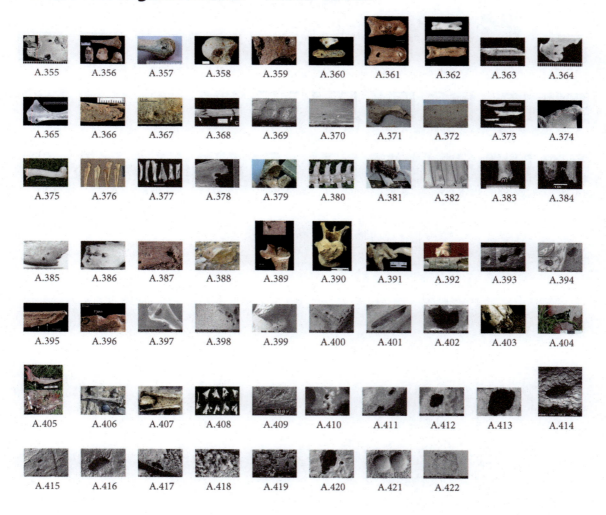

A.355   A.356   A.357   A.358   A.359   A.360   A.361   A.362   A.363   A.364

A.365   A.366   A.367   A.368   A.369   A.370   A.371   A.372   A.373   A.374

A.375   A.376   A.377   A.378   A.379   A.380   A.381   A.382   A.383   A.384

A.385   A.386   A.387   A.388   A.389   A.390   A.391   A.392   A.393   A.394

A.395   A.396   A.397   A.398   A.399   A.400   A.401   A.402   A.403   A.404

A.405   A.406   A.407   A.408   A.409   A.410   A.411   A.412   A.413   A.414

A.415   A.416   A.417   A.418   A.419   A.420   A.421   A.422

4-3 Pits Perforat-Organic-Plant – A.423–A.458

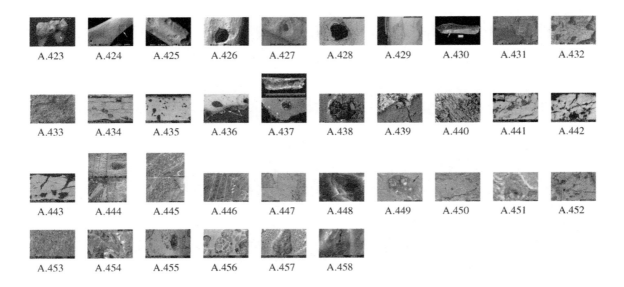

A.423   A.424   A.425   A.426   A.427   A.428   A.429   A.430   A.431   A.432

A.433   A.434   A.435   A.436   A.437   A.438   A.439   A.440   A.441   A.442

A.443   A.444   A.445   A.446   A.447   A.448   A.449   A.450   A.451   A.452

A.453   A.454   A.455   A.456   A.457   A.458

4-4 Pits Perforat-Organic-Microorg – A.459–A.502

A.459   A.460   A.461   A.462   A.463   A.464   A.465   A.466   A.467   A.468

A.469   A.470   A.471   A.472   A.473   A.474   A.475   A.476   A.477   A.478

A.479   A.480   A.481   A.482   A.483   A.484   A.485   A.486   A.487   A.488

A.489   A.490   A.491   A.492   A.493   A.494   A.495   A.496   A.497   A.498

A.499   A.500   A.501   A.502

5 Discolor-Stains – A.503–A.540

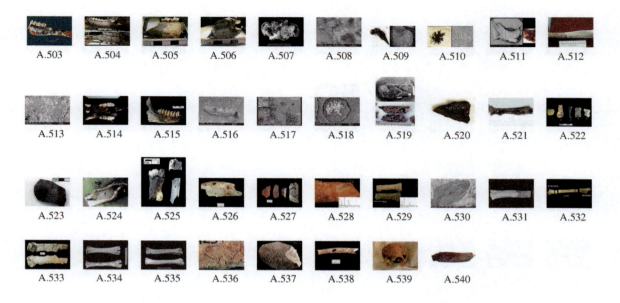

A.503 A.504 A.505 A.506 A.507 A.508 A.509 A.510 A.511 A.512

A.513 A.514 A.515 A.516 A.517 A.518 A.519 A.520 A.521 A.522

A.523 A.524 A.525 A.526 A.527 A.528 A.529 A.530 A.531 A.532

A.533 A.534 A.535 A.536 A.537 A.538 A.539 A.540

6-1 Rounding-Inorg – A.541–A.587

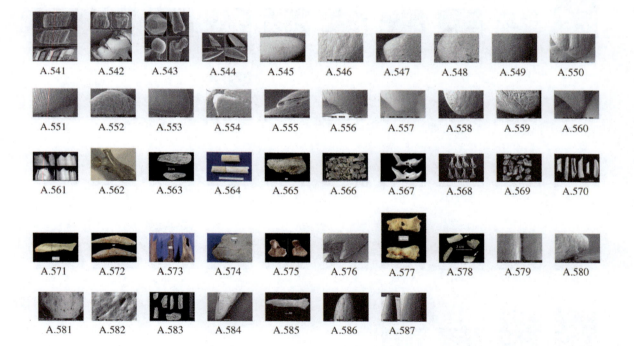

A.541 A.542 A.543 A.544 A.545 A.546 A.547 A.548 A.549 A.550

A.551 A.552 A.553 A.554 A.555 A.556 A.557 A.558 A.559 A.560

A.561 A.562 A.563 A.564 A.565 A.566 A.567 A.568 A.569 A.570

A.571 A.572 A.573 A.574 A.575 A.576 A.577 A.578 A.579 A.580

A.581 A.582 A.583 A.584 A.585 A.586 A.587

6-2 Rounding-Org – A.588–A.638

A.588 A.589 A.590 A.591 A.592 A.593 A.594 A.595 A.596 A.597

A.598 A.599 A.600 A.601 A.602 A.603 A.604 A.605 A.606 A.607

A.608 A.609 A.610 A.611 A.612 A.613 A.614 A.615 A.616 A.617

A.618 A.619 A.620 A.621 A.622 A.623 A.624 A.625 A.626 A.627

A.628 A.629 A.630 A.631 A.632 A.633 A.634 A.635 A.636 A.637

A.638

7-1 Flaking-Cracking-Inorg – A.639–A.743

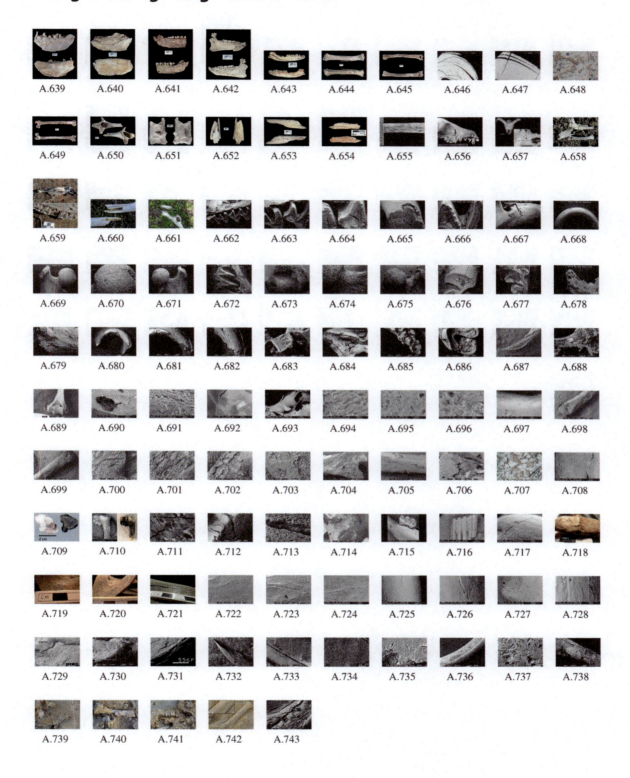

A.639	A.640	A.641	A.642	A.643	A.644	A.645	A.646	A.647	A.648
A.649	A.650	A.651	A.652	A.653	A.654	A.655	A.656	A.657	A.658
A.659	A.660	A.661	A.662	A.663	A.664	A.665	A.666	A.667	A.668
A.669	A.670	A.671	A.672	A.673	A.674	A.675	A.676	A.677	A.678
A.679	A.680	A.681	A.682	A.683	A.684	A.685	A.686	A.687	A.688
A.689	A.690	A.691	A.692	A.693	A.694	A.695	A.696	A.697	A.698
A.699	A.700	A.701	A.702	A.703	A.704	A.705	A.706	A.707	A.708
A.709	A.710	A.711	A.712	A.713	A.714	A.715	A.716	A.717	A.718
A.719	A.720	A.721	A.722	A.723	A.724	A.725	A.726	A.727	A.728
A.729	A.730	A.731	A.732	A.733	A.734	A.735	A.736	A.737	A.738
A.739	A.740	A.741	A.742	A.743					

7-2 Flaking-Cracking-Org – A.744–A.785

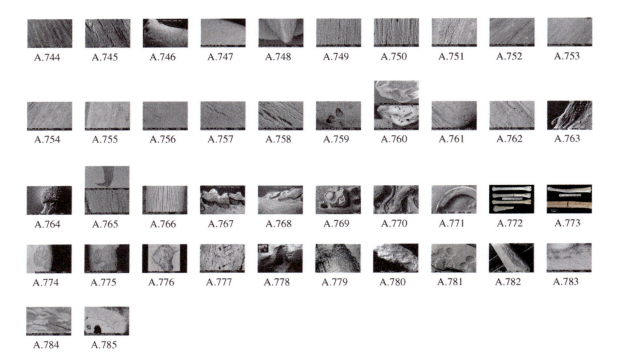

A.744 A.745 A.746 A.747 A.748 A.749 A.750 A.751 A.752 A.753

A.754 A.755 A.756 A.757 A.758 A.759 A.760 A.761 A.762 A.763

A.764 A.765 A.766 A.767 A.768 A.769 A.770 A.771 A.772 A.773

A.774 A.775 A.776 A.777 A.778 A.779 A.780 A.781 A.782 A.783

A.784 A.785

8-1 Corrosion-Inorg – A.786–A.815

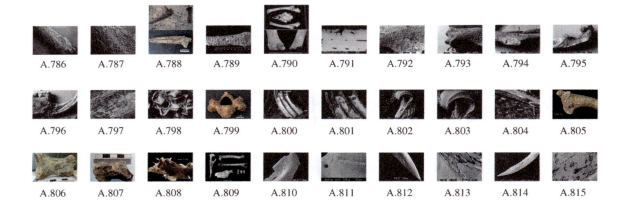

A.786 A.787 A.788 A.789 A.790 A.791 A.792 A.793 A.794 A.795

A.796 A.797 A.798 A.799 A.800 A.801 A.802 A.803 A.804 A.805

A.806 A.807 A.808 A.809 A.810 A.811 A.812 A.813 A.814 A.815

8-2 Corrosion-Org-Animal – A.816–A.878

A.816 A.817 A.818 A.819 A.820 A.821 A.822 A.823 A.824 A.825

A.826 A.827 A.828 A.829 A.830 A.831 A.832 A.833 A.834 A.835

A.836 A.837 A.838 A.839 A.840 A.841 A.842 A.843 A.844 A.845

A.846 A.847 A.848 A.849 A.850 A.851 A.852 A.853 A.854 A.855

A.856 A.857 A.858 A.859 A.860 A.861 A.862 A.863 A.864 A.865

A.866 A.867 A.868 A.869 A.870 A.871 A.872 A.873 A.874 A.875

A.876 A.877 A.878

8-3 Corrosion-Org-Plants – A.879–A.904

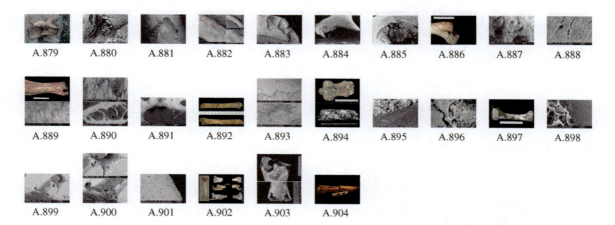

A.879 A.880 A.881 A.882 A.883 A.884 A.885 A.886 A.887 A.888

A.889 A.890 A.891 A.892 A.893 A.894 A.895 A.896 A.897 A.898

A.899 A.900 A.901 A.902 A.903 A.904

8-3 Corrosion-Org-Microorg – A.905–A.950

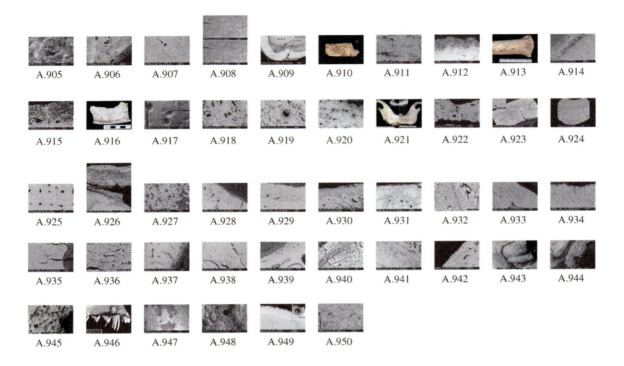

A.905 A.906 A.907 A.908 A.909 A.910 A.911 A.912 A.913 A.914

A.915 A.916 A.917 A.918 A.919 A.920 A.921 A.922 A.923 A.924

A.925 A.926 A.927 A.928 A.929 A.930 A.931 A.932 A.933 A.934

A.935 A.936 A.937 A.938 A.939 A.940 A.941 A.942 A.943 A.944

A.945 A.946 A.947 A.948 A.949 A.950

9-1 Brekg-Deform-Inorg – A.951–A.1000

A.951 A.952 A.953 A.954 A.955 A.956 A.957 A.958 A.959 A.960

A.961 A.962 A.963 A.964 A.965 A.966 A.967 A.968 A.969 A.970

A.971 A.972 A.973 A.974 A.975 A.976 A.977 A.978 A.979 A.980

A.981 A.982 A.983 A.984 A.985 A.986 A.987 A.988 A.989 A.990

A.991 A.992 A.993 A.994 A.995 A.996 A.997 A.998 A.999 A.1000

9-2 Brekg-Deform-Org – A.1001–A.1050

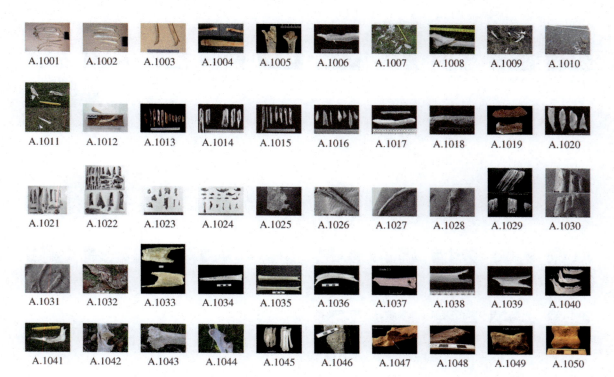

A.1001 A.1002 A.1003 A.1004 A.1005 A.1006 A.1007 A.1008 A.1009 A.1010

A.1011 A.1012 A.1013 A.1014 A.1015 A.1016 A.1017 A.1018 A.1019 A.1020

A.1021 A.1022 A.1023 A.1024 A.1025 A.1026 A.1027 A.1028 A.1029 A.1030

A.1031 A.1032 A.1033 A.1034 A.1035 A.1036 A.1037 A.1038 A.1039 A.1040

A.1041 A.1042 A.1043 A.1044 A.1045 A.1046 A.1047 A.1048 A.1049 A.1050

10 Disartic-Completeness – A.1051–A.1080

A.1051 A.1052 A.1053 A.1054 A.1055 A.1056 A.1057 A.1058 A.1059 A.1060

A.1061 A.1062 A.1063 A.1064 A.1065 A.1066 A.1067 A.1068 A.1069 A.1070

A.1071 A.1072 A.1073 A.1074 A.1075 A.1076 A.1077 A.1078 A.1079 A.1080

Text Figures

 Fig. 1.1

 Fig. 1.2

 Fig. 2.1

 Fig. 2.2

 Fig. 2.3

 Fig. 2.4

 Fig. 2.5

 Fig. 2.6

 Fig. 3.1

 Fig. 3.2

 Fig. 3.2

 Fig. 3.3

 Fig. 3.4

 Fig. 3.5

 Fig. 3.6

 Fig. 3.7

 Fig. 3.8

 Fig. 4.1

 Fig. 4.2

 Fig. 4.3

Fig. 4.4

 Fig. 4.5

 Fig. 4.6

 Fig. 4.7

 Fig. 4.8

 Fig. 4.9

 Fig. 5.1

 Fig. 5.2

 Fig. 5.3

 Fig. 6.1

 Fig. 6.2

 Fig. 6.3

 Fig. 6.4

 Fig. 6.5

 Fig. 6.6

 Fig. 7.1

 Fig. 8.1

 Fig. 8.2

 Fig. 8.3

 Fig. 8.4

Fig. 8.5

 Fig. 8.6

 Fig. 9.1

 Fig. 9.2

 Fig. 9.3

 Fig. 9.4

 Fig. 10.1

 Fig. 10.2

 Fig. 10.3

 Fig. 11.1

 Fig. 11.2a

 Fig. 11.2b

 Fig. 11.3

Index

Note: Page numbers followed by *f* and *t* indicate figures and tables, respectively

Printed by Printforce, the Netherlands